CW00501017

Combinatorics of Finite Sets

Ian Anderson
University of Glasgow

DOVER PUBLICATIONS, INC.
Mineola, New York

Copyright

Bibliographical Note

This Dover edition, first published in 2002, is a corrected republication of the work as published by Oxford University Press, Oxford, England, and New York, in 1989 (first publication: 1987).

Library of Congress Cataloging-in-Publication Data

Anderson, Ian, Ph. D.
Combinatorics of finite sets / Ian Anderson.
p. cm.
". . . a corrected republication of the work as published by Oxford University Press, Oxford, England, and New York, in 1989 (first publication: 1987)"—T.p. verso.
Includes index.
ISBN 0-486-42257-7 (pbk.)
1. Set theory. 2. Combinatorial analysis. I. Title.

QA248 .A657 2002
511.3'22—dc21

2002019228

Manufactured in the United States by Courier Corporation
42257703
www.doverpublications.com

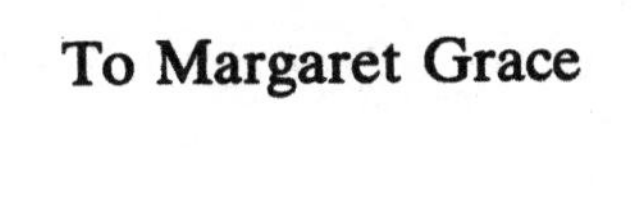

To Margaret Grace

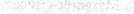

Preface

The past quarter century has seen the remarkable rise of combinatorics as a distinctive and important area of mathematics. Combinatorial topics have found a place in many university degree courses, and, since the founding of the *Journal of Combinatorial Theory* in 1966, there has been a flood of publications in the combinatorial area. Some branches of combinatorics are already well established, with their own unified body of theory and applications: examples are graph theory, with its topological flavour, coding theory and design theory, with their algebraic connections, and enumeration theory, concerned with the techniques of counting. These areas are already supplied with many textbooks at both undergraduate and postgraduate levels. The purpose of the present work, however, is to advertise another area of combinatorics, where a body of theory, at one time very scattered and *ad hoc*, is gradually being moulded into an elegant unity. Without attempting to be exhaustive, the book is intended to be a readable introduction to what is, for the author, a fascinating subject.

The origins of the theory can perhaps be traced back to 1928, when Sperner published a simple theorem which has had repercussions far beyond his wildest dreams. Sperner's theorem simply asserts that if you want to find as many subsets of an n-element set as possible, subject to the condition that no subset is contained in another, then you cannot do better than to choose all the subsets of size $[\frac{1}{2}n]$. This theorem has been reproved and generalized to such an extent that it has given rise to a whole branch of the theory of partially ordered sets (or posets) called Sperner theory. Although we shall go some way down this path, we shall not restrict ourselves to this area. Instead, we shall use Sperner's theorem as a springboard and a signpost, discovering where the ideas involved in its various proofs lead to. We shall be led to consider the structure of the set of subsets of a finite set viewed as a poset, its chain decompositions and its antichains, its rank levels, and the inclusion relations between these levels. This will take us, for example, to the famous Kruskal–Katona theorem which answers the following question: given r subsets of S_t all of size k, what is the least possible number of sets of size $k-1$ contained in them? The answer to this question involves a nice interplay between two different orderings, namely the partial ordering of the subsets of S by inclusion, and the total ordering of the subsets of S of a given size by what we shall call the squashed ordering, which is a variation on the more familiar lexicographic or alphabetical ordering.

Anyone writing a book in this area is faced right at the start by a fundamental problem. Many results for the poset of subsets of a set can be extended to more general posets. Sometimes one of the available proofs for sets easily extends to more general posets; sometimes, however, the nicest proof for sets does not generalize. So should the results be presented in their most general form (thereby sometimes losing out on clarity) or should they be presented for subsets of a set (thereby losing out on generality but perhaps gaining in clarity)? An example of this problem arises in connection with the Kruskal–Katona theorem. Several proofs are available, but a more difficult proof due to Clements and Lindström establishes the result in the more general context of the poset of divisors of a number (or subsets of a multiset). Since we include the Clements–Lindström theorem in Chapter 9, there is strictly speaking no need to include a separate proof of the Kruskal–Katona theorem. However, in the simpler context of sets, the Kruskal–Katona theorem has such an elegant theory surrounding it that it would be almost criminal to omit the simpler case. So we present both proofs. On many other occasions we prove results in more than one way because the different proofs illustrate different ideas and different techniques. On the whole, I have taken the view that I should present results in their simplest forms, concentrating mainly on sets and multisets. Accordingly, the reader will not find a discussion of, say, geometric lattices, although their 'prototype', the poset of partitions of a set, is discussed. In a few places we look at posets more generally; the final chapter, for example, discusses extensions of the theorem of Dilworth concerning chain decompositions of a general poset.

In searching out the material for this book I was greatly helped by several survey articles, two of which deserve special mention: the first is that by Greene and Kleitman (1978), and the second is the more recent one by D. B. West (1982). As will be seen from a glance at the extensive list of references at the end of the book, a number of more recent results have been included. Inevitably some of the results presented here will have been improved upon by the time this book appears in print, but in a sense this does not matter for the aim of the book is not to provide an exhaustive survey but to present some of the ideas and techniques which go to make up the subject. Inevitably, also, my choice of material will not meet with the approval of all, but a number of interesting results not in the text have been included in the exercises at the end of each chapter with hints or outlines of their solutions at the end of the book.

A number of people have helped and encouraged me in the writing of this book. In particular, I gladly acknowledge the helpful comments of Professor George Clements and Dr Hazel Perfect. I should also like to thank the Oxford University Press for encouraging me to write, and the University of Glasgow for granting me a period of study leave during which the final compilation of the book was accomplished.

I.A.

Glasgow
December 1985

[illegible]

... ideas and techniques which go to make up the subject. Certainly [illegible] also the study of material will not [illegible] the subject in detail, but [illegible] numbers [illegible] have been included in the exercises at [illegible] of each chapter with hints or solutions [illegible] at the end of the book.

A number of [illegible] have helped [illegible] encouraged me in the writing of this book. In particular [illegible] acknowledge [illegible] Professor [illegible] and Dr [illegible]. I should [illegible] to thank the [illegible] University [illegible] for encouraging me to [illegible] and the [illegible] the final completion of the [illegible] accomplished.

A. [illegible]

[illegible]
[illegible] 1985

Contents

Some notation used in the text

$\binom{n}{r}$	binomial coefficient (n-choose-r)
$\Delta\mathscr{A}$	the shadow of $\mathscr{A}$
$\Delta^{(k)}\mathscr{A}$	the shadow of $\mathscr{A}$ at level k
$\nabla\mathscr{A}$	the shade of $\mathscr{A}$
$C\mathscr{A}$	the compression of $\mathscr{A}$
$\mathscr{A}'$	the set of complements of members of $\mathscr{A}$
$(k)S$	the set of k-subsets of S
N_i	the number of members of rank i
$<_s$	the squashed ordering
$F_k(n)$	the first n k-sets (or k-vectors)
$\mathfrak{A}(P)$	the set of antichains of the poset P
$\varnothing$	the empty set
$\tau(m)$	the number of divisors of m
$d_k(P)$	the size of the largest k-union in P

1 Introduction and Sperner's theorem

1.1 A simple intersection result

The main theme of this book is the study of collections $\mathscr{A}$ of subsets of a finite set S, where $\mathscr{A}$ is described in terms of intersection, union, or inclusion conditions. An amazing richness and variety of results will be discovered, developed, and extended in various directions. Although our main initial theme will be a study of a theorem of Sperner which could be said to be the inspiration of all that follows, we get into training by first of all asking what must surely be one of the simplest questions possible.

Problem Let $\mathscr{A}$ be a collection of subsets of an n-element set S (or an *n-set* S) such that $A_i \cap A_j \neq \varnothing$ for each pair i, j. How big can $|\mathscr{A}|$ be? The answer, and more besides, is given by the following theorem.

Theorem 1.1.1 If $\mathscr{A}$ is a collection of distinct subsets of the n-set S such that $A_i \cap A_j \neq \varnothing$ for all $A_i, A_j \in \mathscr{A}$, then $|\mathscr{A}| \leq 2^{n-1}$. Further, if $|\mathscr{A}| < 2^{n-1}$, $\mathscr{A}$ can be extended to a collection of 2^{n-1} subsets also satisfying the given intersection property.

Proof If $A \in \mathscr{A}$ then the complement $A' = S - A$ is certainly not in $\mathscr{A}$, since $A \cap A' = \varnothing$. So we immediately obtain $|\mathscr{A}| \leq \frac{1}{2}2^n = 2^{n-1}$. This bound cannot be improved upon since the collection of all subsets of $\{1, \ldots, n\}$ containing 1 satisfies the intersection condition and has 2^{n-1} members.

Now suppose $|\mathscr{A}| < 2^{n-1}$. Then there must be a subset A with $A \notin \mathscr{A}$ and also $A' \notin \mathscr{A}$. We can then add A to the collection $\mathscr{A}$ unless there exists $B \in \mathscr{A}$ such that $A \cap B = \varnothing$. But then $B \subseteq A'$ and so we could add A' to $\mathscr{A}$. If the resulting collection has fewer than 2^{n-1} members, repeat the process. □

This example pinpoints some key questions. Given a property involving union, intersection, and inclusion, how large can a collection $\mathscr{A}$ of subsets of S be if $\mathscr{A}$ satisfies the property? Can we characterize those collections which maximize $|\mathscr{A}|$? These are the sort of questions which we shall study.

1.2 Sperner's theorem

We now consider the property: if $A_i, A_j \in \mathscr{A}$, then $A_i \nsubseteq A_j$. A collection of subsets of S with this property is called a collection of *incomparable* sets, or an *antichain*, or sometimes a *clutter*. It is an antichain in the sense that its property is the other extreme from that of a chain in which every pair of sets is comparable.

Theorem 1.2.1 (Sperner 1928) Let $\mathscr{A}$ be an antichain of subsets of an n-set S. Then

$$|\mathscr{A}| \leqslant \binom{n}{[n/2]}.$$

This result is clearly best possible since the subsets of size $[n/2]$ form an antichain. The original proof given by Sperner will be analysed in Chapter 2, but we start here by giving a simple elegant proof due to Lubell which is pregnant with generalizations and extensions. Altogether we shall give three different proofs, not just because they exist, but because each in its own way presents us with ideas which can be developed to suit a wider range of ordered structures.

Proof of Theorem 1.2.1 (Lubell 1966) First of all note that there are $n!$ permutations of the elements of S. We shall say that a permutation π of the elements of S *begins with* A if the first $|A|$ members of π are the elements of A in some order. Now the number of permutations beginning with A must be $|A|!(n-|A|)!$. Also, no permutation can begin with two different sets in $\mathscr{A}$, since one of these sets would contain the other; therefore permutations beginning with different sets in $\mathscr{A}$ are distinct. Thus

$$\sum_{A \in \mathscr{A}} |A|!(n-|A|)! \leqslant n!$$

If we let p_k denote the number of members of $\mathscr{A}$ of size k, we have

$$\sum_k k!(n-k)!p_k \leqslant n!$$

whence

$$\sum_k \frac{p_k}{\binom{n}{k}} \leqslant 1. \tag{1.1}$$

Thus

$$|\mathscr{A}| = \sum_k p_k = \binom{n}{[n/2]} \sum_k \frac{p_k}{\binom{n}{[n/2]}} \leqslant \binom{n}{[n/2]} \sum_k \frac{p_k}{\binom{n}{k}} \leqslant \binom{n}{[n/2]}. \quad (1.2)$$

□

Note that Lubell's proof actually gives a stronger result than Sperner's theorem. The inequality (1.1) is called a *LYM inequality* after Lubell, Yamamoto (1954), and Meschalkin (1963), whose work includes similar results. Instead of giving an upper bound for $\sum_k p_k$, the inequality gives an upper bound for the weighted sum $\sum_k p_k / \binom{n}{k}$. Note that $p_k / \binom{n}{k}$ is the proportion of all those subsets of S of size k which are in $\mathscr{A}$; the LYM inequality asserts that the sum of these proportions is at most unity. The special case where $p_k = \binom{n}{k}$ if $k = [n/2]$, $p_k = 0$ otherwise, shows that the bound can be attained.

We have thus shown that the maximum size of an antichain of subsets of an n-set S is $\binom{n}{[n/2]}$. Can we identify all the antichains which are as big as this? The first inequality in (1.2) shows that we can attain the bound $\binom{n}{[n/2]}$ only if $p_k = 0$ whenever $\binom{n}{k} < \binom{n}{[n/2]}$. Therefore, if n is even, there is only one maximum-sized antichain, namely the collection of all $n/2$-subsets. If n is odd, all sets in a maximum-sized antichain must be of size $\frac{1}{2}(n-1)$ or $\frac{1}{2}(n+1)$. We now show that there can in fact be no mixture of sizes; a maximum-sized antichain consists either of all subsets of size $\frac{1}{2}(n-1)$ or of all subsets of size $\frac{1}{2}(n+1)$.

Theorem 1.2.2 If n is even, the only antichain consisting of $\binom{n}{[n/2]}$ subsets of the n-set S is made up of all the $n/2$-subsets of S. If n is odd, an antichain of size $\binom{n}{[n/2]}$ consists of either all the $\frac{1}{2}(n-1)$-subsets or all the $\frac{1}{2}(n+1)$-subsets.

Proof (Lovász 1979) The case of even n has been dealt with. Suppose now that $n = 2m+1$ and that $\mathscr{A}$ is an antichain of size $\binom{n}{m}$. Note that in Lubell's proof the only way of finishing up with equality in (1.2) is to have equality at each stage, so in particular every permutation must contribute a member of $\mathcal{A}$ with which it begins.

Now $\mathcal{A}$ must consist only of sets of size m or $m+1$. Suppose that X, Y are subsets of size m, $m+1$ respectively and that $X \subset Y$. If $X = \{x_1, \ldots, x_m\}$ and $Y = \{x_1, \ldots, x_m, x_{m+1}\}$ then since any permutation beginning with $x_1, \ldots, x_{m+1}$ must begin with a member of $\mathcal{A}$, X or Y must be in $\mathcal{A}$.

Our aim is to prove that $\mathcal{A}$ consists of all m-sets or all $(m+1)$-sets. Suppose that $\mathcal{A}$ contains some but not all of the $(m+1)$-sets. Then we can find sets E, F such that $|E| = |F| = m+1$, $E \in \mathcal{A}$, $F \notin \mathcal{A}$. By relabelling the elements of S if necessary we can suppose that $E = \{x_1, \ldots, x_{m+1}\}$ and $F = \{x_i, \ldots, x_{m+i}\}$ for some i. Since $E \in \mathcal{A}$ and $F \notin \mathcal{A}$ there must be a largest integer $j < i$ with $\{x_j, \ldots, x_{m+j}\} \in \mathcal{A}$. Then $E^* = \{x_j, \ldots, x_{m+j}\} \in \mathcal{A}$ and $F^* = \{x_{j+1}, \ldots, x_{m+j+1}\} \notin \mathcal{A}$. We now have an impossible situation. Since $E^* \cap F^* \subset E^*$ where $E^* \in \mathcal{A}$, we must have $E^* \cap F^* \notin \mathcal{A}$. However, $E^* \cap F^* \subset F^*$ where $|E^* \cap F^*| = m$ and $|F^*| = m+1$, so by an earlier part of the proof one of $E^* \cap F^*$ and F^* must be in $\mathcal{A}$. This contradiction shows that our assumption must have been false and so $\mathcal{A}$ indeed consists only of sets all of the same size.

□

1.3 A theorem of Bollobás

As another example of how the permutation approach of Lubell's proof can be used to obtain elegant proofs of results obtained originally by more complicated arguments, we now give a generalization of Sperner's theorem due to Bollobás (1965). The result was also independently proved by Katona (1974), Tarjan (1975), and Griggs, Stahl, and Trotter (1984). This repeated discovery of results by authors working independently is a frequent occurrence in this area of mathematics!

Theorem 1.3.1 (Bollobás 1965) Let $A_1, \ldots, A_m$, $B_1, \ldots, B_m$ be subsets of an n-set S such that $A_i \cap B_j = \varnothing$ if and only if $i = j$. Let $a_i = |A_i|$ and $b_i = |B_i|$. Then

$$\sum_{i=1}^{m} \frac{1}{\binom{a_i + b_i}{a_i}} \leqslant 1.$$

Proof Consider each of the $n!$ permutations of the elements of S, and say that a permutation π *contains* A *followed by* B if all the elements of A occur in π before all the elements of B. If a particular permutation π contains A_i followed by B_i and also contains A_j followed by B_j, then $A_i \cap B_j = \varnothing$ (if A_i ends before B_j begins) or $A_j \cap B_i = \varnothing$ (if A_i ends after B_j begins), and either of these contradicts the hypotheses. So, for each permutation π, there is at most one i for which π contains A_i followed by B_i. However, given i, the number of permutations containing A_i followed by B_i can be found as follows. Choose the $a_i + b_i$ positions to be filled by the elements of A_i and B_i; this can be done in $\binom{n}{a_i + b_i}$ ways. Then place the a_i members of A_i in some order in the first a_i of the chosen positions and then the b_i members of B_i in some order in the remaining b_i positions; this can all be done in $a_i!b_i!$ ways. Finally order the remaining $n - a_i - b_i$ elements of S and place then in the remaining places of the permutation; this can be done in $(n - a_i - b_i)!$ ways. Thus the number of permutations π containing A_i followed by B_i is

$$\binom{n}{a_i + b_i} a_i!b_i!(n - a_i - b_i)! = \frac{n!}{\binom{a_i + b_i}{a_i}}.$$

Summing over all i we now obtain

$$\sum_i \frac{n!}{\binom{a_i + b_i}{a_i}} \leqslant n!$$

as required. □

Note that Sperner's theorem follows on taking $B_i = A_i'$, the complement of A_i, for the condition $A_i \cap B_i = \varnothing$ becomes $A_i \cap A_i' = \varnothing$, the condition $A_i \cap B_j \neq \varnothing$ becomes $A_i \cap A_j' \neq \varnothing$, i.e. $A_i \nsubseteq A_j$, and the conclusion yields

$$\sum_k \frac{p_k}{\binom{n}{k}} = \sum_i \frac{1}{\binom{n}{a_i}} = \sum_i \frac{1}{\binom{a_i + b_i}{a_i}} \leqslant 1.$$

Theorem 1.3.1 has been generalized in a number of ways. Frankl (1982) and Kalai (1984) weakened the condition to $A_i \cap A_j \neq \varnothing$ for $1 \leqslant i < j \leqslant m$ and obtained the same conclusion. Lovász (1977) generalized the theorem to subspaces of a linear space. As a recent application, we now apply Theorem 1.3.1 to the following generalization of the Sperner situation. Suppose we are given m chains of subsets of an n-set S which are incomparable in the sense that no member of one chain is contained in a member of any other chain. How large can m be? In the case where all the chains have $k+1$ members, let $f(n, k)$ denote the largest possible value of m. Then Sperner's theorem corresponds to the case $k=0$ and asserts that $f(n, 0)=\binom{n}{[n/2]}$.

Theorem 1.3.2 (Griggs *et al.* 1984) Let $f(n, k)$ denote the largest value of m for which it is possible to find m chains of $k+1$ distinct subsets of an n-set S such that no member of any chain is a subset of a member of any other chain. Then

$$f(n, k)=\binom{n-k}{[(n-k)/2]}.$$

Proof Suppose that we have m chains

$$A_{i,0} \subset A_{i,1} \subset \ldots \subset A_{i,k} \quad (i=1, \ldots, m)$$

satisfying the conditions of the theorem. In Theorem 1.3.1 take $A_i=A_{i,0}$ and $B_i=S-A_{i,k}$. Then $a_i=|A_{i,0}|$ and $b_i=n-|A_{i,k}|$. It is clear that $|A_{i,k}| \geqslant a_i+k$, so $b_i \leqslant n-k-a_i$; thus

$$\binom{a_i+b_i}{a_i} \leqslant \binom{n-k}{a_i} \leqslant \binom{n-k}{[(n-k)/2]}.$$

Now we certainly have $A_i \cap B_i=\varnothing$. Also, if we had $A_i \cap B_j=\varnothing$ for some $i \neq j$ we would then have $A_{i,0} \subset A_{j,k}$, contradicting the hypotheses. So the sets A_i and B_i satisfy the conditions of Theorem 1.3.1, and we have

$$m=\sum_{i=1}^{m} 1 \leqslant \sum_{i=1}^{m} \binom{n-k}{[(n-k)/2]} \bigg/ \binom{a_i+b_i}{a_i} \leqslant \binom{n-k}{[(n-k)/2]}.$$

Thus

$$f(n, k) \leqslant \binom{n-k}{[(n-k)/2]}.$$

To complete the proof we exhibit $\binom{n-k}{[(n-k)/2]}$ chains with the required properties. Consider the $[(n-k)/2]$-subsets X of

$\{k+1, \ldots, n\}$. There are $\binom{n-k}{[(n-k)/2]}$ such subsets. For each such X take the chain

$$X \subset X \cup \{1\} \subset X \cup \{1, 2\} \subset \ldots \subset X \cup \{1, 2, \ldots, k\}.$$

These chains have the required properties. □

Further generalizations of Sperner's theorem will be discussed later, particularly in Chapter 8 and in the study of the Littlewood–Offord problem in Chapter 11.

Exercises 1

1.1 Prove that if $A_1, \ldots, A_m$ are distinct subsets of an n-set S such that for each pair i, j, $A_i \cup A_j \neq S$ then $m \leqslant 2^{n-1}$.

1.2 Can we have equality in Theorem 1.1.1 without all the A_i having a common element?

1.3 Show that if $\mathscr{A}$ is an antichain of subsets of an n-set with $|A| \leqslant h \leqslant \frac{1}{2}n$ for all $A \in \mathscr{A}$, then $|\mathscr{A}| \leqslant \binom{n}{h}$.

1.4 How many antichains of subsets of S are there if (a) $|S| = 2$, (b) $|S| = 3$? (This will be followed up in Chapter 3.)

1.5 Show that the number of pairs X, Y of distinct subsets of an n-set S with $X \subset Y$ is $3^n - 2^n$.

1.6 A collection $\mathscr{B}$ of subsets of an n-set S is called a *cross-cut* if every subset of S is comparable with (i.e. contains or is contained in) at least one member of $\mathscr{B}$. Suppose that $\mathscr{B}$ is a minimal cross-cut (i.e. $\mathscr{B}$ is a cross-cut but no proper subset of $\mathscr{B}$ is a cross-cut). Show that $|\mathscr{B}| \leqslant \binom{n}{[n/2]}$.

1.7 Let $x_1, \ldots, x_n$ be real numbers, $|x_i| \geqslant 1$ for each i, and let I be any unit interval on the real line. Show that the number of linear combinations $\sum_{i=1}^{n} \varepsilon_i x_i$ with $\varepsilon_i = 0$ or 1 lying inside I is at most $\binom{n}{[n/2]}$. (Hint: First explain why we may assume that each $x_i > 0$. Then associate with each sum $\sum(2\varepsilon_i - 1)x_i$ the corresponding set of indices i for which $(2\varepsilon_i - 1)x_i > 0$.) (Erdös 1945)

1.8 Let $A_1, \ldots, A_m$, $B_1, \ldots, B_m$ be subsets of an n-set S such that

$A_i \subseteq B_i$ for each i and $A_i \nsubseteq B_j$ whenever $i \neq j$. Show that

$$\sum_{i=1}^{m} \frac{1}{\binom{n-|B_i - A_i|}{|A_i|}} \leqslant 1.$$

1.9 Let $\mathscr{A} = \{A_1, \ldots, A_m\}$ be a collection of subsets of an n-set S and let $a_i = |A_i|$. Call a subset B_i of A_i an *own-subset* of A_i if $B_i \nsubseteq A_j$ for all $j \neq i$. Prove that if every member of $\mathscr{A}$ has an own-subset then

$$\sum_{i=1}^{m} \frac{1}{\binom{n-a_i+b_i}{b_i}} \leqslant 1$$

where $b_i = |B_i|$.

(Tuza 1984)

1.10 Let $\mathscr{A} = \{A_1, \ldots, A_m\}$ and let $r(\mathscr{A}) = \min\{|B| : B \cap A_i \neq \varnothing$ for each $i\}$. Call $\mathscr{A}$ *critical* if $r(\mathscr{A} - \{A_i\}) < r(\mathscr{A})$ for all i. Show that if $\mathscr{A}$ is critical with $r(\mathscr{A}) = t+1$, then

$$\sum_{i=1}^{m} \frac{1}{\binom{|A_i|+t}{t}} \leqslant 1.$$

Deduce that, if $|A_i| = k$ for each i and $\mathscr{A}$ is critical with $r(\mathscr{A}) = t+1$, then $\mathscr{A}$ can have at most $\binom{k+t}{t}$ members.

(Tuza 1984)

1.11 $\mathscr{A} = \{A_1, \ldots, A_m\}$ is a *completely separating system* of subsets of $S = \{x_1, \ldots, x_n\}$ if for any two elements $x_i \neq x_j$ of S there exist sets A_k and A_h in $\mathscr{A}$ such that $x_i \in A_k$, $x_j \notin A_k$, $x_i \notin A_h$, $x_j \in A_h$. The dual $\mathscr{A}^*$ of $\mathscr{A}$ is the collection $\mathscr{A}^* = \{X_1, \ldots, X_n\}$ of subsets of $\{1, \ldots, m\}$ given by $X_i = \{k : x_i \in A_k\}$. Show that $\mathscr{A}$ is completely separating if and only if A^* is an antichain.

(Cai 1984)

1.12 (i) If $\mathscr{A}$ is an antichain of subsets of S, define the *blocker* $b(\mathscr{A})$ of $\mathscr{A}$ to be the collection of all minimal subsets of S which intersect every member of $\mathscr{A}$. Clearly $b(\mathscr{A})$ is an antichain. Show that $b(b(\mathscr{A})) = \mathscr{A}$.

(ii) Similarly define the *antiblocker* $c(\mathscr{A})$ of $\mathscr{A}$ to be the collection of all maximal subsets of S having at most one element in common with each set in $\mathscr{A}$. Give an example to show that the assertion that $c(c(\mathscr{A})) = \mathscr{A}$ is not always true.

1.13 An antichain $\mathscr{A}$ is a *Menger antichain* if the maximum number of disjoint sets in $\mathscr{A}$ is equal to the minimum size of a member of the blocker $b(\mathscr{A})$. Verify that

$$\{a, b, c\} \ \{c, d, e\} \ \{b, d, f\} \ \{a, e, f\} \ \{c, f\} \ \{b, e\} \ \{a, d\}$$

form a Menger antichain $\mathscr{A}$ but that $b(\mathscr{A})$ is not a Menger antichain.

1.14 Menger's theorem in graph theory asserts that if x and y are non-adjacent vertices of a graph then the maximum number of vertex-disjoint paths from x to y is equal to the minimum number of vertices whose removal from the graph separates x from y. Re-express this result in terms of Menger antichains. (For further details see, for example, Woodall (1978) or Seymour (1977).)

2 Normalized matchings and rank numbers

2.1 Sperner's proof

The aim of this chapter is to present Sperner's original proof of his theorem and to study a property of sets, called the normalized matching property, which arises in that proof. In doing so we shall discuss one of the basic theorems of combinatorics, the so-called marriage theorem, and we shall finish by having a look at the lattice of partitions of a set, about which some interesting conjectures have both risen and fallen.

The basic idea behind Sperner's proof is the following. Suppose that $\mathscr{A}$ is an antichain of subsets of an n-set S, and suppose that $\mathscr{A}$ has p_i members of size i, $0 \leq i \leq n$. If there is an $i < \frac{1}{2}(n-1)$ for which $p_i > 0$, we consider the smallest such i and show how to replace the p_i sets of size i by p_i sets of size $i+1$, in the process preserving the antichain property. By repeating this process we 'push up' all the sets in $\mathscr{A}$ of size $<\frac{1}{2}(n-1)$ replacing them by sets of size $n/2$ if n is even and size $\frac{1}{2}(n-1)$ if n is odd. We then consider sets of size greater than $n/2$ and push them down. In this way we finish up with an antichain $\mathscr{A}_1$ of the same size as $\mathscr{A}$ but with all sets having size $[n/2]$. The theorem will then follow immediately.

The entire process described above depends upon the following simple lemma about the *shade* and the *shadow* of a collection of k-sets.

Definition Let $\mathscr{B}$ be a collection of k-subsets of an n-set S, $k < n$. The collection

$$\nabla\mathscr{B} = \{D \subseteq S : |D| = k+1, B \subset D \text{ for some } B \in \mathscr{B}\}$$

is called the *shade* of $\mathscr{B}$. Thus $\nabla\mathscr{B}$ consists of all subsets of S which can be obtained by adding an element to a set in $\mathscr{B}$. Similarly the *shadow* $\Delta\mathscr{B}$ of $\mathscr{B}$ is defined to be

$$\Delta\mathscr{B} = \{D \subseteq S : |D| = k-1, D \subset B \text{ for some } B \in \mathscr{B}\}.$$

Thus the shadow of $\mathscr{B}$ consists of all subsets of S which can be obtained by removing an element from a set in $\mathscr{B}$.

Lemma 2.1.1 (Sperner) Let $\mathscr{B}$ be a collection of k-subsets of an n-set S. Then

$$|\nabla\mathscr{B}| \geqslant \frac{n-k}{k+1}|\mathscr{B}| \qquad \text{if } k<n$$

and

$$|\Delta\mathscr{B}| \geqslant \frac{k}{n-k+1}|\mathscr{B}| \quad \text{if } k>0.$$

Proof Consider the pairs (B, D) with $B \in \mathscr{B}$, $D \in \nabla\mathscr{B}$, $B \subset D$. We use a common combinatorial technique and count these pairs in two different ways. For each B there are $n-k$ elements of S not in B which can be added to B to obtain a set D; therefore the number of pairs (B, D) is $(n-k)|\mathscr{B}|$. However, each D is of size $k+1$ and so has $k+1$ subsets of size k (not all of which are, of course, necessarily in $\mathscr{B}$). Therefore the number of pairs (B, D) is at most $(k+1)|\nabla\mathscr{B}|$. Thus

$$(n-k)|\mathscr{B}| \leqslant (k+1)|\nabla\mathscr{B}|$$

as required.

The proof of the second part of the lemma is similar and is left as an exercise. □

Corollary 2.1.2 If $k \leqslant \frac{1}{2}(n-1)$, $|\nabla\mathscr{B}| \geqslant |\mathscr{B}|$. If $k \geqslant \frac{1}{2}(n+1)$, $|\Delta\mathscr{B}| \geqslant |\mathscr{B}|$. □

Proof of Sperner's theorem Suppose that the antichain $\mathscr{A}$ has p_i members of size i, and suppose that the smallest value of i for which $p_i>0$ is $i_0<\frac{1}{2}(n-1)$. By the corollary above we can find p_{i_0} sets of size i_0+1 containing them, none of which can be in $\mathscr{A}$. Therefore we can replace the p_{i_0} sets in $\mathscr{A}$ of size i_0 by an equal number of sets from their shade. We can repeat this process until an antichain is obtained with the same number of sets as $\mathscr{A}$ but with no member of size $\leqslant \frac{1}{2}(n-1)$. We then proceed in the same way as above, replacing all members of $\mathscr{A}$ of size $\geqslant \frac{1}{2}(n+1)$ by an equal number of sets of size $[n/2]$. The resulting antichain $\mathscr{A}_1$ consists entirely of sets of size $[n/2]$, so we have finally $|\mathscr{A}| = |\mathscr{A}_1| \leqslant \binom{n}{[n/2]}$. □

Note that the conclusion of Lemma 2.1.1 can be written in the form

$$\frac{|\nabla \mathscr{B}|}{\binom{n}{k+1}} \geqslant \frac{|\mathscr{B}|}{\binom{n}{k}}.$$

The lemma therefore asserts that the proportion of sets of size $k+1$ in $\nabla \mathscr{B}$ is at least as big as the proportion of sets of size k which are in $\mathscr{B}$. This property of sets is known as the *normalized matching property*. We shall see later that the analogous property holds for divisors of an integer, where inclusion is replaced by divisibility and size by the number of prime factors. It will also be shown that the normalized matching property is intimately related to the LYM property discussed in Chapter 1.

2.2 Systems of distinct representatives

The normalized matching property is reminiscent of a fundamental theorem in combinatorics due to König and P. Hall which is often known by its popular title of the 'marriage problem'. Given a collection of subsets $A_1, \ldots, A_r$ of an n-set S, we define a *system of distinct representatives* (s.d.r.) for $A_1, \ldots, A_r$ to be a system of distinct elements $a_1, \ldots, a_r$ of S such that $a_i \in A_i$ for each i. For example, the sets

$$\{1, 2, 3\} \ \{2, 4\} \ \{1, 4\} \ \{2, 5\}$$

possess an s.d.r. since we can select 3, 2, 1, 5 respectively to 'represent' the sets. However, the sets

$$\{1,2,3\} \ \{2,4\} \ \{1,3\} \ \{1,2\} \ \{2,3\}$$

cannot possess an s.d.r. since, for example, the four sets $\{1,2,3\}$, $\{1,3\}$, $\{1,2\}$, $\{2,3\}$ have only three distinct elements in their union, which is not enough to represent four sets. It is clear that if an s.d.r. exists then, for each k, any k of the sets must have at least k elements in their union. Perhaps surprisingly, this obviously necessary condition turns out to be sufficient.

Before stating the theorem, we remark that the title 'marriage problem' arises from the following interpretation. Imagine that r men are each asked to make a list of the ladies they would be willing to marry. Then (assuming that the ladies have no say in the matter) it is possible to marry each man to a lady on his list if and only if the lists possess an s.d.r.

Theorem 2.2.1 (König 1931, P. Hall 1935) The sets $A_1, \ldots, A_r$ possess an s.d.r. if and only if, for each $m \leq r$, the union of any m of the sets A_i contains at least m elements.

Proof (Halmos and Vaughan 1950) Since the condition is clearly necessary, we prove the 'if' part, and proceed by induction on r. The result is clearly true for $r = 1$, so we assume it to be true for $r = k \geq 1$ and prove it for $r = k + 1$. So let $A_1, \ldots, A_{k+1}$ be given, and put $S = \bigcup_{i=1}^{k+1} A_i$. We consider two cases.

(a) Suppose that any $r \leq k$ sets A_i contain at least $r + 1$ elements in their union. Choose any $a_1 \in A_1$ to represent A_1. Then any r of the remaining k sets contain in their union at least r elements other than a_1 and so by induction hypothesis $A_2, \ldots, A_{k+1}$ possess an s.d.r. which does not contain a_1. We therefore obtain an s.d.r. for all the A_i.

(b) Suppose that there is a value of $r \leq k$ for which there are r sets A_i with exactly r elements in their union. Suppose that these sets are labelled $A_1, \ldots, A_r$. By induction hypothesis, they possess an s.d.r., say $a_1 \in A_1, \ldots, a_r \in A_r$. Now consider the remaining sets $A_{r+1}, \ldots, A_{k+1}$. They will possess an s.d.r. of elements of $S - \{a_1, \ldots, a_r\}$ unless there exist h of them, for some h, which contain less than h elements of $S - \{a_1, \ldots, a_r\}$ in their union. However, in that case these h sets, along with $A_1, \ldots, A_r$, would constitute a collection of $r + h$ sets with less than $r + h$ elements in their union, contradicting the hypotheses of the theorem. So these remaining sets $A_{r+1}, \ldots, A_{k+1}$ do possess an s.d.r. from $S - \{a_1, \ldots, a_r\}$ which, along with $a_1, \ldots, a_r$, constitute an s.d.r for all the sets A_i. □

We shall make use of this theorem while studying the normalized matching property in the next section. For a quantitative version of the theorem, concerned with how many s.d.r.s exist for a given collection of sets, see Exercises 19 and 20 at the end of the chapter.

2.3 LYM inequalities and the normalized matching property

It turns out that the normalized matching property and the LYM property are closely related; this section will be devoted to a study of that relation, describing results which are largely due to Kleitman. To set the scene for the discussion we extend the context from that of subsets of a set to a general partially ordered set with a rank function.

Definition A *partially ordered set* (or *poset*) $P=(S, \leqslant)$ is a set S on which an order relation $\leqslant$ is defined, satisfying

(i) $x \leqslant x$ for all $x \in S$

(ii) if $x \leqslant y$ and $y \leqslant x$ then $x=y$

(iii) if $x \leqslant y$ and $y \leqslant z$ then $x \leqslant z$.

Examples are

(a) the set of subsets of a set under inclusion

(b) the set of divisors of a number under divisibility

(c) the set of partitions of a set under refinement.

We say that x and y are *comparable* if $x \leqslant y$ or $y \leqslant x$. If $x \leqslant y$ and $x \neq y$ we write $x<y$. If $x<y$ and there is no z such that $x<z<y$ we say that y *covers* x. If there is a unique element z in S such that $z \leqslant x$ for all $x \in S$, we call z the *zero* element of the poset and denote it by 0. Any x with no y such that $y<x$ is called a *minimal* element, and any x with no y such that $x<y$ is called a *maximal* element.

If $x_1<x_2<\ldots<x_n$ we say that $x_1, \ldots, x_n$ form a *chain*. A *saturated* chain is a chain $x_1<\ldots<x_n$ such that x_{i+1} covers x_i for each $i<n$. Some posets have the property that, for any $x<y$, all saturated chains from x to y have the same number of elements, this number just depending on x and y. In particular, all saturated chains from 0 to x will have the same size. If we define the *length* of a chain to be 1 less than its cardinality, we can then define the *rank* $r(x)$ of an element x to be the length of all saturated chains from 0 to x. For example, in the poset of subsets of an n-set S, the zero element is the empty set $\varnothing$ and the rank of each subset is just the cardinality of the set.

Associated with a poset P with a rank function r are the *rank numbers* or *Whitney numbers* $N_0, N_1, \ldots$. defined by

$$N_i = \text{number of elements of } P \text{ of rank } i.$$

For example, in the poset of subsets of an n-set, $N_i=\binom{n}{i}$.

Not the same as a saturated chain, a *maximal* chain in a ranked poset is a chain containing one element of each rank. It is clear that a maximal chain must be saturated, but not vice versa.

Finally among these definitions, we record that an *antichain* in a poset P is a set of elements of P, no two of which are comparable.

In all that follows, we assume that all posets under discussion are finite, i.e. have a finite number of elements.

Theorem 2.3.1 (Kleitman 1974) Let P be a poset with a rank function and with rank numbers N_i. Then the following three conditions are equivalent.

(1) *P has the LYM property*, i.e. if $\mathscr{A}$ is an antichain in P with p_i members of rank i, then

$$\sum_i \frac{p_i}{N_i} \leqslant 1.$$

(2) *P has the normalized matching property*, i.e. if $\mathscr{B}$ is a set of members of P of rank k and if $\nabla\mathscr{B}$, the shade of $\mathscr{B}$, is the set of all elements of P of rank $k+1$ which are comparable with at least one member of $\mathscr{B}$, then

$$\frac{|\nabla\mathscr{B}|}{N_{k+1}} \geqslant \frac{|\mathscr{B}|}{N_k}.$$

(3) *There is a regular covering of P by chains*, i.e. there exists a non-empty collection $\mathscr{C}$ of maximal chains in P such that, for each k, every element of rank k occurs in the same number of chains of $\mathscr{C}$.

Proof **(1)⇒(2)** Suppose that P has the LYM property, and let $\mathscr{B}$ be a set of elements of P of rank k. Let $\mathscr{D}$ denote the set of all elements of P of rank $k+1$, excluding the elements of $\nabla\mathscr{B}$. Then $\mathscr{B} \cup \mathscr{D}$ is an antichain, and the LYM inequality gives

$$\frac{|\mathscr{B}|}{N_k} + \frac{N_{k+1} - |\nabla\mathscr{B}|}{N_{k+1}} \leqslant 1,$$

i.e.

$$\frac{|\mathscr{B}|}{N_k} \leqslant \frac{|\nabla\mathscr{B}|}{N_{k+1}}.$$

(2)⇒(3) Suppose that P has the normalized matching property. Define a new poset P' by the following construction. For each j, take $\prod_{i \neq j} N_i$ copies of each element of P of rank j, and then say that in P' each copy of x is less than each copy of y precisely when $x < y$ in P. The poset P' has $\prod_i N_i$ elements of each rank.

Now consider any r elements of P' of rank k. They arise from at least $r/\prod_{i \neq k} N_i$ distinct elements of rank k in P, and by the normalized matching property of P these are comparable with at least

$$\frac{N_{k+1}}{N_k} \frac{r}{\prod_{i \neq k} N_i} = \frac{r}{\prod_{i \neq k+1} N_i}$$

elements of rank $k+1$ in P. These r elements of P' are therefore comparable with at least r elements of P' of rank $k+1$. It follows from the marriage theorem (Theorem 2.2.1) that the elements of P' of rank k can be paired with comparable distinct elements of rank

$k+1$. Repeating this pairing at each rank level and putting the matchings together, we obtain a collection of $\prod_i N_i$ maximal chains such that each element of P of rank k lies in exactly $\prod_{i\neq k} N_i$ chains.

(3) ⇒ (1) Let $\mathscr{C}$ be a collection of maximal chains forming a regular covering. Each element of rank k occurs in precisely $|\mathscr{C}|/N_k$ of the chains. If $\mathscr{A}$ is an antichain with p_k elements of rank k, then these p_k elements together make an appearance in $(p_k/N_k)\,|\mathscr{C}|$ chains. Considering each k in turn and noting that, since $\mathscr{A}$ is an antichain, the chains so obtained for different values of k must be distinct, we obtain

$$\sum_k \frac{p_k}{N_k}|\mathscr{C}| \leq |\mathscr{C}|$$

whence

$$\sum_k \frac{p_k}{N_k} \leq 1. \qquad \square$$

We now present a far-reaching consequence of this important theorem. One of the earliest extensions of Sperner's theorem was obtained by Erdös, who proved that if $\mathscr{A}$ is a collection of subsets A_i of an n-set S such that no $k+1$ of the sets form a chain of length k, then $|\mathscr{A}|$ cannot exceed the sum of the k largest binomial coefficients $\binom{n}{i}$. Clearly, the case $k=1$ reduces to Sperner's theorem. We shall now see that this is an immediate consequence of the following result of Kleitman, which exploits the existence of a regular covering of chains.

Theorem 2.3.2 (Kleitman 1974). Let P be a poset with rank function r and with the LYM property, and let $\mathscr{C}$ be any regular covering of P by chains. Then, if λ is a real-valued function defined on the elements of P, and R is any subset of P,

$$\sum_{x\in R} \frac{\lambda(x)}{N_{r(x)}} \leq \max_{C\in\mathscr{C}} \sum_{x\in C\cap R} \lambda(x).$$

Proof. For any chain $C\in\mathscr{C}$ consider $f(C)=\sum_{x\in C\cap R}\lambda(x)$. We have

$$\sum_{C\in\mathscr{C}} f(C) = \sum_{C\in\mathscr{C}}\sum_{x\in C\cap R} \lambda(x) = \sum_{x\in R}\lambda(x)\sum_{\substack{C\\ x\in C}} 1 = \sum_{x\in R}\lambda(x)\frac{|\mathscr{C}|}{N_{r(x)}}$$

since each of the $N_{r(x)}$ elements of P of rank $r(x)$ occurs in the same number of the $|\mathscr{C}|$ chains. Thus the *average* value of $f(C)$ taken over all chains C in $\mathscr{C}$ is $\sum_{x\in R}\{\lambda(x)/N_{r(x)}\}$. Thus the *maximum* value of $f(C)$ must be at least that value. $\square$

Taking $\lambda(x) = N_{r(x)}$ we immediately obtain the following corollary.

Corollary 2.3.3 Under the hypotheses of Theorem 2.3.2

$$|R| \leq \max_{C \in \mathscr{C}} \sum_{x \in C \cap R} N_{r(x)}. \qquad \square$$

Consider now the special case where P is the set of subsets of an n-set S and R is a collection of subsets of S no $k+1$ of which form a chain. Then, if $\mathscr{C}$ is any regular covering of P by chains and if $C \in \mathscr{C}$, then $C \cap R$ can contain at most k sets; thus $\sum_{x \in C \cap R} N_{r(x)}$ can be at most the sum of the k largest binomial coefficients. We have thus proved the following theorem.

Theorem 2.3.4 (Erdös 1945) If $\mathscr{A}$ is a collection of subsets of an n-set such that no $k+1$ members of $\mathscr{A}$ form a chain, then

$$|\mathscr{A}| \leq \sum_{i=0}^{k-1} \binom{n}{[(n+i)/2]}. \qquad \square$$

We close this section with an important general consequence of Corollary 2.3.3. If R is an antichain in the ranked poset P, then each $C \cap R$ can contain at most one element and so $|R|$ can be no greater than the largest rank number.

Definition A ranked poset P is said to have the *Sperner property* if the size of the largest antichain in P is the largest rank number.

We have proved the following corollary.

Corollary 2.3.5 If the poset P has any of the three equivalent properties described in Theorem 2.3.1, then P has the Sperner property. $\square$

2.4 Rank numbers: some examples

In the poset of subsets of an n-set, the rank numbers N_k are given by

$$N_k = \binom{n}{k}.$$

These rank numbers have the property that $\binom{n}{k-1} < \binom{n}{k}$ if $k \leq \frac{1}{2}n$

and $\binom{n}{k} > \binom{n}{k+1}$ if $k \geq \frac{1}{2}n$. Thus the rank numbers first increase with rank and then decrease. This property is known as unimodality.

Definition A sequence $N_0, N_1, \ldots, N_n$ is *unimodal* if there exists j such that

$$N_0 \leq N_1 \leq \ldots \leq N_j; \qquad N_j \geq \ldots \geq N_n.$$

The poset of subsets of an n-set $S = \{x_1, \ldots, x_n\}$ can be considered as the poset of divisors of a square-free number $m = p_1 \ldots p_n$ (where the p_i are distinct primes) under division, by means of the correspondence

$$A \subseteq S \leftrightarrow p_1^{h_1} \ldots p_n^{h_n}$$

where $h_i = 0$ if $x_i \notin A$ and $h_i = 1$ if $x_i \in A$. For example, the subset $\{x_2, x_4, x_5\}$ of $S = \{x_1, \ldots, x_5\}$ corresponds to the divisor $p_2p_4p_5$ of $m = p_1 \ldots p_5$. Under this interpretation the rank of a divisor of m is the number of prime factors of m.

When this system is generalized the divisors of $m = p_1^{k_1} \ldots p_n^{k_n}$, where the p_i are distinct primes and the k_i may exceed unity, are numbers of the form $p_1^{h_1} \ldots p_n^{h_n}$ where $0 \leq h_i \leq k_i$ for each i. Under divisibility the set of divisors of m form a poset with rank function r given by $r(p_1^{h_1} \ldots p_n^{h_n}) = h_1 + \ldots + h_n =$ total number of prime factors of m counted according to multiplicity.

Example $m = p^2q^3r$ where p,q,r are distinct primes. The divisors of rank 4 are pq^2r, p^2qr, p^2q^2, q^3r, pq^3. The rank numbers are

$$N_0 = 1, \quad N_1 = 3, \quad N_2 = 5, \quad N_3 = 6, \quad N_4 = 5, \quad N_5 = 3, \quad N_6 = 1.$$

Note that these form a unimodal sequence as in the square-free case. That this is true for the divisors of any number will become obvious in the next chapter.

Sometimes the divisors of a non-square-free number are described in terms of *multisets*. The idea here is that the divisor p^2q^2r of p^2q^3r can be thought of as a 'set' $\{p, p, q, q, r\}$ where each distinct element has a multiplicity, namely the number of times it occurs in the factorization. Corresponding to the number $m = p_1^{k_1} \ldots p_n^{k_n}$ is the multiset

$$\{\underbrace{p_1, p_1, \ldots, p_1}_{k_1}, \underbrace{p_2, \ldots, p_2}_{k_2}, \ldots, \underbrace{p_n, \ldots, p_n}_{k_n}\}.$$

Posets of multisets will be discussed more fully in the next two chapters. It will be found that the LYM property holds.

Partition Another familiar poset with a rank function is the poset Π_n of all partitions of a set $\{1, \ldots, n\}$ ordered by refinement. A *partition* of S is a disjoint union of non-empty sets whose union is S. Each partition of S consists of disjoint *blocks*. An order relation $\leqslant$ can be defined on the set of partitions of S by writing $\pi_1 \leqslant \pi_2$ if every block in partition π_1 is wholly contained in a block of π_2 (i.e. if every block of π_2 is a union of blocks of π_1), in which case we call π_1 a *refinement* of π_2.

Example $S = \{1, 2, 3, 4\}$. The partitions of S include

$$\pi_1 = \{1, 3\} \cup \{2\} \cup \{4\}$$
$$\pi_2 = \{1\} \cup \{2, 3, 4\}$$
$$\pi_3 = \{1\} \cup \{2\} \cup \{3, 4\}.$$

Here $\pi_3 \leqslant \pi_2$ but $\pi_1 \not\leqslant \pi_2$.

A rank function r can be defined on Π_n by $r(\pi) = n - |\pi|$, where $|\pi|$ is the number of blocks in π. Therefore in the above example $r(\pi_1) = 1$, $r(\pi_2) = 2$, $r(\pi_3) = 1$. The partition with rank 0 is $\{1\} \cup \{2\} \cup \cdots \cup \{n\}$. The reader should check that in this example the rank numbers are $N_0 = 1$, $N_1 = 6$, $N_2 = 7$, $N_3 = 1$.

The rank numbers associated with partitions are known as *Stirling numbers* of the second kind. We have

$$\begin{aligned} N_k &= \text{number of partitions of an } n\text{-set of rank } k \\ &= \text{number of partitions of an } n\text{-set into } n-k \text{ blocks} \\ &= S(n, n-k) \end{aligned}$$

where, in general, the Stirling number $S(n, m)$ is the number of partitions of an n-set into m blocks. We shall prove that, for fixed n, the numbers $S(n, m)$, $m = 0, \ldots, n-1$, form a unimodal sequence.

Theorem 2.4.1 (i) $S(0, 0) = 1$; $S(n, 0) = 0$ if $n \geqslant 1$
(ii) $S(n+1, k) = S(n, k-1) + kS(n, k)$

(iii) $S(n+1, k) = \sum_{j=k-1}^{n} \binom{n}{j} S(j, k-1).$

Proof (i) $S(0, 0) = 1$ can be taken as definition. $S(n, 0) = 0$ for $n \geqslant 1$ is obvious.

(ii) Consider the $S(n+1, k)$ partitions of $\{x_1, \ldots, x_{n+1}\}$ into k blocks. In any given such partition, either $\{x_{n+1}\}$ is a block or x_{n+1} is in a block of size $\geqslant 2$. The first possibility arises in $S(n, k-1)$ ways. However, a partition with x_{n+1} in a block of size $\geqslant 2$ can be obtained from a partition of $\{x_1, \ldots, x_n\}$ into k blocks by adding x_{n+1} to any of the k blocks; thus there are $kS(n, k)$ such possibilities. Thus $S(n+1, k) = S(n, k-1) + kS(n, k)$.

(iii) A partition of $\{x_1, \ldots, x_n, x_{n+1}\}$ into k blocks can be obtained by first choosing some j, $k-1 \leqslant j \leqslant n$, then choosing $n-j$ of the elements $x_1, \ldots, x_n$ to form a block of size $n-j+1$ with x_{n+1}, and then partitioning the remaining j elements into $k-1$ blocks. Thus

$$S(n+1, k) = \sum_{j=k-1}^{n} \binom{n}{n-j} S(j, k-1) = \sum_{j=k-1}^{n} \binom{n}{j} S(j, k-1).$$

(Note that the summation can equally well be written from $j=1$ to $j=n$.) □

Theorem 2.4.2 For each n the numbers $S(n, 0), S(n, 1), \ldots$ form a unimodal sequence. There is a number $k(n)$ such that

$$S(n, r-1) < S(n, r) \quad \text{if } r \leqslant k(n),$$
$$S(n, k(n)) \geqslant S(n, k(n)+1),$$
$$S(n, r) > S(n, r+1) \quad \text{if } r > k(n).$$

Further, $k(n) = k(n-1)$ or $k(n) = k(n-1) + 1$.

Proof We proceed by induction on n, assuming that the theorem is true for n and considering the case $n+1$.

First of all, suppose that $k \leqslant k(n)$. Then, by Theorem 2.4.1(ii),

$$\begin{aligned} &S(n+1, k) - S(n+1, k-1) \\ &\quad = S(n, k-1) + kS(n, k) - S(n, k-2) - (k-1)S(n, k-1) \\ &\quad = S(n, k-1) - S(n, k-2) + k(S(n, k) - S(n, k-1)) + S(n, k-1) \\ &\quad > 0 \text{ by the induction hypothesis.} \end{aligned}$$

Next suppose that $k \geqslant k(n) + 2$. Then, by Theorem 2.4.1(iii),

$$S(n+1, k) - S(n+1, k-1) = \sum_{j=1}^{n} \binom{n}{j} (S(j, k-1) - S(j, k-2)) < 0$$

since $k-2 \geqslant k(n) \geqslant k(j)$ for each $j \leqslant n$.

We have therefore proved that $S(n+1, k)$ increases strictly up to $k = k(n)$ and decreases strictly from $k = k(n)+1$. It follows that the numbers $S(n, k)$ are unimodal with peak at $k(n+1) = k(n)$ or $k(n)+1$. □

How large is $k(n)$? It was shown by Harper (1967) that $k(n) \sim n/\log n$. More recently, Canfield (1978*b*) and Jichang and Kleitman (1984) have shown that, for sufficiently large n, $k(n)$ lies around the root of the equation

$$(x + \tfrac{3}{2})\log x = n - 2.$$

We now turn our attention to the normalized matching property. Does Π_n possess it? It turns out that the answer is 'no' if $n \geqslant 20$, and a proof of this is outlined in the exercises. It was thought for many years, nevertheless, that the partition poset had the Sperner property, i.e. that the maximum size of an antichain in Π_n is equal to the largest Stirling number $S(n, k(n))$. This conjecture was publicised by Rota (1967) and much work was done on the problem. However, thanks to the work of Canfield (1978*a*) it is now known that the conjecture is false for sufficiently large n. Since Canfield's paper, Shearer (1979) and Jichang and Kleitman (1984) have reduced the estimate for the smallest value of n for which the conjecture fails. We finish this chapter by giving an outline of Shearer's argument.

First, though, we introduce some notation. A partition of $\{1, \ldots, n\}$ into k_1 blocks of size 1, k_2 blocks of size 2, ..., k_r blocks of size r, where necessarily $\sum_i ik_i = n$, is called a partition of *type*

$$1^{k_1}2^{k_2} \ldots r^{k_r}.$$

Consider now the set $\mathscr{C}$ of all partitions of $\{1, \ldots, n\}$ of type $m^{h_1}(2m)^{h_2}(3m)^{h_3}$ into $k(n)+1$ blocks where

$$h_1 + h_2 + h_3 = k(n) + 1$$

and

$$mh_1 + 2mh_2 + 3mh_3 = n.$$

Choose the h_i as follows:

$$h_1 = \frac{1}{3}\left\{5(k(n)+1) - \frac{2n}{m} - \theta\right\}$$

$$h_2 = \frac{1}{3}\left\{\frac{n}{m} - (k(n)+1) + 2\theta\right\}$$

$$h_3 = \frac{1}{3}\left\{\frac{n}{m} - (k(n)+1) - \theta\right\},$$

where θ is chosen as 0, 1 or -1 so as to make each h_i an integer. The value of m remains to be chosen.

Now if we have a partition into $k(n)$ blocks which is related under the ordering $\leqslant$ to a partition in $\mathscr{C}$, it must be of one of the following types:

(i) $m^{h_1-2}(2m)^{h_2+1}(3m)^{h_3}$
(ii) $m^{h_1-1}(2m)^{h_2-1}(3m)^{h_3+1}$
(iii) $m^{h_1-1}(2m)^{h_2}(3m)^{h_3-1}(4m)^1$
(iv) $m^{h_1}(2m)^{h_2-2}(3m)^{h_3}(4m)^1$
(v) $m^{h_1}(2m)^{h_2-1}(3m)^{h_3-1}(5m)^1$
(vi) $m^{h_1}(2m)^{h_2}(3m)^{h_3-2}(6m)^1$.

If there are altogether less than $|\mathscr{C}|$ partitions of these types, then the partitions in $\mathscr{C}$ and the remaining partitions with $k(n)$ blocks will form an antichain of size greater than $S(n, k(n))$ as required. So we have to show that, for sufficiently large n, there are less than $|\mathscr{C}|$ partitions of types (i)–(vi).

Now the number of partitions of type (i) is, by Exercise 2.15,

$$\frac{n!}{(m!)^{h_1-2}((2m)!)^{h_2+1}((3m)!)^{h_3}(h_1-2)!(h_2+1)!h_3!}$$

and the number of partitions in $\mathscr{C}$ is

$$\frac{n!}{(m!)^{h_1}((2m)!)^{h_2}((3m)!)^{h_3}h_1!h_2!h_3!}$$

so that the ratio of these two numbers is

$$\frac{h_1(h_1-1)(m!)^2}{(h_2+1)((2m)!)}=\frac{h_1(h_1-1)}{(h_2+1)\binom{2m}{m}}.$$

If we carry out a similar calculation for each of the other types (ii)–(vi) we find that we have to prove that

$$\frac{h_1(h_1-1)}{(h_2+1)\binom{2m}{m}}+\frac{h_1h_2}{(h_3+1)\binom{3m}{m}}+\frac{h_1h_3}{\binom{4m}{m}}$$
$$+\frac{h_2(h_2-1)}{\binom{4m}{2m}}+\frac{h_2h_3}{\binom{5m}{m}}+\frac{h_3(h_3-1)}{\binom{6m}{3m}}<1. \qquad (2.1)$$

Now take $m=[n/\alpha k(n)]$. Since $k(n)\approx n/\log n$ we have $m\approx \log n/\alpha$. We choose α a little greater than unity to ensure that each h_i is positive. We now have

$$h_1\approx\tfrac{1}{3}(5-2\alpha)\frac{n}{\log n} \qquad h_2\approx h_3\approx\tfrac{1}{3}(\alpha-1)\frac{n}{\log n}.$$

We can now use Stirling's approximation $n! \approx (2\pi n)^{\frac{1}{2}} n^n e^{-n}$ to show that each of the six terms on the left-hand side of (2.1) tends to zero as $n \to \infty$. Their sum is therefore eventually less than unity as required. Detailed numerical work shows that if $\alpha = 1.06$ then it is sufficient to take $n > 4 \times 10^9$. (Jichang and Kleitman (1984) reduced this estimate to 3.4×10^6.)

Exercises 2

2.1 Complete the proof of Lemma 2.1.1.

2.2 Let $\mathscr{A}$ be an antichain of subsets of an n-set S, and suppose that $\mathscr{A}$ contains $\alpha_i \binom{n}{i}$ sets of size i, $1 \leqslant i \leqslant n$. Let $\mathscr{B} = \{B \subseteq S : A \subseteq B$ for some $A \in \mathscr{A}\}$, and suppose that $\mathscr{B}$ has $\beta_i \binom{n}{i}$ sets of size i. Show that $\beta_{i+1} \geqslant \beta_i + \alpha_{i+1}$ for each $i < n$.

2.3 Let $\mathscr{A}$ be a collection of subsets of an n-set S, such that $A \in \mathscr{A}$, $A \subset B \Rightarrow B \in \mathscr{A}$. Show that the average size of the sets in $\mathscr{A}$ is at least $\frac{1}{2}n$.

2.4 Show that if $k < \frac{1}{2}n$ then all the k-subsets of an n-set S can be paired with distinct $(k+1)$-subsets containing them.

2.5 Let $|S| = mn$, $S = A_1 \cup \ldots \cup A_n = B_1 \cup \ldots \cup B_n$ where $|A_i| = |B_i| = m$ for each i. Show that there exists a permutation α of $1, \ldots, n$ such that $A_i \cap B_{\alpha(i)} \neq \varnothing$ for each i. Deduce that S can be split into m disjoint sets of n elements each of which is an s.d.r. for the As and for the Bs.

2.6 Show that if P is a *regular* poset in the sense that, for each k, each element of rank k covers the same number of elements of rank $k-1$ and is covered by the same number of elements of rank $k+1$, then P has the normalized matching property. (Baker 1969)

2.7 Obtain the rank numbers for the poset of divisors of 60.

2.8 Prove that, if $\mathscr{A}$ is a collection of subsets of an n-set such that there are no sets $A, B \in \mathscr{A}$ with $B \subset A$ and $|A - B| < k$, then $|\mathscr{A}|$ is at most the largest sum of the form $\Sigma_i \binom{n}{a + ki}$. (Katona 1972*a*)

2.9 Prove that, if $\mathscr{A}$ is a collection of subsets of an n-set such that

$A_i, A_j \in \mathscr{A}, A_i \subset A_j \Rightarrow |A_j - A_i| < r$, then $|\mathscr{A}|$ is at most the sum of the r largest binomial coefficients $\binom{n}{i}$. Give two different proofs: (a) imitate the proof of Sperner's theorem given in this chapter; (b) use Theorem 2.3.2 or its corollary. (Erdös 1945)

2.10 Let $\mathscr{A}_1, \ldots, \mathscr{A}_t$ be t collections of subsets of $S = \{1, \ldots, n\}$ such that if $i \neq j$ then there is no set in $\mathscr{A}_i$ which is a proper subset of a set in $\mathscr{A}_j$. Use Theorem 2.3.2 to prove that

$$|\mathscr{A}_1| + \ldots + |\mathscr{A}_t| \leqslant \max\left(2^n, t\binom{n}{[n/2]}\right).$$

(Daykin, Frankl, Greene and Hilton 1981)

2.11 Verify, by listing all possible partitions, that $S(5, 4) = 10$ and $S(5, 3) = 25$.

2.12 Prove that (a) $S(n, 2) = 2^{n-1} - 1$ and (b) $S(n, 3) = \frac{1}{2}(3^{n-1} - 2^n + 1)$.

2.13 Show that $k^{n-k} \leqslant S(n, k) \leqslant \binom{n-1}{k-1} k^{n-k}$ whenever $k \leqslant n$.

2.14 Prove that Π_n does *not* have the normalised matching property as follows. Consider n even. Let X be the set of all partitions of the n-set S into blocks of size $\frac{1}{2}n$; then $|X| = \frac{1}{2}\binom{n}{n/2}$. Verify that each partition in X covers $2S(n/2, 2)$ partitions into three blocks, covering $\binom{n}{n/2} S(n/2, 2)$ such partitions altogether. The partitions of X and the remaining partitions into three blocks form an antichain, so if Π_n has the LYM property, then the LYM inequality implies $S(n, 3) \leqslant 2S(n, 2)S(n/2, 2)$. Use Exercise 2.12 to show this false if $n \geqslant 20$. (Spencer 1974)

2.15 Prove that the number of partitions of $\{1, \ldots, n\}$ of type $1^{k_1} 2^{k_2} \ldots r^{k_r}$ is

$$\frac{n!}{\prod_{m=1}^{r} (m!)^{k_m} (k_m)!}$$

2.16 This exercise is to construct a ranked poset which does not have the Sperner property. A graph G has n vertices, a, b, $c_1, \ldots, c_{n-2}$; a and b are adjacent, and each of a and b is adjacent to each of the c_i, but no two c_i are adjacent. Let P be the set of partitions of the vertex set V into blocks which are such that the subgraphs induced by the vertices in the blocks are connected. Then P is a poset under refinement. If π is a partition define $r(\pi)$ to be the number of blocks in π; then r is a rank function. Verify that the elements of rank k in P are as follows: (i) partitions in which $k-1$ of the c_i are blocks on their own, the other $(n-k-1)$ c_i being with a and b in another block; (ii) partitions in which $k-2$ of the c_i, form blocks on their own, a is in a block with j c_is and b is in a block with $n-k-j$ c_i's for some j. Verify also that there are $\binom{n-2}{k-1}$ partitions of type (i) and

$$\binom{n-2}{k-2}\sum_{j=0}^{n-k}\binom{n-k}{j}=2^{n-k}\binom{n-2}{k-2}$$

partitions of type (ii). Next note that we can obtain an antichain by taking all the partitions of type (i) for some k_1 and all those of type (ii) for some $k_2<k_1$. Now the maximum value of $\binom{n-2}{k-1}$ occurs when k is approximately $\frac{1}{2}n$, and the maximum value of $2^{n-k}\binom{n-2}{k-2}$ occurs approximately when $k=\frac{1}{3}n$. Choosing k_1 and k_2 appropriately we obtain an antichain of size approximately

$$\binom{n-2}{\frac{1}{2}(n-2)}+2^{2n/3}\binom{n-2}{\frac{1}{3}n-2}$$

However,

$$N_k=\binom{n-2}{k-1}+2^{n-k}\binom{n-2}{k-2}$$

so that the maximum N_k is attained when k is approximately $\frac{1}{3}n$, the value of N_k being smaller than the size of the antichain. In particular, choose $n=12$; in this case $\max N_k=N_5=15\,570$ whereas the antichain obtained by choosing $k_1=6$ and $k_2=5$ has 15 612 members. (Dilworth and Greene 1971)

2.17 Show that if $|X|=n$ and $Y=\{1,\ldots,k\}$ then the number of surjections ('onto' maps) $f:X\to Y$ is $k!S(n,k)$.

2.18 Use Exercise 2.17 to show that if $[k]_i=k(k-1)(k-2)\ldots(k-i+1)$ then $k^n=\sum_i S(n,i)[k]_i$. Illustrate with $n=3$.

2.19 Suppose that $A_1,\ldots,A_r$ possess an s.d.r.. Show that there exists j such that, for any $a \in A_j$, there is an s.d.r. in which A_j is represented by a.

2.20 Show that, if $A_1, \ldots, A_r$ possess an s.d.r. and $|A_i| \geq k$ for each i, then there are at least $k!$ different s.d.r.'s if $r \geq k$, and at least $k!/(k-r)!$ s.d.r.s if $k > r$. (M. Hall 1948)

3 Symmetric chains

3.1 Symmetric chain decompositions

The problem of generalizing Sperner's theorem on subsets of a set (or divisors of a square-free number) to multisets (or non-square-free numbers) was posed in 1949 as a prize problem by the Dutch Wiskundig Genootschap. The solution, due to de Bruijn, Tengbergen, and Kruyswijk (1951), is not only very simple but exhibits a structural aspect of the poset of divisors of a number which turns out to be of great usefulness. The aim of this chapter is to describe this structural property and to give a number of varied applications.

The basic idea in their solution was to decompose the poset into chains with certain symmetry properties. If P is a ranked poset with rank function r, we say that the elements $x_1, \ldots, x_h$ of P form a *symmetric chain* if

(i) x_{i+1} covers x_i for each $i<h$
(ii) $r(x_1)+r(x_h)=r(P)$

where $r(P)$ is the largest rank in P. Note that (i) requires that the chains are saturated and (ii) requires that each chain finishes as far above the middle rank as it started below, i.e. it is 'symmetrically positioned' about the middle rank. In the case of subsets of an n-set S, $A_1, \ldots, A_h$ form a symmetric chain if

(i) $A_i \subset A_{i+1}$ and $|A_{i+1}|=1+|A_i|$ for each $i<h$
(ii) $|A_1|+|A_h|=n$.

In the case of divisors of a number m, the divisors $d_1, \ldots, d_h$ form a symmetric chain if

(i) d_i divides d_{i+1} and d_{i+1}/d_i is prime for each $i<h$
(ii) $r(d_1)+r(d_h)=r(m)$,

where $r(d_i)$ is the number of prime factors of d_i counted according to multiplicity. de Bruijn *et al.* showed that the set of divisors (subsets) of a number m(set S) can be partitioned into symmetric chains. Here are two examples to illustrate the idea.

Example 3.1.1 Subsets of $\{1,2,3,4\}$.

$$
\begin{array}{l}
\qquad\qquad\qquad\quad \{3,4\} \\
\qquad\qquad\qquad\quad \{2,4\} \\
\qquad\qquad \{4\} \subset \{1,4\} \subset \{1,2,4\} \\
\qquad\qquad \{3\} \subset \{1,3\} \subset \{1,3,4\} \\
\qquad\qquad \{2\} \subset \{2,3\} \subset \{2,3,4\} \\
\varnothing \subset \{1\} \subset \{1,2\} \subset \{1,2,3\} \subset \{1,2,3,4\}.
\end{array}
$$

Example 3.1.2 Divisors of $180 = 2^2 3^2 5$.

$$
\begin{array}{cccccc}
 & & 2 \cdot 5 & 2^2 \cdot 5 & & \\
 & & 3^2 & 3^2 \cdot 5 & & \\
 & 5 & 3 \cdot 5 & 2 \cdot 3 \cdot 5 & 2^2 \cdot 3 \cdot 5 & \\
 & 3 & 2 \cdot 3 & 2 \cdot 3^2 & 2 \cdot 3^2 \cdot 5 & \\
1 & 2 & 2^2 & 2^2 \cdot 3 & 2^2 \cdot 3^2 & 2^2 \cdot 3^2 \cdot 5
\end{array}
$$

In terms of multisets, Example 3.1.2 partitions the set of subsets (or more strictly submultisets) of the multiset $\{2, 2, 3, 3, 5\}$ as follows:

$$
\begin{array}{l}
\qquad\qquad\quad \{2,5\} \subset \{2,2,5\} \\
\qquad\qquad\quad \{3,3\} \subset \{3,3,5\} \\
\qquad \{5\} \subset \{3,5\} \subset \{2,3,5\} \subset \{2,2,3,5\} \\
\qquad \{3\} \subset \{2,3\} \subset \{2,3,3\} \subset \{2,3,3,5\} \\
\varnothing \subset \{2\} \subset \{2,2\} \subset \{2,2,3\} \subset \{2,2,3,3\} \subset \{2,2,3,3,5\}.
\end{array}
$$

Theorem 3.1.1 (de Bruijn *et al.* 1951) The set of divisors of a number (or subsets of a set) can be expressed as a disjoint union of symmetric chains.

Proof We shall use the notation of divisors of m, and proceed by induction on the number n of distinct primes dividing m. If $m = p^\alpha$ then $1, p, p^2, \ldots, p^\alpha$ is a symmetric chain containing all the divisors of m; thus the theorem is true for $n = 1$. Suppose now that the theorem is true for $n = k$, and consider a number with $k + 1$ distinct prime factors. Such a number can be written as $m = m_1 p^\alpha$ where m_1 has k distinct prime factors and where $p \nmid m_1$. We show how to construct symmetric chains for m from symmetric chains for m_1. Let $d_1, \ldots, d_h$ be one of the symmetric chains in the chain decomposition of m_1, and consider all the divisors of m of the form $d_i p^\beta$, $1 \leqslant i \leqslant h$, $0 \leqslant \beta \leqslant \alpha$. We put these divisors of m into symmetric chains by the following procedure. Write them in a rectangular array as shown, and then peel off chains as indicated:

$$\begin{array}{lllll}
d_1 \; d_2 & \dots \; d_{h-2} & d_{h-1} & d_h & \\
\hline
d_1p \; d_2p & \dots \; d_{h-2}p & d_{h-1}p & \vert \; d_hp \\
\hline
d_1p^2 \; d_2p^2 & \dots \; d_{h-2}p^2 & \vert \; d_{h-1}p^2 & \vert \; d_hp^2 \\
\hline
\vdots & \vert \; \vdots & \vert \; \vdots & \vert \; \vdots \\
d_1p^{\alpha-1} \; d_2p^{\alpha-1} & \vert \; d_{h-2}p^{\alpha-1} & \vert \; d_{h-1}p^{\alpha-1} & \vert \; d_hp^{\alpha-1} \\
d_1p^{\alpha} \; d_2p^{\alpha} & \vert \; d_{h-2}p^{\alpha} & \vert \; d_{h-1}p^{\alpha} & \vert \; d_hp^{\alpha}
\end{array}$$

The outer layer gives the first chain

$$d_1, d_2, \dots, d_{h-1}, d_h, d_hp, d_hp^2, \dots, d_hp^{\alpha}$$

which clearly satisfies condition (i) and is symmetric since

$$r(d_1) + r(d_hp^{\alpha}) = r(d_1) + r(d_h) + \alpha = r(m_1) + r(p^{\alpha}) = r(m).$$

Similarly the second layer gives a symmetric chain, and so on. Finally, every divisor of m will be obtained in this way from some symmetric chain for m_1. □

Example 3.1.3 We can use Example 3.1.1 and the construction given in the proof of the theorem to obtain a symmetric chain decomposition of the set of subsets of $\{1, 2, 3, 4, 5\}$. For example, starting with the chain $\{3\} \subset \{1, 3\} \subset \{1, 3, 4\}$ we consider the rectangular array

$$\begin{array}{ccc}
\{3\} \; \{1, 3\} & & \{1, 3, 4\} \\
\hline
\{3, 5\} \; \{1, 3, 5\} & \vert & \{1, 3, 4, 5\}
\end{array}$$

and obtain the chains

$$\{3\} \subset \{1, 3\} \subset \{1, 3, 4\} \subset \{1, 3, 4, 5\} \quad \text{and} \quad \{3, 5\} \subset \{1, 3, 5\}.$$

Having introduced the idea of symmetric chains, we can now give our third proof of Sperner's theorem.

Proof of Theorem 1.2.1 Partition the poset of subsets of the n-set S into symmetric chains. Now the number of chains must be $\binom{n}{[n/2]}$ since each chain must contain exactly one subset of size

$[n/2]$. However, if $\mathscr{A}$ is an antichain then $\mathscr{A}$ can contain at most one member of each chain; so $|\mathscr{A}| \leqslant \binom{n}{[n/2]}$. □

3.1.1 Rank numbers for multisets

Consider a symmetric chain decomposition of $m = p_1^{k_1} \ldots p_r^{k_r}$ and put $K = k_1 + \ldots + k_r$. Since each chain starting at rank h ends at rank $K - h$, we must have $N_h = N_{K-h}$. Also, if $h < \frac{1}{2}K$ then all chains containing a divisor of rank h must contain a divisor of rank $h+1$, and so we must have $N_{h+1} \geqslant N_h$. Thus we have the following theorem.

Theorem 3.1.2 If $N_0, \ldots, N_K$ are the rank numbers for the poset of divisors of $m = p_1^{k_1} \ldots p_r^{k_r}$, where $K = \sum_i k_i$, then

$$\text{(i) } N_0 \leqslant N_1 \leqslant \ldots \leqslant N_{[K/2]}; \qquad N_{[K/2]} \geqslant \ldots \geqslant N_K$$

and

$$\text{(ii) } N_h = N_{K-h} \quad \text{for each } h.$$ □

In the next chapter we shall find out where we have strict inequalities in (i). Meanwhile we note that the N_i form a unimodal sequence and that the largest value taken by any N_i is $N_{[K/2]}$. The proof of Theorem 1.2.1 given above therefore generalizes directly to give the following theorem.

Theorem 3.1.3 If $\mathscr{A}$ is an antichain in the poset of divisors of $m = p_1^{k_1} \ldots p_r^{k_r}$, then $|\mathscr{A}| \leqslant N_{[K/2]}$ where $K = \sum_i k_i$. □

Thus the poset of divisors of m possesses the Sperner property.

3.2 Dilworth's theorem

Before looking further at symmetric chains it is of interest to compare Theorem 3.1.1 with the theorem of Dilworth concerning chains in a general poset. It is clear that if a poset can be expressed as a union of w chains then any antichain in P can have at most w members, one from each chain. Thus

$$\text{maximum size of an antichain in } P \leqslant c(P) \tag{3.1}$$

where $c(P)$ denotes the minimum number of chains into which P can be decomposed. Dilworth's theorem asserts that there is in fact equality; $c(P)$ *is* the largest size of an antichain in P. If there is a symmetric chain decomposition, then we acquire some extra information, namely that $c(P)$ is the largest rank number, but we emphasise that Dilworth's theorem holds in any poset, even one without any rank function.

There are now many different proofs of Dilworth's theorem in the literature. The proof we now present is one of the simplest, and is due to Perles (1963).

Theorem 3.2.1 (Dilworth 1950) In any poset P, the maximum size of an antichain is equal to the minimum number of chains in a chain decomposition of P.

Proof (Perles 1963) In view of (3.1), we have to show that the minimum number of chains in a chain decomposition of P is less than or equal to the size of a maximum antichain. We use induction, and assume that the required result is true for all posets Q with $|Q|<|P|$. Let m be the maximum size of an antichain in P. We now consider two cases.

a(i) Suppose first that there is an antichain of size m which is neither the set of all maximal elements nor the set of all minimal elements of P. Let W be such an antichain, and define

$$P^+ = \{x \in P : x \geqslant y \text{ for some } y \in W\}$$
$$P^- = \{x \in P : x \leqslant y \text{ for some } y \in W\}.$$

Since there is at least one minimal element not in W, $P^+ \neq P$. Similarly $P^- \neq P$. Thus $|P^+|<|P|$ and $|P^-|<|P|$, so we shall be able to apply the induction hypothesis to them. Observe that $P^+ \cup P^- = P$, for if there were an element of P not in $P^+ \cup P^-$ then we could add it to W to obtain a larger antichain. Also, $P^+ \cap P^- = W$. Now, as already mentioned, P^+ and P^- are, by induction, the union of m chains, say $P^+ = \bigcup_{i=1}^m C_i$ and $P^- = \bigcup_{i=1}^m D_i$. The elements of W, being the minimal elements of P^+ and the maximal elements of P^-, are the minimal elements of the chains C_i and the maximal elements of the chains D_i. We can therefore glue the chains C_i and D_i together in pairs to form m chains which form a chain decomposition of P.

(ii) Next suppose that there are at most two maximum-sized antichains, one or both of the set of all maximal elements and the set of all minimal elements. Choose a maximal element a and a minimal element b with $b \leqslant a$. Then $P - \{a, b\}$ contains an antichain of size $m-1$ but none of size m, so that by the induction hypothesis it can be decomposed into $m-1$ chains. These chains, together with $\{a, b\}$, give a decomposition of P into m chains. □

3.2.1 The marriage problem

Dilworth's theorem will be considered further in Chapter 13. Meanwhile we close this section by showing how it can be used to give an alternative proof of the marriage theorem (Theorem 2.2.1).

Proof Let $A_1, \ldots, A_m$ be $m \leq n$ subsets of $S = \{x_1, \ldots, x_n\}$ such that, for each $k \leq m$, any k sets A_i contain at least k elements in their union. Define a poset as follows: take objects $u_1, \ldots, u_n$ and $v_1, \ldots, v_m$ and define an order relation by

$$u_i \leq v_j \quad \text{if and only if } x_i \in A_j$$
$$u_i \nleq u_j \quad \text{if } i \neq j$$
$$v_i \nleq v_j \quad \text{if } i \neq j.$$

It can be seen that u_i is really representing x_i and v_i is representing A_i. Now $\{u_1, \ldots, u_n\}$ is an antichain in P, and we now show that it is a maximum-sized antichain. Suppose, after a suitable relabelling if necessary, that $\{u_1, \ldots, u_r, v_1, \ldots, v_s\}$ is an antichain. Then none of $x_1, \ldots, x_r$ is in any of $A_1, \ldots, A_s$, so the only possible elements of $A_1, \ldots, A_s$ are $x_{r+1}, \ldots, x_n$. Thus the s sets $A_1, \ldots, A_s$ have $\leq n-r$ elements in their union. However, they have $\geq s$ elements in their union, so we must have $s \leq n-r$, i.e. $r+s \leq n$. Therefore n is indeed the maximum size of an antichain. It now follows from Dilworth's theorem that P can be decomposed into n chains. Each chain must contain exactly one u_i. Since the v_j are all in different chains, there corresponds to each v_j a u_j in the same chain with $u_j \leq v_j$; thus to each A_j there corresponds an x_j with $x_j \in A_j$. □

3.3 Symmetric chains for sets

Having looked at symmetric chains in a more general setting, we are now going to return to the particular case of subsets of a set and examine the structure of the chains more closely. Suppose we wish to find a symmetric chain decomposition of the set of subsets of a given set. The construction given in the proof of Theorem 3.1.1 is a step-by-step construction, altering the chains at each step. This makes for a tedious construction if the given set is large. Is there a neater way of writing down the chains without having to find the chains for all the smaller sets in the process? A positive answer to this question has been given by Leeb (pers. comm.) and, independently, by Greene and Kleitman (1976*b*).

Corresponding to each subset A of $S = \{x_1, \ldots, x_n\}$ we associate a sequence of n parentheses. If $x_i \in A$ we put a right parenthesis in the ith place; if $x_i \notin A$ we put a left parenthesis in the ith place. For example, if $S = \{1, \ldots, 9\}$ and $A = \{2, 3, 4, 7, 8\}$ we represent A by

$$(\;) \;) \;) \; (\; (\;) \;) \; (\; .$$

Having done this we can now 'close' the brackets as far as possible. In the example given, we obtain

$$\underbrace{(\;)} \;) \;) \; \underbrace{(\; \underbrace{(\;)} \;)} \; (\; .$$

Elements of S which correspond to right parentheses which are paired in the closing process are called the *basic elements* of A. In the example 2, 7 and 8 are the basic elements of A. The unpaired right parentheses represent non-basic elements of A. Note that all unpaired left parentheses must occur to the right of all the unpaired right parentheses, for otherwise further closing of brackets would be possible. These unpaired left parentheses represent unpaired elements of the complement A'.

Now define A^- to be the set of basic elements of A, and A^+ to be the union of A with the set of unpaired elements of A'. Then

$$\begin{aligned} |A^+| &= |A| + |A'| - \text{number of paired elements of } A' \\ &= |S| - \text{number of paired elements of } A \\ &= |S| - |A^-| \end{aligned}$$

so that $|A^-| + |A^+| = |S|$.

Now form a chain containing A by starting with A^-, the set of basic elements of A, and adding one by one the non-basic elements of A in order, and then adding one by one the unpaired elements of A' in order. The chain therefore starts with A^-, contains A, and ends with A^+. Since $|A^-| + |A^+| = |S|$, the chain is a *symmetric* chain containing A. In the example above, the basic elements of A are 2, 7, 8, and the elements 3, 4, 9 are added in turn to give the chain

$$\{2, 7, 8\} \subset \{2, 3, 7, 8\} \subset \{2, 3, 4, 7, 8\} \subset \{2, 3, 4, 7, 8, 9\}. \quad (3.2)$$

The reader could check that this is one of the chains which the proof of Theorem 3.1.1 would construct!

In terms of parentheses, the basic elements determine all the closed parentheses uniquely and leave all the unpaired parentheses to be fitted in. The chains are constructed by first taking all the unpaired parentheses as left ones (giving A^-), then changing the first of them to a right parenthesis, then the second, and so on. The chain (3.2) is, in terms of parentheses,

```
( ) ( ( ( ( ) ) (
( ) ) ( ( ( ) ) (
( ) ) ) ( ( ) ) (
( ) ) ) ( ( ) ) ) .
```

Observe how the closed parts remain unchanged while the remaining parts change step by step from left to right parentheses.

If two subsets of S have the same basic elements, they will have the same closed parts and hence will give rise to (and be contained in) the same chain. The set of all chains obtained in this way forms a symmetric chain decomposition of the set of subsets of S; every set lies in a chain, and the chains are disjoint since they are each determined by basic elements and each subset has a unique set of basic elements.

We can now see that the chains formed in this way are precisely those obtained in Theorem 3.1.1. For small values of n we could find the chains explicitly and verify this; therefore we proceed with an induction argument. Suppose that the chains of Theorem 3.1.1 for an n-set are the same as those of the parenthesis construction, and consider how the chains for an $(n+1)$-set are formed. The first new chain in the de Bruijn construction is obtained by extending an existing chain by adjoining x_{n+1} to its final set:

$$\begin{array}{cccc} A_1 & A_2 & \cdots & A_h \\ \hline A_1 \cup \{x_{n+1}\} & A_2 \cup \{x_{n+1}\} & \cdots & \big| \; A_h \cup \{x_{n+1}\} \end{array}$$

This corresponds to appending '(' to each string of parentheses representing the sets $A_1, \ldots, A_h$ and then changing it to ')' to represent $A_h \cup \{x_{n+1}\}$. The resulting chain is precisely a chain of subsets of $\{x_1, \ldots, x_{n+1}\}$ with the same basic elements as those of the original chain. The second new chain formed from $A_1 \subset \ldots \subset A_h$ is obtained by appending x_{n+1} to each of $A_1, \ldots, A_{h-1}$, i.e. appending ')' to the strings of parentheses representing $A_1, \ldots, A_{h-1}$. The fact that A_h is omitted corresponds to making the last non-basic element, say x_j, of A_h no longer available. This is achieved by forming a closed pair of brackets, '(' at the jth place and ')' at the $(n+1)$th place;

therefore the resulting chain consists of all subsets of $\{x_1, \ldots, x_{n+1}\}$ whose basic elements are precisely those of the original chain together with x_{n+1}. Thus again the new chain consists of subsets with the same basic elements.

The chains possess some elementary properties which are worth recording.

Theorem 3.3.1 Let $A_1 \subset \ldots \subset A_r$ be one of the symmetric chains of subsets of the n-set S constructed as above.

(i) If $A_{i+1} = A_i \cup \{x_{m_i}\}$, then $m_i < m_{i+1}$ for each i, and the m_i are alternatively odd and even.

(ii) The number of basic elements in each of the A_i is $\frac{1}{2}(n+1-r)$.

(iii) For each i, there is a set B_i such that $A_{i-1} \subset B_i \subset A_{i+1}$ and with B_i in a *shorter* chain.

Proof (i) That $m_i < m_{i+1}$ is immediate from the construction. Note also that any block of closed parentheses must contain an even number of elements, so that $m_{i+1} = m_i + (\text{even number}) + 1$.

(ii) We have

$$\begin{aligned} n &= (\text{number of paired elements}) \\ &\quad + (\text{number of unpaired elements}) \\ &= 2(\text{number of basic elements}) \\ &\quad + (\text{number of unpaired elements}). \end{aligned}$$

Now the chain is constructed by adding one unpaired element at a time, so that the number of unpaired elements must be $r-1$. Therefore

$$n = 2(\text{number of basic elements}) + r - 1.$$

(iii) The sequences of parentheses representing A_{i-1} and A_{i+1} differ in two places. We have

$$\begin{aligned} A_{i-1}&: \ \ldots \ (\ \mathrm{XX} \ (\ \ldots \\ A_i&: \ \ldots \ ) \ \mathrm{XX} \ (\ \ldots \\ A_{i+1}&: \ \ldots \ ) \ \mathrm{XX} \ ) \ \ldots \end{aligned}$$

where XX denotes a (possibly empty) closed part. Consider the subset B_i represented by

$$B_i: \ \ldots \ (\ \mathrm{XX} \ ) \ \ldots .$$

It clearly satisfies $A_{i-1} \subset B_i \subset A_{i+1}$. Also, since a further bracketing is now possible in the sequence for B_i it follows that B_i has more basic elements than A_i, and so, by part (ii), the chain in which it lies is shorter.

□

There are yet more equivalent ways of constructing the same symmetric chain decompositions as those described above, and we briefly mention two of these. If we look at, for example, the 2-subsets of $\{1, \ldots, 4\}$ in lexicographic order (alphabetical, with 1 in place of a, 2 in place of b, etc.), and assign to each the first available 3-set containing it we obtain

$$\begin{array}{l}\{1,2\}-\{1,2,3\}\\ \{1,3\}-\{1,3,4\}\\ \{1,4\}-\{1,2,4\}\\ \{2,3\}-\{2,3,4\}\\ \{2,4\}-\text{none left}\\ \{3,4\}- \end{array}.$$

Note that this is precisely the pairing between 2-subsets and 3-subsets in the symmetric chain decomposition of Example 3.1. Aigner (1973) has shown that this always holds, so that the chains can be built up rank by rank.

Another approach is due to White and Williamson (1977). To find out which set covers the set $\{a_1, \ldots, a_h\}$ in its chain, work out $a_i - 2i$ for each i (taking $a_0 = 0$) and define t to be the largest interger i, $0 \leq i \leq h$, for which $a_i - 2i$ is minimal; then the set covering $\{a_1, \ldots, a_h\}$ in its chain is $\{a_1, \ldots, a_h, 1 + a_t\}$. For example, to find the set covering $\{2, 3, 7, 8\}$, work out $0 - 2.0 = 0$, $2 - 2.1 = 0$, $3 - 2.2 = -1$, $7 - 2.3 = 1$, and $8 - 2.4 = 0$; take $t = 2$ to obtain the set $\{2, 3, 4, 7, 8\}$ since $1 + a_2 = 1 + 3 = 4$.

3.4 Applications

We now give some applications of the symmetric chain structure.

3.4.1 The number of antichains

Given an n-set S, how many antichains of subsets of S are there? We shall accept as antichains two rather dubious candidates, namely the empty antichain which contains no subsets and the antichain consisting of only the empty set. This is to enable us to formulate an equivalent version of the question posed above.

We show that antichains of subsets of S are in one–one correspondence with increasing functions from $\mathscr{P}(S)$, the set of subsets of S, to

$\{0, 1\}$. Here we say that a function $f: \mathscr{P}(S) \to \{0, 1\}$ is *increasing* if $f(X) \leqslant f(Y)$ whenever $X \subseteq Y$. Given an antichain $\mathscr{A}$, we can define $f: \mathscr{P}(S) \to \{0, 1\}$ by $f(B) = 0$ if B is a subset of a member of $\mathscr{A}$ and $f(B) = 1$ otherwise. To check that this f is increasing we need only check that if $B \subset D$ and $f(D) = 0$ then $f(B) = 0$. However, if $f(D) = 0$ then $D \subseteq A$ for some $A \in \mathscr{A}$; then $B \subseteq A$ also, so $f(B) = 0$.

It is clear that under this correspondence the sets in the antichain are precisely the maximal sets on which f takes the value zero. Conversely, given an increasing function $f: \mathscr{P}(S) \to \{0, 1\}$, the maximal sets on which f takes the value zero will form an antichain. Thus the required correspondence has been exhibited. Note that the empty antichain corresponds to the function f which takes the value unity everywhere.

The number of antichains is therefore the number of ways of defining an increasing function $f: \mathscr{P}(S) \to \{0, 1\}$. We therefore consider the problem of constructing as many increasing functions as we can. Suppose that we decompose the set of subsets of S into symmetric chains, and that we attempt to construct f by defining it on the sets in one chain at a time. More precisely, suppose that we have succeeded in defining an increasing function f consistently on all chains of size less than k and that we now attempt to extend the definition of f to a given chain of size k.

If $k = 2$, any chain of size k is of the form $X \subset Y$. Here there are at most three possible choices of values for $f(X)$ and $f(Y)$: we could have $f(X) = f(Y) = 0$, or $f(X) = f(Y) = 1$, or $f(X) = 0$ and $f(Y) = 1$. Certainly the choice $f(X) = 1$, $f(Y) = 0$ is not allowed. Thus there are at most three possibilities.

If $k > 2$, consider any two sets X, Y in the chain under consideration with $|Y| = |X| + 2$. We then have

$$. . X \subset Z \subset Y . . .$$

for some set Z. Now by Theorem 3.3.1(iii) there is a set U, $X \subset U \subset Y$, with U in a shorter chain, and hence with $f(U)$ already defined. It follows that if the chain

$$X_1 \subset X_2 \subset . . . \subset X_k$$

is being considered, then, for each $i = 2, . . . , k - 1$, there is a set U_i such that $X_{i-1} \subset U_i \subset X_{i+1}$ and with $f(U_i)$ already defined. If $f(U_i) = 0$ for all i, $2 \leqslant i \leqslant k - 1$, then we must take $f(X_{i-1}) = 0$ for each $i \leqslant k - 1$ and we are possibly free to define f on X_{k-1} and X_k. As before, there are at most three possible choices here, $f(X_{k-1}) = 1$, $f(X_k) = 0$ not being allowed. So again there are at most three possible ways of extending f. If $f(U_i)=1$ for all i, $2 \leqslant i \leqslant k-1$, then similarly there is only possible freedom of choice at $f(X_1)$ and $f(X_2)$, and again there are at most three possible ways of extending f. Finally, if f is not constant on the U_i, let j be the first i at which $f(U_i)=1$. Then $f(X_i)=1$ for all $i>j$ and $f(X_i)=0$ for all $i<j-1$, so we are only free to define f on X_{j-1} and X_j. As before there are at most three possible choices.

Thus as each chain is considered in turn, there are at most three ways of extending the definition of f consistently to that chain. The total number of possible increasing functions f is therefore at most 3^m where m is the number of chains. Since $m=\binom{n}{[n/2]}$ we therefore have the following theorem.

Theorem 3.4.1 (Hansel 1966) The number of antichains of subsets of an n-set is at most

$$3^{\binom{n}{[n/2]}}. \qquad \square$$

By using a much more complicated argument based on the same ideas, Kleitman and Markowsky (1974) later reduced this upper bound to

$$2^{(1+O(\log n/n))\binom{n}{[n/2]}}.$$

More recently, Korshunov (1981) has published a complicated asymptotic formula for the number of antichains. In the case of even n, the estimate is

$$2^{\binom{n}{n/2}} \exp\left\{\binom{n}{n/2-1}(2^{-n/2}+n^2 2^{-n-5}-n\times 2^{-n-4})\right\}.$$

Even in the case of $n=6$ this gives remarkably good results: 7 996 118 instead of the exact value 7 828 354. The first few exact values are tabulated below.

n	Number of antichains
1	3
2	6
3	20
4	168
5	7581
6	7 828 354
7	2 414 682 040 998

3.4.2 A storage problem

The following problem arises in the mathematical theory of information storage and retrieval. Given a collection $\mathscr{B}$ of subsets of an n-set S, find the shortest sequence T of elements of S such that every set in $\mathscr{B}$ appears at least once as a subsequence of consecutive terms of T.

In the case where $\mathscr{B}$ is the set of all subsets of S, the problem is to find as short a sequence as possible containing every subset as a subsequence of consecutive terms. If $n = 5$, for example, and $S = \{1, 2, 3, 4, 5\}$, the sequence

$$1\ 2\ 3\ 4\ 5\ 1\ 2\ 4\ 1\ 3\ 5\ 2\ 4$$

satisfies the requirements.

A trivial upper bound for the length of the shortest sequence is obtained by considering the sequence obtained simply by writing down each subset one after the other. Since there are 2^n subsets of average size $\frac{1}{2}n$, the length of the resulting sequence is $\frac{1}{2}n2^n$. However, an improvement on this can be achieved by exploiting the existence of a symmetric chain decomposition of the poset of subsets.

Theorem 3.4.2 (Lipski 1978) If S is an n-set, there exists a sequence of length (asymptotically) $\leqslant (2/\pi)2^n$ which contains every subset of S as a subsequence of consecutive terms.

Proof. We consider first n even, say $n = 2k$, and we write $S = T \cup R$ where $|T| = |R| = k$. Then both T and R have symmetric chain decompositions of their posets of subsets into $m = \binom{k}{[k/2]}$ symmetric chains: $\mathscr{P}(T) = \phi_1 \cup \ldots \cup \phi_m$, $\mathscr{P}(R) = \psi_1 \cup \ldots \cup \psi_m$. Corresponding to the chain ϕ_i

$$\phi_i : \{x_1, \ldots, x_j\} \subset \{x_1, \ldots, x_j, x_{j+1}\} \subset \ldots \subset \{x_1, \ldots, x_h\} \quad (h \geqslant j)$$

we associate the sequence $U_i = x_1x_2 \ldots x_h$. Then *every* subset of T occurs as an *initial* part of one of the sequences $U_1, \ldots, U_m$. Similarly let $V_1, \ldots, V_m$ be sequences corresponding to the chains $\psi_1, \ldots, \psi_m$ such that every subset of R occurs as an initial part of some V_i. If we let $\bar{V}_i$ denote the sequence obtained by writing V_i in reverse order, then every subset of R occurs as a *final* part of one of the $\bar{V}_i$. Now consider the sequence

$$\bar{V}_1U_1\bar{V}_1U_2 \ldots \bar{V}_1U_m\bar{V}_2U_1\bar{V}_2U_2 \ldots \bar{V}_2U_m \ldots \bar{V}_mU_1 \ldots \bar{V}_mU_m.$$

If A is any subset of S we can write $A = E \cup F$ where $E \subseteq T$ and $F \subseteq R$. Now E occurs as the initial part of some U_e and F occurs as the final part of some $\bar{V}_f$: thus A occurs in the above sequence as part of $\bar{V}_f U_e$. Thus the sequence contains every subset of S as required. The length of the sequence is

$$\leqslant 2m^2(k+1) = 2(k+1)\binom{k}{[k/2]}^2.$$

However,

$$\binom{k}{[k/2]} \approx \left(\frac{2}{k\pi}\right)^{1/2} 2^k$$

by Stirling's approximation, so that

$$2m^2(k+1) \approx \frac{4}{k\pi} \cdot 2^{2k} \cdot k = \frac{4}{\pi} 2^n.$$

This estimate for the length of the sequence can be halved by using Exercise 3.10, thus proving the theorem.
The proof for n odd is similar. □

3.4.3 A probability result related to Sperner's theorem

Another application of the symmetric chain structure leads to an elegant probabilistic version of Sperner's theorem. We investigate the following problem. Suppose that two subsets A, B are chosen independently and at random from the set of subsets of an n-set S, and suppose that the underlying probabilities of each being chosen are known. Then what can be said about the probability that $A \subseteq B$? We shall show that, no matter what the underlying probability distribution is, the probability that $A \subseteq B$ is at least $1/\binom{n}{[n/2]}$.

Lemma 3.4.3 Let $p_1, \ldots, p_m$ be a probability distribution on an m-element set T. Then the probability that two elements of T chosen at random are identical is $\geqslant 1/m$, irrespective of the p_i.

Proof The required probability is $\sum_{i=1}^m p_i^2$. Now we clearly have

$$\sum_{i=1}^{m} \left(p_i - \frac{1}{m}\right)^2 \geqslant 0$$

so

$$\sum_{i=1}^{m} p_i^2 - \frac{2}{m}\sum_{i=1}^{m} p_i + m \cdot \frac{1}{m^2} \geqslant 0$$

whence

$$\sum_{i=1}^{m} p_i^2 \geqslant \frac{2}{m}\sum_{i=1}^{m} p_i - \frac{1}{m} = \frac{2}{m} - \frac{1}{m} = \frac{1}{m}.$$

□

Lemma 3.4.4 If $\{q(x)\}$ is a probability distribution on a poset P and P is partitioned into m chains, then the probability that two elements of P chosen at random lie in the same chain is $\geqslant 1/m$, no matter what the values of q are.

Proof In Lemma 3.4.3 take T to be the set of chains. The required probability is the probability of choosing two chains at random and obtaining the same chain twice. The probability p_i of choosing the ith chain is given by $p_i = \Sigma\, q(x)$ where the summation is over all x in the ith chain. The result is immediate from Lemma 3.4.3. □

Note that the probability that two elements chosen at random lie in the same chain can be written as

$$\sum_{x \in P} q(x)\Bigg(\sum_{\substack{y \\ x,\, y \text{ in} \\ \text{same chain}}} q(y)\Bigg) = \sum_x q^2(x) + 2 \sum_{\substack{x<y \\ x,\, y \text{ in} \\ \text{same chain}}} q(x)q(y).$$

Therefore the result of Lemma 3.4.4 can be written as

$$\sum_x q^2(x) + 2 \sum_{\substack{x<y \\ x,\, y \text{ in} \\ \text{same chain}}} q(x)q(y) \geqslant \frac{1}{m}.$$

Suppose now that we can find two chain decompositions of P into m chains with the property that no pair of elements of P in the same chain in one of the decompositions lies in the same chain in the second decomposition. (Such a pair of chain decompositions are said to be *orthogonal*.) We can then obtain an inequality as above for both the chain decompositions, and on adding them together and dividing by 2 we obtain

$$\sum_x q^2(x) + \sum{}^{*} q(x)q(y) \geqslant \frac{1}{m}$$

where Σ^* denotes the sum over all x, y such that $x<y$ and x, y occur in the same chain in one of the two decompositions. Since

$$\sum{}^{*} q(x)q(y) \leqslant \sum_{x<y} q(x)q(y)$$

we therefore have

$$\sum_x q^2(x) + \sum_{x<y} q(x)q(y) \geqslant \frac{1}{m}. \tag{3.3}$$

We can now prove the following theorem.

Theorem 3.4.5 (Baumert, McEliece, Rodemich, and Rumsey 1980) If two sets A, B are chosen independently according to a probability distribution defined on the subsets of an n-set S, then the probability that $A \subseteq B$ is at least $1\Big/\binom{n}{[n/2]}$.

Proof Take $m = \binom{n}{[n/2]}$. The required probability is

$$\sum_A q^2(A) + \sum_{A \subset B} q(A)q(B)$$

where $q(A)$ is the probability of choosing A. Thus, by (3.3), the probability is $\geq 1/m$—provided of course that an orthogonal pair of chain decompositions exists! This is now proved. □

Theorem 3.4.6 (Shearer and Kleitman 1979) If $n \geq 2$, there exist two orthogonal chain decompositions of the poset of subsets of an n-set into $\binom{n}{[n/2]}$ chains.

Proof If $n = 2$ or 3 it is easy to find such decompositions (see Exercise 3.11). Therefore we now suppose that $n \geq 4$.

First of all take the symmetric chain decomposition constructed earlier in this chapter. Then obtain another chain decomposition by replacing each set by its complement. Certainly $\varnothing$ and S will lie in the same chain in both decompositions, but we show that no other pair of sets have that property. If A, B with $A \subset B$ lie in the same chain in both decompositions then A and B must have the same basic elements and their complements A' and B' must also have the same basic elements. Also, the string of parentheses representing A' is obtained from the string representing A by replacing all left parentheses by right ones and vice versa. We now study the strings for A and B and show that if A and B have the same basic elements, then A' and B' cannot have the same basic elements except when A and B are $\varnothing$ and S.

Suppose first that A is not the smallest set in its chain. Then with A and B in the same chain and $A \subset B$, we must have the following strings of parentheses:

$$A\colon\ C_0\ )\ C_1 \ldots\ )\ C_i\ (\ C_{i+1} \ldots$$
$$B\colon\ C_0\ )\ C_1 \ldots\ )\ C_i\ )\ C_{i+1} \ldots$$

where the C_i respresent (possibly empty) closed parts. Since A is not the smallest set in its chain, there is at least one ')' before C_i. Now the strings representing A' and B' are obtained by interchanging left and right parentheses, and it is clear that this interchange applied to the string for A will result in the reversed bracket between C_i and C_{i+1} closing with a previous bracket whereas this closing will not take place in the case of B. Thus A' and B' do not have the same closed parts.

A similar argument applies if B is not the last set in its chain. We are left with the case when A and B are the first and last members of their chain. We then have

$$A\colon C_0 \ (\ C_1 \ (\ \ldots$$
$$B\colon C_0 \ ) \ C_1 \ ) \ \ldots$$

If there is a non-empty C_i to the left of a '(' in A then on reversal there will be a closing of brackets in A which will not occur in B. If there is a non-empty C_i to the right of a ')' in B then there will, on reversal, be a closing of brackets in B which will not occur in A. Thus, unless all C_i are empty (in which ase $A = \varnothing$ and $B = S$), A' and B' will have different closed parts and hence different basic elements.

The two symmetric chain decompositions are therefore almost orthogonal. To make them orthogonal, we remove $\varnothing$ from its chain in the first decomposition and add it to any chain which does not contain any of the $n-1$ sets other than S or itself in its chain in the other decomposition. This is possible since there are more than n chains if $n > 3$. □

3.5 Nested chains

In obtaining a generalization of Sperner's theorem, Lih (1980) considered collections $\mathscr{A}$ of subsets of an n-set S which are not only antichains but which have the further property that every member of $\mathscr{A}$ has a non-empty intersection with a given k-element subset T of S. The case $k = n$, i.e. $T = S$, clearly reduces to Sperner. Lih showed that $\mathscr{A}$ has the Sperner property, so that

$$|\mathscr{A}| \leqslant \max_i \{\text{no. of subsets of size } i \text{ which intersect } T\}$$
$$= \max_i \left\{ \binom{n}{i} - \binom{n-k}{i} \right\}$$

since the number of i-subsets of S which do *not* intersect T is $\binom{n-k}{i}$.

It is not hard to check that the maximum occurs when $i = \lceil \frac{1}{2}n \rceil$ (see

Exercise 3.12), so we finally obtain

$$|\mathscr{A}| \leq \binom{n}{[n/2]} - \binom{n-k}{[n/2]}.$$

An alternative approach to proving that $\mathscr{A}$ is Sperner has been given by Griggs (1982). By relabelling the elements of $S = \{x_1, \ldots, x_n\}$ if necessary, we can suppose that $T = \{x_1, \ldots, x_k\}$. Let $C(n, k)$ denote the collection of all subsets of S which intersect T. Then Griggs showed that, although $C(n, k)$ need not have a decomposition into symmetric chains, it does nevertheless have the next best thing, a decomposition into *nested* chains.

Definition Two chains in a ranked poset P are *nested* if they are saturated and if the chain containing the element of least rank in their union also contains the element of greatest rank. Also, P is called a *nested chain order* if P can be partitioned into pairwise nested chains.

Example 3.5.1 Let $S = \{1, 2, 3, 4\}$ and $T = \{1, 2\}$. Then $C(4, 2)$ consists of all subsets of S which contain 1 or 2. A decomposition into pairwise nested chains is shown:

$$\{1\} \subset \{1,2\} \subset \{1,2,3\} \subset \{1,2,3,4\}$$
$$\{2\} \subset \{2,3\} \subset \{2,3,4\}$$
$$\{1,3\} \subset \{1,3,4\}$$
$$\{1,4\} \subset \{1,2,4\}$$
$$\{2,4\}$$

Note how these nested chains are related to the symmetric chains exhibited in Example 3.1.1. This gives us a hint of the proof of the following theorem.

Theorem 3.5.1 (Griggs 1982) $C(n, k)$ is a nested chain order (and so has the Sperner property).

Proof Consider the symmetric chain decomposition given by de Bruijn *et al.* (1951). If the first member of a chain intersects T then every member of that chain will intersect T, and so the whole chain is in $C(n, k)$. If none of the members of a chain intersect T, then ignore that chain. Consider next a chain in which the first member does not intersect T, but at least one other member of the chain does intersect T. Since each member of a chain is obtained by adding an element to the previous one, the elements being added in the order of the x_i, and since $T = \{x_1, \ldots, x_k\}$, it follows that if an element of T is added at some stage in the chain, it must be added at the very first stage. Thus every member of the chain except the first is in $C(n, k)$. In conclusion, therefore, $C(n, k)$ is the union of chains each of which is one of the symmetric chains or is a symmetric chain minus its first member. Thus $C(n, k)$ is a nested chain order. □

Corollary 3.5.2 Any antichain $\mathscr{A}$ in $C(n, k)$ has at most

$$\binom{n}{[n/2]} - \binom{n-k}{[n/2]}$$

members.

Proof $\mathscr{A}$ can contain at most one member of each of the nested chains but the number of chains is the number of subsets of size $[\frac{1}{2}n]$ which intersect $\{1, \ldots, k\}$, i.e. is $\binom{n}{[n/2]} - \binom{n-k}{[n/2]}$. □

3.6 Posets with symmetric chain decompositions

A ranked poset is called a *symmetric chain order* if it possesses a decomposition into symmetric chains. In this section we look for some partial answers to the following question: which posets are symmetric chain orders? Our first result is really just a reworking of Theorem 3.1.1.

Definition If P, Q are posets with rank functions r, r' respectively, then the *direct product* $P \times Q = \{(p, q) : p \in P, q \in Q\}$ is a poset with $(p_1, q_1) \leqslant (p_2, q_2) \Leftrightarrow p_1 \leqslant p_2$ in P and $q_1 \leqslant q_2$ in Q. $P \times Q$ has rank function ρ defined by $\rho(p, q) = r(p) + r'(q)$.

Example 3.6.1 The poset of divisors of $p^a q^b$ (where p, q are distinct primes) is the direct product of the two chains $\{1, p, p^2, \ldots, p^a\}$ and $\{1, q, \ldots, q^b\}$.

Theorem 3.6.1 If P and Q are symmetric chain orders, then so is $P \times Q$.

Proof Let $P = C_1 \cup \ldots \cup C_m$ and $Q = D_1 \cup \ldots \cup D_n$ be symmetric chain decompositions of P and Q. Choose any C_i and D_j, say

$$C_i: c_0 < c_1 < \ldots < c_k$$

and

$$D_j\colon d_0 < d_1 < \ldots < d_h,$$

where

$$r(c_0) = r, \qquad r(c_k) = R, \qquad r + R = r(P)$$

and

$$r'(d_0) = s, \qquad r'(d_h) = S, \qquad s + S = r'(Q).$$

We imitate the proof of Theorem 3.1.1 and consider the array

$$\begin{array}{lllll}
E_0 \to c_0d_0\ c_1d_0 \ldots & c_{k-2}d_0 & c_{k-1}d_0 & c_kd_0 \\
E_1 \to c_0d_1\ c_1d_1 \ldots & c_{k-2}d_1 & c_{k-1}d_1 & c_kd_1 \\
E_2 \to c_0d_2 \ldots & & \vdots & \vdots \\
 & & c_{k-1}d_h & c_kd_h
\end{array}$$

from which we read off chains E_0, E_1, etc. We show that these chains are symmetric (they certainly cover $P \times Q$). Consider

$$E_j\colon (c_0, d_j) < (c_1, d_j) < \ldots < (c_{k-j}, d_j) < \ldots < (c_{k-j}, d_h)$$

where the rank of (c_0, d_j) is $r + j + s$ and the rank of (c_{k-j}, d_h) is $r + k - j + s + h = R + S - j$, so that

$$\begin{aligned}\text{rank of } (c_0, d_j) + \text{rank of } (c_{k-j}, d_h) &= r + R + s + S = r(P) + r'(Q) \\ &= \rho(P \times Q).\end{aligned}$$

This completes the proof. □

We now turn our attention to the posets $L(m, n)$ defined by

$$L(m, n) = \{(x_1, \ldots, x_m) : x_i \text{ integers}, 0 \leqslant x_1 \leqslant x_2 \leqslant \ldots \leqslant x_m \leqslant n\}$$

with order relation $\leqslant$ defined by

$$\boldsymbol{x} = (x_1, \ldots, x_m) \leqslant \boldsymbol{y} = (y_1, \ldots, y_m) \Leftrightarrow x_i \leqslant y_i \text{ for each } i.$$

The posets $L(m, n)$ are ranked posets, with rank given by $r(\boldsymbol{x}) = x_1 + \ldots + x_m$. As an example, consider $L(3, 2)$. The element of maximum rank in $L(3, 2)$ is $(2, 2, 2)$ with rank 6. There are 10 elements altogether, and they can be placed into two symmetric chains as follows:

$$(0, 0, 2) < (0, 1, 2) < (0, 2, 2)$$

$$(0, 0, 0) < (0, 0, 1) < (0, 1, 1) < (1, 1, 1) < (1, 1, 2) < (1, 2, 2) < (2, 2, 2)$$

It has been conjectured that each $L(m, n)$ is a symmetric chain order. Now there are two necessary conditions on the rank numbers of a poset which have to be satisfied; they are

(i) the rank numbers N_i are unimodal,

(ii) $N_i = N_{r-i}$ for each i,

where $r = r(P)$ is the largest rank in the poset P. Since $(x_1, \ldots, x_m)$ has rank j if and only if $(n - x_m, \ldots, n - x_1)$ has rank $mn - j$, condition (ii) is certainly satisfied. It is also known that the rank numbers satisfy (i).

At present, the conjecture has been confirmed only for $L(3, n)$ by Lindström (1980) and for $L(4, n)$ by West (1980). We give the proof for the first of these.

Theorem 3.6.2 (Lindström 1980) $L(3, n)$ is a symmetric chain order.

Proof We shall consider the cases of n odd and n even separately. First we consider n odd, $n = 2k + 1$. The case $k = 0$ corresponds to $L(3, 1)$ which is trivially symmetric as it consists of one chain $(0, 0, 0) < (0, 0, 1) < (0, 1, 1) < (1, 1, 1)$. We therefore use an induction argument, assuming that $L(3, 2k - 1)$ is symmetric ($k \geqslant 1$) and considering $L(3, n)$ where $n = 2k + 1$. Let $S(n)$ denote the set of all (a, b, c) in $L(3, n)$ with $a = 0$ or $c = n$ (or both). The remaining elements of $L(3, n)$ have $1 \leqslant a \leqslant b \leqslant c \leqslant n - 1$ and form a poset isomorphic to $L(3, n - 2) = L(3, 2k - 1)$ under the correspondence $(a, b, c) \in L(3, n) \leftrightarrow (a - 1, b - 1, c - 1) \in L(3, n - 2)$. By the induction hypothesis, $L(3, n) - S(n)$ has a symmetric chain decomposition. We then show that $S(n)$ has a symmetric chain decomposition as well, by exhibiting $k + 1$ chains C_i, $i = 0, \ldots, k$, which partition $S(n)$:

$$\begin{aligned}
C_i\colon (0, i, i) &< (0, i, i + 1) < \ldots < (0, i, n - i) < (0, i + 1, n - i) \\
&< (0, i + 1, n - i + 1) < (0, i + 2, n - i + 1) < \ldots \\
&< (0, 2i, n) < (0, 2i + 1, n) < (1, 2i + 1, n) \\
&< (1, 2i + 2, n) < (2, 2i + 2, n) < \ldots \\
&< (n - 2i - 1, n, n) < (n - 2i, n, n).
\end{aligned}$$

These chains C_i are symmetric and partition $S(n)$.

We next consider n even. We let $S(n)$ be as before, and we obtain $S^*(n)$ by adjoining to $S(n)$ all elements $(1, 2i + 1, n - 1)$, $i = 0, \ldots, k - 1$, where $n = 2k$. It can now be checked that $S^*(n)$ can be partitioned into chains C and C_i^*, $i = 0, \ldots, k - 1$, defined by

$$C\colon (0, 0, n) < (0, 1, n) < \ldots < (0, n, n)$$

and

$$C_i^*: (0, i, i) < (0, i, i+1) < \ldots < (0, i, n-i-1) < (0, i+1, n-i-1)$$
$$< (0, i+1, n-i) < (0, i+2, n-i) < (0, i+2, n-i+1) < \ldots$$
$$< (0, 2i+1, n-1) < (1, 2i+1, n-1) < (1, 2i+1, n)$$
$$< (1, 2i+2, n) < (2, 2i+2, n) < (2, 2i+3, n) < \ldots$$
$$< (n-2i-1, n, n) < (n-2i, n, n).$$

Each chain C_i^* contains an element $(1, 2i+1, n-1)$ of $S^*(n) - S(n)$; this element is in $T(n) = \{(a, b, c) \in L(3, n);\ a = 1 \text{ or } c = n-1 \text{ or both}\}$. We can also check that $T(n) - \{(1, 2i+1, n-1)\}$ is the union of symmetric chains D_i, $i = 1, \ldots, k-1$, given by

$$D_i: (1, i, i) < (1, i, i+1) < \ldots < (1, i, n-i-1) < (1, i+1, n-i-1)$$
$$< (1, i+1, n-i) < \ldots < (1, 2i, n-1) < (2, 2i, n-1)$$
$$< (2, 2i+1, n-1) < (3, 2i+1, n-1) < \ldots$$
$$< (n-2i, n-1, n-1) < (n-2i+1, n-1, n-1).$$

Now $S(n) \cup T(n)$ contains all elements (a, b, c) in $L(3, n)$ with $a = 0$ or 1 or $c = n-1$ or n. Therefore $L(3, n) - (S(n) \cup T(n))$ is isomorphic to $L(3, n-4)$ under the correspondence $(a, b, c) \leftrightarrow (a-2, b-2, c-2)$. Thus we can use an induction argument, obtain a symmetric chain decomposition of $L(3, n) - (S(n) \cup T(n))$ from the induction hypothesis, and combine it with the symmetric chain decomposition obtained above for $S(n) \cup T(n)$. □

The proof for $L(4, n)$ given by West (1980) is more complicated. Although it is not known whether $L(m, n)$ is symmetric in general, it has recently been shown by Stanley (1980) that $L(m, n)$ is always Sperner. His proof uses deep algebraic methods.

3.6.1 Symmetric chains and the normalized matching property

We now show that, along with the conditions (i) and (ii) on the rank numbers which are necessary for a symmetric chain decomposition to exist, a sufficient condition which ensures that P is symmetric is, remarkably, that P has the normalized matching property. Before we prove this, we have to describe a generalization of the marriage problem to the context of *common* systems of distinct representatives.

Definition If $A_1, \ldots, A_m$ and $B_1, \ldots, B_m$ are two families of sets, a *common system of distinct representatives* (c.s.d.r.) is a set of m

elements which is simultaneously an s.d.r. for the A_i and for the B_i (in some order).

Example 3.6.2 The sets $A_1 = \{1, 2, 3\}$, $A_2 = \{1, 4\}$, $A_3 = \{3, 4\}$ and the sets $B_1 = \{3, 4\}$, $B_2 = \{2, 4\}$, $B_3 = \{3\}$ have a c.s.d.r. consisting of 2, 3, 4, for we can represent A_1 and B_2 by 2, A_3 and B_3 by 3, and A_2 and B_1 by 4.

Theorem 3.6.3 (Ford and Fulkerson 1958) The families $A_1, \ldots, A_m$ and $B_1, \ldots, B_m$ of subsets of the n-set S possess a c.s.d.r. if and only if

$$\left|\left(\bigcup_{i\in E} A_i\right) \cap \left(\bigcup_{j\in F} B_j\right)\right| \geqslant |E| + |F| - m \qquad (3.4)$$

for all subsets E, F of $\{1, \ldots, m\}$.

Proof (Perfect 1969) The key idea behind Perfect's elegant proof is to show first that the existence of a c.s.d.r. is equivalent to the existence of an s.d.r. for a related family of sets. Suppose that $S = \{x_1, \ldots, x_n\}$, let $T = S \cup \{1, \ldots, m\}$, and define $m + n$ subsets Y_i of T by

$$Y_j = A_j \quad (j = 1, \ldots, m)$$

$$Y_x = \{x\} \cup \{i : x \in B_i\} \quad (x \in S).$$

Now any s.d.r. for the Y_i is uniquely described by an injection $\alpha: \{1, \ldots, m\} \to \{1, \ldots, n\}$ and a permutation β of $\{1, \ldots, m\}$, which are given by

$$x_{\alpha(i)} \in A_i \qquad (i = 1, \ldots, m)$$

$$\beta(i) \in Y_{x_{\alpha(i)}} \quad (i = 1, \ldots, m),$$

the remaining Y_{x_i} being represented by x_i. The second of these relations is equivalent to

$$x_{\alpha(i)} \in B_{\beta(i)}$$

so that we obtain $x_{\alpha(i)} \in A_i \cap B_{\beta(i)}$ for each $i = 1, \ldots, m$. Clearly a different s.d.r. for the Y_i would involve a different α or a different β and so would give rise to a different c.s.d.r.. Conversely, it is easy to check that any c.s.d.r. yields an s.d.r. for the Y_i.

Now by the marriage theorem the Y_i possesses an s.d.r. if and only if $|\bigcup_{i\in K} Y_i| \geqslant |K|$ whenever $K \subseteq \{1, \ldots, m\} \cup \{x_1, \ldots, x_n\}$. If we write $K = I \cup J$ where $I \subseteq \{1, \ldots, m\}$ and $J \subseteq \{x_1, \ldots, x_n\}$, the condition is

$$\left|\left(\bigcup_{i\in I} A_i\right) \cup \left(\bigcup_{j=J} Y_j\right)\right| \geqslant |I| + |J| \tag{3.5}$$

whenever $I \subseteq \{1, \ldots, m\}$ and $J \subseteq \{x_1, \ldots, x_n\}$. However,

$$\left(\bigcup_{i\in I} A_i\right) \cap \left(\bigcup_{j\in J} Y_j\right) = \left(\bigcup_{i\in I} A_i\right) \cap J$$

and

$$\bigcup_{j\in J} Y_j = J \cup \{i : B_i \cap J \neq \varnothing\}$$

so that, since $|S \cap T| + |S \cup T| = |S| + |T|$, expression (3.5) can be rewritten as

$$\left|\bigcup_{i\in I} A_i\right| + \left|\bigcup_{j\in J} Y_j\right| - \left|\left(\bigcup_{i\in I} A_i\right) \cap J\right| \geqslant |I| + |J|,$$

i.e.

$$\left|\bigcup_{i\in I} A_i \cap (\{x_1, \ldots, x_n\} - J)\right| + |\{i : B_i \cap J \neq \varnothing\}| \geqslant |I|. \tag{3.6}$$

Finally we show that (3.6) is equivalent to the Ford–Fulkerson condition (3.4). Suppose that (3.6) holds for all I, J. Given subsets E, F of $\{1, \ldots, m\}$, define J by $\{x_1, \ldots, x_n\} - J = \cup \{B_j : j \in F\}$ and take $I = E$. J is the set of elements not in any B_j, $j \in F$, so that if $B_j \cap J \neq \varnothing$ then $j \notin F$. Then (3.6) yields

$$\left|\left(\bigcup_{i\in E} A_i\right) \cap \left(\bigcup_{j\in F} B_j\right)\right| + m - |F| \geqslant |E|.$$

Conversely, suppose that (3.4) holds for all E, F. Given I, J, define E and F by $E = I$ and $F = \{i : B_i \cap J = \varnothing\}$. Then

$$\left|\left(\bigcup_{i\in E} A_i\right) \cap \left(\bigcup_{j\in F} B_j\right)\right| \leqslant \left|\bigcup_{i\in E} A_i \cap (\{x_1, \ldots, x_n\} - J)\right|$$

so that (3.4) gives

$$\begin{aligned}\left|\bigcup_{i\in I} A_i \cap (\{x_1, \ldots, x_n\} - J)\right| &\geqslant |E| + |F| - m \\ &= |I| - (m - |F|) \\ &= |I| - |\{i : B_i \cap J \neq \varnothing\}|,\end{aligned}$$

i.e. (3.6) holds. □

We now use Theorem 3.6.3 to obtain a sufficient condition for a poset to be a symmetric chain order.

Theorem 3.6.4 (Anderson 1967b, Griggs 1977) Let P be a ranked poset with normalized matching property whose rank numbers are unimodal and satisfy $N_i = N_{r-i}$ where r is the rank of P. Then P is a symmetric chain order.

Proof We construct a symmetric chain decomposition step by step. Suppose that all elements of P of ranks $t, k<t<r-k$ have been placed in symmetric chains. We then have N_{k+1} pairs (u, v) of elements of P with $r(u)=k+1$, $r(v)=r-k-1$, $u<v$, and u in the same chain as v. Let H denote the set of all such pairs. For each $a_i \in P$ of rank k, let $A_i = \{(u, v) \in H: a_i < u\}$, and for each $b_i \in P$ of rank $r-k$ let $B_i = \{(u, v) \in H: v < b_i\}$. Then we can extend the symmetric chains to include all elements of P of ranks k and $n-k$ if we can find a c.s.d.r. for the sets $A_i, i=1, \ldots, N_k$, and the sets $B_i, i = 1, \ldots, N_k$. We therefore only have to show that the sets A_i and B_i satisfy the condition (3.4) of the Ford–Fulkerson theorem.

Therefore let E, F be any two subsets of $\{1, \ldots, N_k\}$. We assume that $|E|+|F|>N_k$ since otherwise (3.4) is trivial. Now, by the normalized matching property,

$$\left|\bigcup_{i\in E} A_i\right| \geqslant \frac{N_{k+1}}{N_k}|E|$$

$$\left|\bigcup_{j\in F} B_j\right| \geqslant \frac{N_{k+1}}{N_k}|F|$$

so that

$$\begin{aligned}\left|\left(\bigcup_{i\in E} A_i\right)\cap\left(\bigcup_{j\in F} B_j\right)\right| &= \left|\bigcup_{i\in E} A_i\right| + \left|\bigcup_{j\in F} B_j\right| - \left|\left(\bigcup_{i\in E} A_i\right)\cup\left(\bigcup_{j\in F} B_j\right)\right| \\ &\geqslant \frac{N_{k+1}}{N_k}(|E|+|F|) - N_{k+1} \\ &= \frac{N_{k+1}}{N_k}(|E|+|F|-N_k) \\ &\geqslant |E|+|F|-N_k.\end{aligned}$$

Thus the Ford–Fulkerson condition is satisfied, and the chains can be extended to include all elements of ranks k and $n-k$. Repeating the process at each rank level gives a construction of the chains for P. If n is even the whole process begins with the elements of the middle rank in place of the pairs (u, v), and if n is odd the first step is to pair off the elements of rank $\frac{1}{2}(n-1)$ with those of rank $\frac{1}{2}(n+1)$, which is possible since the normalized matching condition reduces to the Hall condition between the two ranks in this case. □

Quantitative results in this area are also possible: here we ask how many symmetric chain decompositions does P have? Recall from Exercise 2.20 that if the sets $A_1, \ldots, A_m$ possess an s.d.r. and if $|A_i| \geqslant t$ for each i, then there are at least $f(m, t)$ different ways of choosing an s.d.r. for the A_i, where

$$f(m, t) = \begin{cases} t! & \text{if } m \geqslant t, \\ t(t-1) \ldots (t-m+1) & \text{if } m < t. \end{cases} \tag{3.7}$$

Using this result we find that the Perfect proof of the Ford–Fulkerson theorem yields the following theorem.

Theorem 3.6.5 (Anderson 1985) Let $A_1, \ldots, A_m$ and $B_1, \ldots, B_m$ be two collections of subsets of an n-set S which satisfy condition (3.4). Suppose also that $|A_i| \geqslant t$ for each i and that every element of S is contained in at least u of the B_i. Let $p = \min(t, 1+u)$. Then there are at least $p!$ different c.s.d.r.s for the A_i and the B_i. (Here two c.s.d.r.s are considered different if at least one of the sets is represented by different elements in the two c.s.d.r.s.)

Proof In Perfect's proof, $|Y_i| \geqslant t$ for each $i = 1, \ldots, m$ and $|Y_x| \geqslant 1+u$ for each $x \in S$, so the result follows from (3.7). □

Corollary 3.6.6 The poset of subsets of an n-set can be partitioned into symmetric chains in at least $g(n)$ ways where

$$g(n) = 3!4! \ldots (\tfrac{1}{2}n+1)!n \quad \text{if } n \text{ is even}$$

and

$$g(n) = 3!4! \ldots \left(\frac{n+1}{2}\right)! \left(\frac{n+1}{2}\right)! n \quad \text{if } n \text{ is odd.}$$

Proof If n is even, then, in the notation of the proof of Theorem 3.6.3, $|A_i| = n-k$ and each element is in $k+1$ sets B_i. Since $\min(t, 1+u) = \min(n-k, k+2) = k+2$, there are at least $(k+2)!$ ways of extending the chains to ranks k and $n-k$. At $k=0$, $\varnothing$ and S can clearly be added to any of n chains. The case of odd n is similar, except that at the start there are at least $\{(n+1)/2\}!$ ways of pairing the sets of ranks $(n-1)/2$ and $(n+1)/2$. □

This lower bound can be increased by a more detailed argument, but it is certainly an improvement on the trivial lower bound of $n!$, which is the number of symmetric chain decompositions obtained using the construction of de Bruijn *et al.* (1951) by considering each of the $n!$ possible orderings of the elements of S.

Exercises 3

3.1 Show that, in a symmetric chain decomposition of the set of subsets of an n-set, there are $\binom{n}{k} - \binom{n}{k-1}$ chains of size $n+1-2k$, and hence $\binom{n}{[(n+i)/2]} - \binom{n}{[(n+i+1)/2]}$ chains of size i.

3.2 Define $\mu(A) = (-1)^{|A|}$. Show that if $|S| = n$ then

$$\left|\sum_{A \in \mathscr{A}} \mu(A)\right| \leqslant \binom{n}{[n/2]}$$

whenever $\mathscr{A}$ is a *convex* set of subsets of S, i.e. if $A \in \mathscr{A}$, $B \in \mathscr{A}$, $A \subset D \subset B \Rightarrow D \in \mathscr{A}$. (de Bruijn *et al.* 1951)

3.3 Use symmetric chains to show that if $\mathscr{A}_1, \ldots, \mathscr{A}_r$ are r disjoint antichains of subsets of an n-set then

$$\left|\bigcup_{i=1}^{r} \mathscr{A}_i\right| \leqslant \sum_{j=1}^{r} \binom{n}{[(n+j)/2]}.$$

(Kleitman 1965)

3.4 Show that, if $A_1, \ldots, A_m$ are distinct k-subsets of an n-set and $k \leqslant h \leqslant n-k$, then there exist distinct h-subsets $B_1, \ldots, B_m$ such that $A_i \cap B_i = \varnothing$ for each $i = 1, \ldots, m$.

3.5 Show that the number of antichains of subsets of an n-set is $\geqslant 2^{\binom{n}{[n/2]}}$.

3.6 Use Dilworth's theorem to show that if $A_1, \ldots, A_n$ are any n sets then we can find $\geqslant n^{1/2}$ of them such that the union of no two is a third.

3.7 (a) Show that the maximum size of a chain in a finite poset P equals the minimum number of disjoint antichains into which P can be decomposed.
(b) Deduce that if r, s are positive integers then a poset of $rs+1$ elements possesses a chain of size $r+1$ or an antichain of size $s+1$ (or both).
(c) Deduce that, given a sequence of $rs+1$ real numbers, there is either an increasing subsequence of length $r+1$ or a decreasing subsequence of length $s+1$. (Mirsky 1971)

3.8 Show that a subset A of $S = \{x_1, \ldots, x_n\}$ is a set of basic elements if and only if, for each i, at most half of $x_1, \ldots, x_i$ are in A. A *zigzag* path between two points with integer components is a path made up of straight line segments joining a point (x, y) to either $(x+1, y+1)$ or $(x+1, y-1)$. Given a set of basic elements, construct a zigzag path from $(0,0)$ as follows: if x_1 is a basic element go from $(0,0)$ to $(1,-1)$, but if it is not a basic element go to $(1,1)$; at the $(i+1)$th stage, go from the point (i, v_i) so far reached to $(i+1, v_i-1)$ if x_i is basic, but to $(i+1, v_i+1)$ if x_i is not basic. Hence show that the number of zigzag paths from $(0,0)$ to a point on $x=n$ which do not go below the x-axis is $\binom{n}{[n/2]}$. Show also that the number of zigzag paths from $(0,0)$ to a particular (n, p) is $\binom{n}{k} - \binom{n}{k-1}$ where $k = \frac{1}{2}(n-p)$ is integral.

(Greene and Kleitman 1976*b*)

3.9 Show that $(2/n\pi)^{1/2}\, 2^n$ is a lower bound for the storage problem of Theorem 3.3.2.

3.10 Obtain the upper bound of Theorem 3.4.2 by showing that the average size of the greatest member of a chain in a symmetric chain decomposition of the set of subsets of an n-set is $\frac{1}{2}n + O(n^{1/2})$.

3.11 Find orthogonal chain decompositions as required in the proof of Theorem 3.4.6 for $n=2$ and $n=3$.

3.12 Show that, if n and k are fixed, $d_m = \binom{n}{m} - \binom{n-k}{m}$ takes its maximum value when $m = [\frac{1}{2}n]$.

(Lih 1980)

3.13 By imitating the first proof of Sperner's theorem show that if $\mathscr{A}$ is an antichain in $C(n, k)$ then

$$\sum_{A \in \mathscr{A}} \frac{1}{\binom{n-1}{|A|-1}} |A \cap T| \leqslant k.$$

Also extend this result to the poset $C(n, k, h)$, consisting of those members of $C(n, k)$ which intersect T in $\geqslant h$ elements.

(Griggs 1982)

3.14 Show that $L(m, n)$ has $\binom{m+n}{n}$ members.

3.15 An *intersecting* family is a collection of subsets of a set no two of which are disjoint. A *maximal* intersecting family is one which cannot be extended to a larger one. Show that the number of maximal intersecting families of subsets of an n-set is at most the number of antichains of an $(n-1)$-set, as follows.

(i) For any maximal intersecting family $\mathscr{A}$ define $g(\mathscr{A}) = \Big\{A \subseteq \{1, \ldots, n-1\} : A = B \cap \{1, \ldots, n-1\}$ for some $B \in \mathscr{A}$, and there does not exist $C \in \mathscr{A}$ such that $C \cap \{1, \ldots, n-1\} \subset A\Big\}$.

Prove that $g(\mathscr{A})$ is an antichain.

(ii) Prove that if $A \subseteq \{1, \ldots, n-1\}$ then

(a) $A \cup \{n\} \in \mathscr{A}$ if and only if $B \subseteq A$ for some $B \in g(\mathscr{A})$

(b) $A \in \mathscr{A}$ if and only if $B \subseteq A$ for some $B \in g(\mathscr{A})$ and there is no $C \in g(\mathscr{A})$ with $C \cap A = \varnothing$.

(iii) Observe that (a) and (b) show that g is one–one.

(Erdös and Hindman 1984)

4 Rank numbers for multisets

4.1 Unimodality and log concavity

In this chapter we shall take a closer look at the rank numbers in the poset of divisors of a number (or subsets of a multiset). We shall use the notation of divisors of a number and throughout the chapter we shall assume that

$$m = p_1^{k_1} p_2^{k_2} \dots p_r^{k_r} \tag{4.1}$$

where the p_i are distinct primes and $k_i \geqslant 1$ for each i. The rank of m is then $r(m) = k_1 + \dots + k_r$, and $N_i(m)$ is the number of divisors of m of rank i. We have already seen from the symmetric chain decomposition of the poset that, if $r(m) = K$ and $\lambda = [\frac{1}{2}K]$,

$$N_i(m) = N_{K-i}(m) \tag{4.2}$$

for each i, and

$$N_0(m) \leqslant N_1(m) \leqslant \dots \leqslant N_\lambda(m); N_\lambda(m) \geqslant \dots \geqslant N_K(m). \tag{4.3}$$

In the case of subsets of an n-set, or square-free m, we have $N_i = \binom{n}{i}$, and $N_i < N_{i+1}$ whenever $i < \frac{1}{2}n$, so that strict inequality occurs throughout (4.3) if n is even and everywhere except at $\binom{n}{\frac{1}{2}(n-1)} = \binom{n}{\frac{1}{2}(n+1)}$ if n is odd. It is natural to ask where strict inequality occurs in (4.3) when m is not square free. This question has been answered by Clements (1984*a*, *b*) and Griggs (1984).

Theorem 4.1.1 Let m be as in (4.1), with $k_1 \geqslant k_2 \geqslant \dots \geqslant k_r$. Let $K = \sum_i k_i$, and put $L = L(m) = \min\{[\frac{1}{2}K], K - k_1\}$. Then

$$N_0 < N_1 < \dots < N_L; N_L = N_{L+1} = \dots = N_{K-L}; N_{K-L} > \dots > N_K.$$

Note that this tells us that if $k_1 \leqslant \frac{1}{2}K$ the rank numbers increase strictly to N_λ and there is a unique maximum rank number if K is even and two equal maximum rank numbers if K is odd. However, if $k_1 > \frac{1}{2}K$

there will be a 'plateau' of $2k_1 - K + 1$ equal maximum rank numbers $N_{K-k_1}, \ldots, N_{k_1}$.

Example 4.1.1 $m = p^4q^3r^2$. Here $K = 9$, $L = \min(4, 5) = 4$, $k_1 < \frac{1}{2}K$. The rank numbers are

$$1, 3, 6, 9, 11, 11, 9, 6, 3, 1.$$

Example 4.1.2 $m = p^5q^2r$. Here $K = 8$, $L = \min(4, 3) = 3$, $k_1 > \frac{1}{2}K$, $2k_1 - K + 1 = 3$, so that there should be a plateau of three maximum rank numbers. The rank numbers are

$$1, 3, 5, 6, 6, 6, 5, 3, 1.$$

Proof of Theorem 4.1.1 (Clements 1984*a*, *b*) We first prove that $N_{i-1} < N_i$ if $i \leqslant L$. The proof is by induction on r, the number of distinct prime factors. The case $r = 1$, in which $m = p^\alpha$, is trivial since then $L = 0$. For the induction step, assume that the result is true for $m' = p_1^{k_1} \ldots p_{r-1}^{k_{r-1}}$, $k_1 \geqslant \ldots \geqslant k_{r-1}$, and consider $m = m'p^{k_r}$ where $k_{r-1} \geqslant k_r$. We then have to prove that, if $i \leqslant L(m)$, $N_{i-1}(m) < N_i(m)$. Now

$$\begin{aligned} N_i(m) - N_{i-1}(m) &= \{N_i(m') + N_{i-1}(m') + \ldots + N_{i-k_r}(m')\} \\ &\quad - \{N_{i-1}(m') + \ldots + N_{i-1-k_r}(m')\} \\ &= N_i(m') - N_{i-1-k_r}(m'), \end{aligned}$$

so we have to show that

$$N_{i-1-k_r}(m') < N_i(m'). \tag{4.4}$$

It will help to have an idea of how large $i - 1 - k_r$ is, so we next show that if $i \leqslant L(m)$ then

$$i - k_r \leqslant L(m'). \tag{4.5}$$

There are two cases to check in proving (4.5). First suppose that $L(m') = [\frac{1}{2}K(m')]$. Then

$$\begin{aligned} i \leqslant L(m) &\Rightarrow i \leqslant \tfrac{1}{2}K(m) \Rightarrow i - k_r \leqslant [\tfrac{1}{2}\{K(m') + k_r\}] - k_r \\ &\Rightarrow i - k_r \leqslant [\tfrac{1}{2}K(m')] = L(m') \end{aligned}$$

as required. Next suppose $L(m') = K(m') - k_1$. Then

$$\begin{aligned} i \leqslant L(m) &\Rightarrow i \leqslant K(m) - k_1 \\ &\Rightarrow i - k_r \leqslant K(m) - k_1 - k_r \\ &\Rightarrow i - k_r \leqslant K(m') - k_1 = L(m'). \end{aligned}$$

So (4.5) is proved. Thus, by the induction hypothesis,

$$N_{i-1-k_r}(m') < N_{i-k_r}(m').$$

To prove (4.4) it is therefore sufficient to show that $N_{i-k_r}(m') \leq N_i(m')$. Since $i - k_r \leq L(m')$, this will be true by (4.2) provided that $i \leq K(m') - (i - k_r)$, i.e. provided that $i \leq \frac{1}{2}\{K(m') + k_r\} = \frac{1}{2}K(m)$. This in turn is true since $i \leq L(m) \leq \frac{1}{2}K(m)$. So (4.4) is proved.

We next show that $N_i(m) = N_{i+1}(m)$ for all i such that $L(m) \leq i < [\frac{1}{2}K(m)]$. Clearly we can assume that $L(m) = K(m) - k_1$. Again, we proceed by induction and assume that $N_i(m') = N_{i+1}(m')$ for all i such that $L(m') \leq i < [\frac{1}{2}K(m')]$. We then have that $N_i(m') \geq N_j(m')$ for all i, j such that $L(m') \leq i \leq j$.

Now if $L(m) \leq i < [\frac{1}{2}K(m)]$ we have

$$i - k_r \geq K(m) - k_1 - k_r = K(m') - k_1 \geq L(m')$$

so that $N_{i-k_r}(m') \geq N_{i+1}(m')$. However, $N_{i+1}(m) - N_i(m) = N_{i+1}(m') - N_{i-k_r}(m')$, so we have $N_{i+1}(m) \leq N_i(m)$. However, $N_i(m) \leq N_{i+1}(m)$ since $i < [\frac{1}{2}K(m)]$, so we conclude that $N_i(m) = N_{i+1}(m)$ as required. □

For completeness we record the number of maximum rank numbers as a corollary.

Corollary 4.1.2 In the notation of Theorem 4.1.1, if $k_1 > \frac{1}{2}K$ there are exactly $2k_1 - K + 1$ equal rank numbers of maximum size. □

4.1.1 Logarithmic concavity

In attempting to prove that a sequence $\{a_n\}$ is unimodal it is sometimes easier to prove a stronger result, namely that the sequence is log concave.

Definition The sequence $\{a_n\}$ of positive numbers is *log concave* if

$$a_{i-1}a_{i+1} \leq a_i^2 \tag{4.6}$$

For each i. Similarly, $\{b_n\}$ is *log convex* if

$$b_{i-1}b_{i+1} \geq b_i^2$$

for each i.

If we write (4.6) as $a_{i+1}/a_i \leq a_i/a_{i-1}$ it becomes clear that if $a_{i+1} \geq a_i$ then $a_i \geq a_{i-1}$, so that $\{a_n\}$ is certainly unimodal. Also note that we can write (4.6) as

$$\tfrac{1}{2}(\log a_{i-1} + \log a_{i+1}) \leq \log a_i$$

so that the sequence $\{\log a_i\}$ is concave in the sense that in the cartesian plane the point $(i, \log a_i)$ lies on or above the straight line segment joining $(i-1, \log a_{i-1})$ and $(i+1, \log a_{i+1})$.

It is trivial to check that, for fixed n, the binomial coefficients $\binom{n}{i}$ are log concave:

$$\binom{n}{i-1}\binom{n}{i+1} \leq \binom{n}{i}^2.$$

It is also fairly straightforward to show that, for fixed n, the Stirling numbers $S(n, k)$ are log concave (see Exercise 4.1) We now show that the rank numbers for multisets are also log concave. This is a special case of results of various authors (see Exercise 4.7) but we give a simple direct proof.

Theorem 4.1.3 The rank numbers for multisets are log concave, i.e. $N_{i-1}N_{i+1} \leq N_i^2$ for each i, $0 < i < K$.

Proof (Anderson 1968) Observe that $N_{i-1}N_{i+1}$ is the number of ordered pairs (a, b) of divisors a, b of m with $r(a) = i-1$, $r(b) = i+1$, and that similarly N_i^2 is the number of ordered pairs (c, d) of divisors c, d of m with $r(c) = r(d) = i$.

Let M be any divisor of m^2. The theorem will be proved if we can show that M can be expressed in the form cd in at least as many ways as in the form ab. To prove this for all M dividing m^2 we use induction on n, the rank of m. The result is trivial if $n = 1$, so we suppose that m has rank $n \geq 2$ and that the result is true for all M arising from an integer of rank less than n. Let M be any divisor of m^2 of rank $2i$.

First suppose that $m \nmid M$. Then there is a prime p which divides m to a higher power than it divides M. Let $m = m_1 p$. Then $M \mid m_1^2$ and $r(m_1) < n$. However, the numbers of representations of M in the forms ab and cd remain unchanged on replacing m by m_1, so that the result follows by induction.

The condition $m \nmid M$ is certainly satisfied if $i < \tfrac{1}{2}n$, since if $m \mid M$ then $r(m) \leq r(M)$, i.e. $n \leq 2i$. Suppose next that $i = \tfrac{1}{2}n$, i.e. $r(m) = r(M)$. If $m \neq M$ then $m \nmid M$ and the result follows as before. If $m = M$ then the numbers of representations of M in the forms ab and cd are precisely $N_{\frac{1}{2}n-1}$ and $N_{\frac{1}{2}n}$, and the result follows since $N_{\frac{1}{2}n-1} \leq N_{\frac{1}{2}n}$.

It remains only to consider those M for which $i > \frac{1}{2}n$. Now there is a one–one correspondence between representations of divisors of m^2 of degree $2i$ and representations of divisors of m^2 of degree $2(n-i)$. For if $ab = cd = M$ then

$$\frac{m}{b} \cdot \frac{m}{a} = \frac{m}{c} \cdot \frac{m}{d} = \frac{m^2}{M},$$

where

$$r\left(\frac{m^2}{M}\right) = 2(n-i).$$

So if the result is true for $i < \frac{1}{2}n$ then it is also true for $i > \frac{1}{2}n$. □

Corollary 4.1.4 If $i < j$ then $N_i N_j \leqslant N_{i+1} N_{j-1}$.

Proof

$$\frac{N_i}{N_{i+1}} \leqslant \frac{N_{i+1}}{N_{i+2}} \leqslant \ldots \leqslant \frac{N_{j-1}}{N_j}.$$ □

4.2 The normalized matching property

We next show that the poset of divisors of m has the normalized matching property. As in the case of log concavity, there are several methods of proof. As well as the proof given here we shall present one in Chapter 12 which treats the poset as a direct product of chains.

Theorem 4.2.1 Let $\mathcal{B}$ be a set of divisors of m of rank i, and let $\nabla\mathcal{B}$ denote the shade of $\mathcal{B}$. Then

$$\frac{|\nabla\mathcal{B}|}{N_{i+1}(m)} \geqslant \frac{|\mathcal{B}|}{N_i(m)}.$$

Proof (Anderson 1968) Let $|\mathcal{B}| = s$ and $|\nabla\mathcal{B}| = t$. We have to prove that $tN_i(m) \geqslant sN_{i+1}(m)$. Now $tN_i(m)$ is the number of ordered pairs (a, v) with $a \in \nabla\mathcal{B}$, $v \mid m$, and $r(v) = i$, while $sN_{i+1}(m)$ is the number of ordered pairs (b, u) with $b \in \mathcal{B}$, $u \mid m$, and $r(u) = i+1$. As in the proof of Theorem 4.1.3 it is sufficient to prove that any divisor M of m^2 of rank $2i+1$ can be expressed in the form av in at least as many ways as in the form bu.

We proceed by induction on the rank n of m. The case $n = 1$ is trivial so we suppose that m has rank $n \geqslant 2$ and that the result is true for divisors of numbers of rank less than n. We let M be a divisor of m^2 of rank $2i+1$. The case $m \nmid M$ is dealt with as in the previous proof: $m = m_1 p$ where M is effectively a divisor of m_1^2. So now suppose that $m \mid M$, say $M = mg$ where $r(g) = 2i - n + 1$. Suppose that the representations of M in the form bu are

$$M = mg = b_1 u_1 = \ldots = b_r u_r.$$

Since $b_i \mid m$ and $u_i \mid m$ it follows that $g \mid u_i$ and $g \mid b_i$, so we can write $b_i = ge_i$, $u_i = gw_i$. Then $m/g = e_i w_i$ and

$$r(e_i) = \frac{1}{2}\left\{r\left(\frac{m}{g}\right) - 1\right\}.$$

If we now consider a symmetric chain decomposition of the poset of divisors of m/g, we see that, since

$$r(e_i) = \frac{1}{2}\left\{r\left(\frac{m}{g}\right) - 1\right\},$$

we can associate with each e_i a multiple f_i of rank one more, namely the next number in the chain containing e_i, to obtain numbers f_i such that $e_i \mid f_i \mid m/g$, $r(f_i) = 1 + r(e_i)$. We then have

$$\frac{m}{g} = f_i z_i \quad \text{where} \quad z_i = w_i \cdot \frac{e_i}{f_i} \,\Big|\, \frac{m}{g}.$$

Thus

$$M = mg = gf_1 \cdot gz_1 = \ldots = gf_r \cdot gz_r$$

where each $gf_i \in \nabla\mathscr{B}$ and each $gz_i \mid m$. This gives r representations of M in the form av as required. □

Corollary 4.2.2 Let $\mathscr{B}$ be a set of divisors of m of rank i, and let $\Delta\mathscr{B}$ denote the shadow of $\mathscr{B}$. Then

$$\frac{|\Delta\mathscr{B}|}{N_{i-1}} \geqslant \frac{|\mathscr{B}|}{N_i}.$$

Proof Let $\mathscr{B}' = \{m/b : b \in \mathscr{B}\}$. Then $|\mathscr{B}'| = |\mathscr{B}|$ and $|\Delta\mathscr{B}| = |\nabla\mathscr{B}'|$. Applying Theorem 4.2.1 to $\mathscr{B}'$, we have

$$\frac{|\nabla\mathscr{B}'|}{N_{n-i+1}} \geqslant \frac{|\mathscr{B}'|}{N_{n-i}}$$

so that

$$\frac{|\Delta\mathscr{B}|}{N_{i-1}} \geqslant \frac{|\mathscr{B}|}{N_i}. \qquad \square$$

We also record the fact that, since the poset of divisors of m has the normalized matching property, it must also have the other properties equivalent to it, as described in Theorem 2.3.1. We thus have the following theorem.

Theorem 4.2.3 If D is an antichain of divisors of m then

$$\sum_{d \in D} \frac{1}{N_{r(d)}} \leqslant 1. \qquad \square$$

As an application of this theorem, we establish a result of Kleitman and Milner (1973) concerning the average rank size in an antichain. The proof given here is essentially due to Grimmett (1973). As in Theorem 4.1.1, we let $L(m)$ denote the largest value of i such that $N_0 < N_1 < \ldots < N_i$ holds.

Theorem 4.2.4 (Kleitman and Milner 1973) Let D be an antichain in the poset of divisors of $m = p_1^{k_1} \ldots p_n^{k_n}$, and suppose that $|D| \geqslant N_k$ for some $k \leqslant L(m)$. Then $\sum_{d \in D} r(d) \geqslant k\,|D|$, i.e. the average rank size of the divisors in D is at least k.

Proof Let q denote the average rank size of the divisors in D, and suppose that $q < k$ with a view to obtaining a contradiction. Let $f(j) = 1/N_j$ for each $j = 0, \ldots, L(m)$. Then, since $N_{j-1}N_{j+1} \leqslant N_j^2$ for each j,

$$f(j-1) + f(j+1) = \frac{1}{N_{j-1}} + \frac{1}{N_{j+1}} \geqslant 2\left(\frac{1}{N_{j-1}N_{j+1}}\right)^{1/2}$$
$$\geqslant \frac{2}{N_j} = 2f(j),$$

so that f is a convex function on $0, 1, \ldots, \sum_i k_i$. By Theorem 4.1.1 it is strictly decreasing on $\{0, \ldots, \sum_i k_i\}$. We can extend f to a convex function $f : [0, \sum_i k_i] \to \mathbb{R}$ by linear interpolation; this extended f is also strictly decreasing on $[0, L(m)]$. If we let d_i denote the number of members of D of rank i, we have

$$\sum_i \frac{d_i}{|D|} = 1,$$

so that, by the convexity of f,

$$f\Big(\sum_i \frac{id_i}{|D|}\Big) \leqslant \sum_i \frac{d_i}{|D|} f(i).$$

The LYM inequality of Theorem 4.2.3 therefore yields

$$\frac{1}{|D|} \geqslant \frac{1}{|D|} \sum_{d \in D} f(r(d)) = \frac{1}{|D|} \sum_i d_i f(i) \geqslant f\Big(\sum_i \frac{id_i}{|D|}\Big)$$
$$= f(q) > f(k) = \frac{1}{N_k}.$$

Thus $|D| < N_k$, giving the required contradiction. □

4.3 The largest size of a rank number

In the case of subsets of an n-set the largest rank number is $\binom{n}{[n/2]}$. Using Stirling's approximation $n! \sim (2\pi n)^{1/2} n^n e^{-n}$ we obtain

$$\binom{n}{[n/2]} \approx \Big(\frac{2}{\pi}\Big)^{1/2} \frac{2^n}{n^{1/2}}$$

If $m = p_1^{k_1} \ldots p_r^{k_r}$, the largest rank number is $N_{[K/2]}(m)$ where $K = k_1 + \ldots + k_r = r(m)$. If we let $\tau(m)$ denote the total number of divisors of m, we have $\tau(m) = N_0(m) + \ldots + N_K(m) = (1+k_1) \ldots (1+k_r)$ since any divisor of m is of the form $p_1^{h_1} \ldots p_r^{h_r}$, where there are $1+k_i$ choices $0, 1, \ldots, k_i$ for h_i. In this section we obtain an estimate for the size of $N_{[K/2]}$ in terms of the k_i. If we write

$$A(m) = \tfrac{1}{3} \sum_i k_i(k_i+2)$$

we find that we can show that the size of the middle-rank number is essentially a multiple of

$$\frac{\tau(m)}{\{A(m)\}^{1/2}}. \tag{4.7}$$

Note that in the square-free case where each $k_i = 1$,

$$\frac{\tau(m)}{\{A(m)\}^{1/2}} = \frac{2^n}{n^{1/2}}.$$

The first lemma below shows that $A(m)$ is closely related to the standard deviation of the distribution of ranks.

Lemma 4.3.1 Let σ^2 be the variance of the distribution of ranks among the divisors of m. Then

$$\sigma^2 = \sum_i \tfrac{1}{12} k_i(k_i + 2).$$

Proof The power of p_i occurring in a divisor of m as in (4.1) is a random variable taking the values $0, 1, \ldots, k_i$ with equal probability. Its average value is therefore $\frac{1}{2}k_i$ and its variance is

$$\begin{aligned}\frac{1}{1+k_i}\sum_{m=0}^{k_i}(m - \tfrac{1}{2}k_i)^2 &= \frac{1}{1+k_i}\left\{\sum_m m^2 - k_i \sum_m m + \tfrac{1}{4}k_i^2(k_i+1)\right\}\\ &= \tfrac{1}{6}k_i(2k_i+1) - \tfrac{1}{2}k_i^2 + \tfrac{1}{4}k_i^2\\ &= \tfrac{1}{12}k_i(k_i+2).\end{aligned}$$

The lemma now follows since the variance of a sum of independent distributions is the sum of the variances. □

Since $A(m) = 4\sigma^2$, we obtain information about $A(m)$ by studying the variance σ^2. Now since

$$\sigma^2 = \frac{1}{\tau(m)}\sum_{i=0}^{K}\left(\frac{K}{2} - i\right)^2 N_i = \frac{1}{2\tau(m)}\sum_{i=0}^{\lambda}(K-2i)^2 N_i \tag{4.8}$$

where $\lambda = [\frac{1}{2}K]$, we have to estimate $\sum_{i=0}^{\lambda}(K-2i)^2 N_i$. To do this, we shall make use of the following lemma.

Lemma 4.3.2 (Reduction formula)

$$\begin{aligned}N_{\lambda-r}^2 \sum_{i=0}^{\lambda-r}(K-2i)^2 N_i &\leq N_{\lambda-r-1}^2 \sum_{i=0}^{\lambda-r-1}(K-2i)^2 N_i + 8N_{\lambda-r-1}^2 \sum_{i=0}^{\lambda-r-1}(K-2i)N_i\\ &\quad + 16N_{\lambda-r-1}^2 \sum_{i=0}^{\lambda-r-1} N_i + 3(2r+5)^2 N_{\lambda-r}^3.\end{aligned}$$

Proof

$$N_{\lambda-r}^2 \sum_{i=0}^{\lambda-r}(K-2i)^2 N_i \leq N_{\lambda-r}^2 \sum_{i=0}^{\lambda-r-2}(K-2i)^2 N_i + 2(2r+5)^2 N_{\lambda-r}^3$$

where the first term on the right is, by two applications of Corollary

4.1.4,

$$\leqslant N_{\lambda-r-1}^2 \sum_{i=0}^{\lambda-r-2} (K-2i)^2 N_{i+2}$$

$$\leqslant N_{\lambda-r-1}^2 \sum_{i=0}^{\lambda-r} (K-2i+4)^2 N_i$$

$$\leqslant N_{\lambda-r-1}^2 \sum_{i=0}^{\lambda-r-1} (K-2i+4)^2 N_i + (2r+5)^2 N_{\lambda-r}^3$$

$$= N_{\lambda-r-1}^2 \sum_{i=0}^{\lambda-r-1} (K-2i)^2 N_i + 8N_{\lambda-r-1}^2 \sum_{i=0}^{\lambda-r-1} (K-2i) N_i$$

$$+ 16 N_{\lambda-r-1}^2 \sum_{i=0}^{\lambda-r-1} N_i + (2r+5)^2 N_{\lambda-r}^3. \qquad \square$$

Lemma 4.3.3 There is a constant $C > 0$ such that

$$\sigma < C \frac{\tau(m)}{N_\lambda(m)}.$$

Proof As in (4.8),

$$2\tau(m) N_\lambda^2(m) \sigma^2 = N_\lambda^2(m) \sum_{i=0}^{\lambda} (K-2i)^2 N_i(m).$$

Using the reduction formula of the previous lemma repeatedly, for $r = 0, 1, \ldots, \lambda - 1$, we obtain

$$2\tau(m) N_\lambda^2(m) \sigma^2 \leqslant 8 \sum_{t=1}^{\lambda} N_{\lambda-t}^2 \left\{ \sum_{i=0}^{\lambda-t} (K-2i) N_i \right\} + 16 \sum_{t=1}^{\lambda} N_{\lambda-t}^2 \left(\sum_{i=0}^{\lambda-t} N_i \right)$$

$$+ 3 \sum_{t=0}^{\lambda-1} (2t+5)^2 N_{\lambda-t}^3. \tag{4.9}$$

However,

$$\sum_{t=0}^{\lambda-1} (2t+5)^2 N_{\lambda-t}^3 \leqslant 25 N_\lambda^3 + \sum_{t=1}^{\lambda-1} N_{\lambda-t} (7t N_{\lambda-t})^2$$

$$\leqslant 25 N_\lambda^3 + 49 \sum_{t=1}^{\lambda-1} N_{\lambda-t} (N_{\lambda-t} + \ldots + N_{\lambda-1})^2$$

$$\leqslant 25 N_\lambda^3 + 49 \sum_{t=1}^{\lambda-1} N_{\lambda-t} \{\tfrac{1}{2}\tau(m)\}^2$$

$$\leqslant 25\tau^2(m) \left(N_\lambda + \sum_{t=1}^{\lambda} N_{\lambda-t} \right) < 25\tau^3(m). \tag{4.10}$$

We also have

$$\sum_{t=1}^{\lambda} N_{\lambda-t}^2\left(\sum_{i=0}^{\lambda-t} N_i\right) \leqslant \tfrac{1}{2}\tau(m)\sum_{t=1}^{\lambda} N_{\lambda-t}^2 \leqslant \tfrac{1}{2}\tau(m)\left(\sum_{t=1}^{\lambda} N_{\lambda-t}\right)^2$$
$$< \tau^3(m). \tag{4.11}$$

Finally, a reduction similar to the above (see Exercise 4.6) leads to

$$N_\lambda \sum_{i=0}^{\lambda} (K-2i)N_i \leqslant 4\tau^2(m),$$

so that we have

$$\sum_{t=1}^{\lambda} N_{\lambda-t}^2\left(\sum_{i=0}^{\lambda-t} (K-2i)N_i\right) \leqslant 4\tau^2(m)\left(\sum_{t=1}^{\lambda} N_{\lambda-t}\right) < 4\tau^3(m). \tag{4.12}$$

Combining (4.9)–(4.12), we therefore obtain

$$2\tau(m)N_\lambda^2(m)\sigma^2 \leqslant (\text{constant})\tau^3(m)$$

whence

$$\sigma \leqslant C\frac{\tau(m)}{N_\lambda(m)}$$

as required. □

Note that, since $\sigma \to \infty$ as $K \to \infty$, $N_\lambda(m)/\tau(m) \to 0$ as $K \to \infty$, and so we can write

$$N_\lambda(m) = o(\tau(m)). \tag{4.13}$$

We now use symmetric chains to obtain a lower bound for σ. In the next lemma we assume that the poset of divisors of m has been decomposed into symmetric chains.

Lemma 4.3.4 Let $H(m)$ denote the number of ways of choosing three divisors of m from the same symmetric chain. Then

$$\sigma^2 > \frac{H(m)}{2\tau(m)}.$$

Proof We shall use the identity

$$\binom{x+2}{3} - \binom{x}{3} = x^2.$$

We have, by (4.8),

$$\begin{aligned}\sigma^2 &\geqslant \frac{1}{2\tau(m)} \sum_{i=0}^{\lambda-1} (K-2i-1)^2 N_i \\ &= \frac{1}{2\tau(m)} \sum_{i=0}^{\lambda-1} \left\{ \binom{K-2i+1}{3} - \binom{K-2i-1}{3} \right\} N_i \\ &= \frac{1}{2\tau(m)} \sum_{i=1}^{\lambda-1} \left\{ \binom{K+1-2i}{3} (N_i - N_{i-1}) \right\} + \binom{K+1}{3} \\ &= \frac{1}{2\tau(m)} H(m)\end{aligned}$$

since, for $i \geqslant 1$, there are $N_i - N_{i-1}$ chains of size $K+1-2i$. □

Lemma 4.3.5 There is a constant $C' > 0$ such that

$$\sigma > C' \frac{\tau(m)}{N_\lambda(m)}.$$

Proof Suppose that the divisors of m are in $s = N_\lambda(m)$ symmetric chains of sizes $x_1, \ldots, x_s$ where $x_1 + \ldots + x_s = \tau(m)$. Then

$$H(m) = \sum_{i=1}^{s} \binom{x_i}{3} = \frac{1}{6} \sum_{i=1}^{s} x_i(x_i - 1)(x_i - 2).$$

Now under the constraint $\sum_i x_i = \tau(m)$, $H(m)$ has its minimum value essentially when all the x_i are equal, i.e. when each $x_i = \tau(m)/s$. It follows from (4.13) that

$$H(m) \geqslant \{\tfrac{1}{6} - o(1)\} s \left(\frac{\tau}{s}\right)^3 = \{\tfrac{1}{6} - o(1)\} \frac{\tau^3(m)}{N_\lambda^2(m)}$$

and the lemma follows from Lemma 4.3.4. □

We can now state our main theorem of this section, justifying the estimate (4.7) for $N_\lambda(m)$.

Theorem 4.3.6 (Anderson 1969) Let $m = p_1^k \ldots p_r^k$, $\lambda = [\frac{1}{2}K]$, $\tau(m) = (1+k_1), \ldots (1+k_r)$, and $A(m) = \frac{1}{3} \sum_i k_i(k_i + 2)$. Then there exist positive constants c_1 and c_2 such that

$$c_1 \frac{\tau(m)}{\{A(m)\}^{1/2}} < N_\lambda(m) < c_2 \frac{\tau(m)}{\{A(m)\}^{1/2}}.$$

Proof The theorem follows immediately from Lemmas 4.3.1, 4.3.3, and 4.3.5. □

Observe that in the case where $k_1 = \ldots = k_n = 1$ and $K = n$,

$$A(m) = \tfrac{1}{3} . 3n = n$$

so that $\tau(m)/\{A(m)\}^{1/2} = 2^n/n^{1/2}$. Thus Theorem 4.3.6 agrees with the estimate for $N_\lambda(m)$ in the square-free case.

By using complex variable methods, an asymptotic formula can be obtained for $N_\lambda(m)$. It has been shown by Anderson (1967*a*) that, under reasonably general conditions,

$$N_\lambda(m) \approx \left(\frac{2}{\pi}\right)^{1/2} \frac{\tau(m)}{\{A(m)\}^{1/2}}.$$

Exercises 4

4.1 Show that, for fixed n, (a) the binomial coefficients $\binom{n}{k}$ and (b) the Stirling numbers $S(n, k)$ are log concave. (In (b), use induction and Theorem 2.4.1(ii).)

4.2 Verify Theorems 4.1.1 and 4.1.2 for $m = p^7q^3r^2$.

4.3 Extend Theorem 2.3.4 to multisets.

4.4 Extend Exercise 2.2 to multisets.

4.5 Show that if $N_{i-1} \leqslant N_i$ and $\mathscr{B}$ is a collection of divisors of m of rank i, then $|\mathscr{B}| - |\Delta\mathscr{B}| \leqslant N_i - N_{i-1}$.

4.6 (a) Prove the reduction formula

$$N_{\lambda-r} \sum_{i=0}^{\lambda-r} (K-2i)N_i \leqslant N_{\lambda-r-1} \sum_{i=0}^{\lambda-r-1} (K-2i)N_i + 2N_{\lambda-r} \sum_{i=0}^{\lambda-r} N_i + 2r\{N_{\lambda-r}^2 + N_{\lambda-r-1}N_{\lambda-r}\}.$$

(b) Use (a) to prove that $N_\lambda \sum_{i=0}^{\lambda} (K-2i)N_i \leqslant 4\tau^2(m)$.

(c) Let $J(m)$ denote the number of ways of choosing two divisors of m from the same chain of a given symmetric chain decomposition. Show that $J(m) \leqslant 2\sum_{i=0}^{\lambda} (K-2i)N_i$, and use (b) to deduce that

$$J(m) < C\frac{\tau^2(m)}{N_\lambda(m)}.$$

(d) Verify that, if m is square-free,

$$J(m) \approx \frac{2}{\pi}\frac{\tau^2(m)}{N_\lambda(m)}. \qquad \text{(Anderson 1968)}$$

4.7 In the product $H = F \times G$ of two ranked posets (defined in Section 3.6) the rank numbers h_k of H are given in terms of the rank numbers f_k, g_k of F, G by

$$h_k = \sum_{p \leqslant k} f_p g_{k-p}.$$

Suppose that the f_k and the g_k are log concave. Show that the h_k are then also log concave as follows.

(i) We have to prove that

$$\sum_{\substack{p \leqslant k-1 \\ m \leqslant k+1}} f_p g_{k-1-p} f_m g_{k+1-m} \leqslant \sum_{\substack{p \leqslant k \\ m \leqslant k}} f_m f_p g_{k-m} g_{k-p}.$$

(ii) If $p \leqslant k$, verify that both of the expressions $f_m f_p - f_{m-1} f_{p+1}$ and $g_{k-m} g_{k-p} - g_{k-m+1} g_{k-p-1}$ are $\geqslant 0$ or $\leqslant 0$ according as $m \leqslant p$ or $m \geqslant p$, so that their product is always $\geqslant 0$. Deduce that

$$f_m g_{k-m+1} f_p g_{k-p-1} + f_{m-1} g_{k-m} f_{p+1} g_{k-p} \leqslant f_m f_p g_{k-m} g_{k-p} + f_{m-1} f_{p+1} g_{k-(m-1)} g_{k-(p+1)}.$$

(iii) Sum over all p and m to prove the inequality of (i).

(iv) Deduce Theorem 4.1.3.

(Hsieh and Kleitman 1973, Hoggar 1974)

5 Intersecting systems and the Erdös–Ko–Rado theorem

5.1 The EKR theorem

The main result to be discussed so far, Sperner's theorem, is concerned with antichains of subsets of a set. We now turn our attention to a special class of antichains which have the extra property that no two members of the antichain are disjoint.

Definition A collection of sets $A_1, \ldots, A_r$ is called an *intersecting family*, or is said to have the *intersection property*, if $A_i \cap A_j \neq \varnothing$ for all i, j. The following result due to Erdös, Ko, and Rado (1961) was published in 1961.

Theorem 5.1.1 (Erdös–Ko–Rado (EKR) theorem) Let $\{A_1, \ldots, A_m\}$ be an antichain of subsets of an n-set S which has the intersection property. Suppose also that there is a number $k \leqslant \frac{1}{2}n$ such that $|A_i| \leqslant k$ for each i. Then

$$m \leqslant \binom{n-1}{k-1}.$$

Before proving this result we observe that we need only prove it in the special case where $|A_i| = k$ for each i, for if some of the sets A_i have fewer than k elements, we can replace each of them by a set of size k containing it (e.g. by the k-set in the same chain of a symmetric chain decomposition), meanwhile preserving the antichain property. The intersection property is also automatically preserved since each set is replaced by one containing it. It follows that we need only prove the following theorem.

Theorem 5.1.2 Let $A_1, \ldots, A_m$ be m k-subsets of an n-set S, $k \leqslant \frac{1}{2}n$, which are pairwise non-disjoint. Then

$$m \leqslant \binom{n-1}{k-1}.$$

Note that the upper bound on m given by this theorem is best possible; it is attained when the A_i are precisely the k-subsets of S which contain a chosen fixed element of S. Note also that the condition $k \leq \frac{1}{2}n$ is essential since if $k > \frac{1}{2}n$ any two k-subsets are non-disjoint, so that m could be as large as $\binom{n}{k}$. The case $k > \frac{1}{2}n$ will be considered in Section 5.3.

The first proof of Sperner's theorem in this book involved a study of all permutations of the elements of S, estimating how often the initial elements of such a permutation can constitute one of the sets A_i. We shall make use of similar ideas in proving Theorem 5.1.2, this time looking at *cyclic* orderings of the elements of S. There are $(n-1)!$ such cyclic permutations. We say that a cyclic permutation *contains* A if the elements of A are consecutive in the permutation.

Proof of Theorem 5.1.2 (Katona 1972*b*) Suppose that $S = \{1, \ldots, n\}$. For each cyclic permutation α of $1, \ldots, n$ and each set A_i, define

$$f(\alpha, A_i) = \begin{cases} 1 & \text{if } \alpha \text{ contains } A_i \\ 0 & \text{otherwise} \end{cases}.$$

We count $\sum_{i,\alpha} f(\alpha, A_i)$ in two different ways. First,

$$\sum_{i,\alpha} f(\alpha, A_i) = \sum_i \sum_\alpha{}' 1$$

where $\sum'$ is over all cyclic permutations α which contain A_i. Now each α has n k-sets of consecutive elements, and, by symmetry, each of the $\binom{n}{k}$ k-subsets of S will occur as consecutive elements equally often when all of the $(n-1)!$ cyclic permutations α are considered. So we must have

$$\sum_\alpha{}' 1 = \frac{n \,.\, (n-1)!}{\binom{n}{k}} = \frac{n!}{\binom{n}{k}}$$

so that

$$\sum_{i,\alpha} f(\alpha, A_i) = \frac{m \,.\, n!}{\binom{n}{k}}. \tag{5.1}$$

However, if we can prove that any one cyclic permutation α can contain at most k of the sets A_i, it will follow that

$$\sum_{i,\alpha} f(\alpha, A_i) = \sum_\alpha \sum_{\substack{i \\ (A_i \text{ in } \alpha)}} 1 \leq \sum_\alpha k = (n-1)!\,k. \tag{5.2}$$

Comparing the two estimates (5.1) and (5.2) we shall then be able to conclude that

$$\frac{m \cdot n!}{\binom{n}{k}} \leqslant k \cdot (n-1)!,$$

i.e.

$$m \leqslant \frac{k}{n}\binom{n}{k} = \binom{n-1}{k-1}.$$

It therefore remains to prove the following lemma.

Lemma 5.1.3 If α is a cyclic permutation of $1, \ldots, n$ and the subsets A_i of $\{1, \ldots, n\}$ are as in Theorem 5.1.2, then α can contain at most k of the sets A_i.

Proof Suppose that A_h appears as consecutive elements $x_1, \ldots, x_k$ of α. Then, owing to the intersection property of the sets A_i, the only sets of k consecutive elements of α which can constitute other sets are the $k-1$ sets A_i beginning with x_i $(2 \leqslant i \leqslant k)$ and the sets B_i ending with x_i $(1 \leqslant i \leqslant k-1)$. But at most one of A_{i+1}, B_i can be in the collection if $1 \leqslant i \leqslant k-1$, and, since $A_1 = B_k$, the collection can therefore contain at most $1 + (k-1) = k$ of these sets. □

The above proof is due to Katona (1972*b*); the original proof was not so elegant. It will be possible to make further use of the ideas of Katona's proof in obtaining more general results in the following sections. Another proof of the EKR theorem will be given in Chapter 7 where it will be shown that it is an easy consequence of the Kruskal–Katona theorem.

It is natural to generalize the intersection property. Suppose we replace the condition $A_i \cap A_j \neq \varnothing$ (i.e. all intersections of two sets are non-empty) by the condition all intersections of h sets are non-empty (where $h \geqslant 2$ is a given integer). For this more general situation the following result holds.

Theorem 5.1.4 (Frankl 1976*a*) Let $\mathscr{A}$ be a collection of k-subsets of $\{1, \ldots, n\}$ such that $A_1 \cap \ldots \cap A_h \neq \varnothing$ for all choices of $A_1, \ldots, A_h \in \mathscr{A}$. Suppose that $kh \leqslant n(h-1)$. Then $|\mathscr{A}| \leqslant \binom{n-1}{k-1}$. □

Clearly the case $h = 2$ reduces to the EKR theorem. The theorem in fact follows from the following result in the same way that Theorem 5.1.2 followed from Lemma 5.1.3.

Lemma 5.1.5 If α is a cyclic permutation of $1, \ldots, n$ and the sets A_i are as in Theorem 5.1.4, then, if $kh \leq (h-1)n$, at most k of the sets A_i can be contained in α. □

5.2 Generalizations of EKR

Just as the original version of Sperner's theorem can be strengthened to a LYM inequality, so can the EKR theorem be strengthened to a LYM version. The first stage in this strengthening is to adapt Lemma 5.1.3.

Lemma 5.2.1 If $\{A_1, \ldots, A_m\}$ is an antichain of intersecting subsets of $S = \{1, \ldots, n\}$, if $|A_i| \leq \frac{1}{2}n$ for each i, and if α is any cyclic permutation of $1, \ldots, n$ which contains A_h, then α contains at most $|A_h|$ of the sets A_i.

Proof. Follow the proof of Lemma 5.1.3. Let A_h appear in α as the consecutive elements $x_i, \ldots, x_k$. Then any other set A_i contained in α must begin at one of $x_2, \ldots, x_k$ or end at one of $x_1, \ldots, x_{k-1}$. The sets need not all be the same size, but the antichain property ensures that no two can begin at the same element of α or end at the same element. As in the proof of Lemma 5.1.3, the antichain can contain at most $k = |A_h|$ of these sets. □

Theorem 5.2.2 (Bollobás 1973). Let $\{A_1, \ldots, A_m\}$ be an intersecting antichain of subsets of an n-set S such that $|A_i| \leq \frac{1}{2}n$ for each i. Then

$$\sum_{i=1}^{m} \frac{1}{\binom{n-1}{|A_i|-1}} \leq 1.$$

Proof (Greene, Katona, and Kleitman 1976) We imitate the proof of Theorem 5.1.2, but this time define

$$f(\alpha, A_i) = \begin{cases} \dfrac{1}{|A_i|} & \text{if } \alpha \text{ contains } A_i \\ 0 & \text{otherwise.} \end{cases}$$

Again we estimate $\sum_{i,\alpha} f(\alpha, A_i)$ in two different ways. First we have

$$\sum_{i,\alpha} f(\alpha, A_i) = \sum_{\alpha} \sum_{\substack{i \\ (A_i \text{ in } \alpha)}} \frac{1}{|A_i|}.$$

Consider the inner sum where α has been fixed. Choose j such that A_j is the smallest of the A_i contained in α. Then, by the lemma, there are at most $|A_j|$ terms in the inner sum, each $\leqslant 1/|A_j|$. Therefore the inner sum is at most

$$|A_j| \cdot \frac{1}{|A_j|} = 1,$$

and we have

$$\sum_{i,\alpha} f(\alpha, A_i) \leqslant \sum_{\alpha} 1 = (n-1)!. \tag{5.3}$$

However,

$$\sum_{i,\alpha} f(\alpha, A_i) = \sum_{i} \frac{1}{|A_i|} \sum_{\substack{\alpha \\ (A_i \text{ in } \alpha)}} 1 = \sum_{i} \frac{1}{|A_i|} \frac{n \cdot (n-1)!}{\binom{n}{|A_i|}}$$

as in the proof of Theorem 5.1.2, so that

$$\sum_{i,\alpha} f(\alpha, A_i) = \sum_{i} \frac{1}{|A_i|} \frac{n!}{\binom{n}{|A_i|}} = \sum_{i} \frac{(n-1)!}{\binom{n-1}{|A_i|-1}}. \tag{5.4}$$

Comparing (5.3) and (5.4) we obtain

$$\sum_{i} \frac{1}{\binom{n-1}{|A_i|-1}} \leqslant 1. \qquad \square$$

Theorem 5.1.1 is clearly a consequence of this result since, if $|A_i| \leqslant k \leqslant \frac{1}{2}n$ for each i, then $\binom{n-1}{k-1} \geqslant \binom{n-1}{|A_i|-1}$ for each i. Note also that Theorem 5.2.2 implies that if $|A_i| < k$ for some i, we have *strict* inequality: $m < \binom{n-1}{k-1}$.

Corollary 5.2.3 (Bollobás 1973) Let $\mathscr{A}$ be an antichain of subsets of an n-set S such that $A \in \mathscr{A} \Rightarrow A' \in \mathscr{A}$. Then

$$\sum_{A \in \mathscr{A}} \frac{1}{\theta(A)} \leqslant 2$$

where

$$\theta(A) = \min\left\{\binom{n-1}{|A|-1}, \binom{n-1}{n-|A|-1}\right\}.$$

Thus

$$|\mathscr{A}| \leq 2\binom{n-1}{[n/2]-1}.$$

Proof We first note that if $\mathscr{A}$ contains any sets of size $\frac{1}{2}n$, these sets must occur in complementary pairs. Form a subcollection $\mathscr{B}$ of $\mathscr{A}$ by putting into $\mathscr{B}$ all those sets $A \in \mathscr{A}$ with $|A| < \frac{1}{2}n$, and one set from each complementary pair of sets of size $\frac{1}{2}n$ in $\mathscr{A}$. Then $\mathscr{B}$ must be an intersecting family, for if $X, Y \in \mathscr{B}$ and $X \cap Y = \varnothing$, then $X \subset Y'$, contradicting the antichain property. Since $\theta(A) = \binom{n-1}{|A|-1}$ for each $A \in \mathscr{B}$, we can apply Theorem 5.2.2 to $\mathscr{B}$ to obtain

$$\sum_{A \in \mathscr{B}} \frac{1}{\theta(A)} \leq 1. \tag{5.5}$$

The remaining members of $\mathscr{A}$ are precisely the complements of the sets in $\mathscr{B}$. Applying Theorem 5.2.2 to their complements, we obtain

$$\sum_{A \in \mathscr{A} - \mathscr{B}} \frac{1}{\binom{n-1}{(n-|A|)-1}} \leq 1,$$

i.e.

$$\sum_{A \in \mathscr{A} - \mathscr{B}} \frac{1}{\theta(A)} \leq 1. \tag{5.6}$$

The required result now follows by adding (5.5) and (5.6). Finally, note that

$$\theta(A) \leq \binom{n-1}{[n/2]-1} \tag{5.7}$$

so that

$$\frac{|\mathscr{A}|}{\binom{n-1}{[n/2]-1}} \leq \sum_{A \in \mathscr{A}} \frac{1}{\theta(A)} \leq 2. \qquad \square$$

Corollary 5.2.4 (Brace and Daykin 1972) Let $\mathscr{A}$ be an antichain of subsets of an n-set S such that if $A, B \in \mathscr{A}$ then $A \cap B \neq \varnothing$ and

$A \cup B \neq S$. Then

$$|\mathscr{A}| \leq \binom{n-1}{[(n-2)/2]}.$$

Proof Let $\mathscr{A}' = \{A' : A \in \mathscr{A}\}$. Then $\mathscr{A} \cup \mathscr{A}'$ is an antichain satisfying the conditions of Corollary 5.2.3 so that

$$\sum_{A \in \mathscr{A} \cup \mathscr{A}'} \frac{1}{\theta(A)} \leq 2.$$

Thus, by (5.7),

$$\frac{2|\mathscr{A}|}{\binom{n-1}{[n/2]-1}} \leq \sum_{A \in \mathscr{A} \cup \mathscr{A}'} \frac{1}{\theta(A)} \leq 2,$$

whence $|\mathscr{A}| \leq \binom{n-1}{[(n-2)/2]}$.

□

The conditions of Corollary 5.2.4 can be rewritten as follows: for all $A, B \in \mathscr{A}$, $A \cap B \neq \varnothing$, $A \cap B' \neq \varnothing$, $A' \cap B \neq \varnothing$, $A' \cap B' \neq \varnothing$, i.e. all four regions defined by A and B in a Venn diagram are non-empty. The collection $\mathscr{A}$ is in this case called *2-independent.* More generally, $\mathscr{A}$ is *k-independent* if, for all choices of k sets $A_1, \ldots, A_k$ in $\mathscr{A}$, all 2^k regions of their Venn diagram are non-empty.

We conclude this section with a few remarks about alternative paths through the theorems. Instead of proving Theorem 5.2.2 by altering the choice of function $f(\alpha, A_i)$, we could derive it from Theorem 5.1.2. This can be achieved by 'moving up' all sets of smallest size, replacing them by sets of size one more, and using the following lemma (see Exercise 7.14).

Lemma 5.2.5 Let $\mathscr{A}$ be a set of k-subsets of an n-set such that $|\mathscr{A}| \leq \binom{n-1}{k-1}$. Then

$$|\nabla \mathscr{A}| \geq \frac{\binom{n-1}{k}}{\binom{n-1}{k-1}} |\mathscr{A}|.$$

However, we cannot prove this lemma until we have met the Kruskal–Katona theorem in Chapter 7. The argument in fact enables a LYM inequality to be derived from Theorem 5.1.4 also.

5.3 Intersecting antichains with large members

The EKR theorem was concerned with intersecting antichains whose members are all of size $\leqslant k \leqslant \frac{1}{2}n$. In this section we remove this restriction on size and show that $|\mathscr{A}|$ is then at most $\binom{n}{[n/2]+1}$. By considering the set of all subsets of size $[n/2]+1$ we see that this bound is the best possible.

Theorem 5.3.1 (Greene *et al.* 1976) Let $\mathscr{A} = \{A_1, \ldots, A_m\}$ be an antichain of subsets of an n-set S such that $A_i \cap A_j \neq \varnothing$ for all i, j. Then

$$\sum_{\substack{A \in \mathscr{A} \\ |A| \leqslant n/2}} \frac{1}{\binom{n}{|A|-1}} + \sum_{\substack{A \in \mathscr{A} \\ |A| > n/2}} \frac{1}{\binom{n}{|A|}} \leqslant 1.$$

Proof The proof follows that of Theorems 5.1.2 and 5.2.2, but involves yet another choice of the weight function $f(\alpha, A_i)$. Define

$$f(\alpha, A_i) = \begin{cases} \dfrac{n-|A_i|+1}{|A_i|} & \text{if } |A_i| \leqslant \frac{1}{2}n \text{ and } \alpha \text{ contains } A_i \\ 1 & \text{if } |A_i| > \frac{1}{2}n \text{ and } \alpha \text{ contains } A_i \\ 0 & \text{otherwise.} \end{cases}$$

Then

$$\begin{aligned} \sum_{i,\alpha} f(\alpha, A_i) &= \sum_i \Big\{ \sum_{\substack{\alpha \\ A_i \text{ in } \alpha}} f(\alpha, A_i) \Big\} \\ &= \sum_{\substack{i \\ |A_i| \leqslant n/2}} \frac{n-|A_i|+1}{|A_i|} \sum_{\substack{\alpha \\ A_i \text{ in } \alpha}} 1 + \sum_{\substack{i \\ |A_i| > n/2}} \sum_{\substack{\alpha \\ A_i \text{ in } \alpha}} 1 \\ &= \sum_{\substack{i \\ |A_i| \leqslant n/2}} \frac{n-|A_i|+1}{|A_i|} \frac{n!}{\binom{n}{|A_i|}} + \sum_{\substack{i \\ |A_i| > n/2}} \frac{n!}{\binom{n}{|A_i|}} \\ &= n! \sum_{|A_i| \leqslant n/2} \frac{1}{\binom{n}{|A_i|-1}} + n! \sum_{|A_i| > n/2} \frac{1}{\binom{n}{|A_i|}}. \end{aligned}$$

However,

$$\sum_{i,\alpha} f(\alpha, A_i) = \sum_{\alpha} \Big\{ \sum_{\substack{i \\ A_i \text{ in } \alpha}} f(\alpha, A_i) \Big\}$$

$$= \sum_{\alpha} \Big\{ \sum_{\substack{|A_i| \leq n/2 \\ A_i \text{ in } \alpha}} \frac{n - |A_i| + 1}{|A_i|} + \sum_{\substack{|A_i| > n/2 \\ A_i \text{ in } \alpha}} 1 \Big\}.$$

If we can show that right-hand side of this expression is

$$\leq \sum_{\alpha} n = n \,.\, (n-1)! = n!$$

then the required result will follow on comparing the two expressions for $\sum_{i,\alpha} f(\alpha, A_i)$. Thus it remains to prove the following lemma.

Lemma 5.3.2 Let the sets A_i be as in Theorem 5.3.1, and let α be any cyclic permutation of the elements of S. Then

$$\sum_{\substack{|A_i| \leq n/2 \\ A_i \text{ in } \alpha}} \frac{n - |A_i| + 1}{|A_i|} + \sum_{\substack{|A_i| > n/2 \\ A_i \text{ in } \alpha}} 1 \leq n.$$

Proof Suppose first that the first sum is empty. We have then to prove that the number of sets A_i contained in α is at most n. However, by the antichain property at most one A_i can begin at any given element of the cyclic permutation α, so the result is trivial.

Next suppose that there is a set A_j of $\mathscr{A}$ contained in α, with $|A_j| = r \leq \frac{1}{2}n$. Choose such an A_j with r as small as possible, and without loss of generality suppose that $A_j = \{1, \ldots, r\}$ and $S = \{1, \ldots, n\}$. Any other A_i contained in α must consist of consecutive elements of α beginning with some s, $2 \leq s \leq r$, or ending with some $s-1$, $2 \leq s \leq r$. By the antichain property, at most one A_i can begin with any such s, and at most one A_i can end with any such $s-1$. However, note that if A_{i_1} begins with s and A_{i_2} ends with $s-1$ then, by the intersection property, at least one of them must contain more than $\frac{1}{2}n$ elements. Since r is the minimum size of an A_i, the sum of the weights of A_{i_1} and A_{i_2} is therefore at most

$$1 + \frac{n-r+1}{r} = \frac{n+1}{r}.$$

The total sum of the weights of the sets contained in α is therefore at most

$$\frac{n-r+1}{r} + (r-1)\frac{n+1}{r} = n. \qquad \square$$

Corollary 5.3.3 (Brace and Daykin 1972; Schonheim 1974b) If $\mathscr{A}$ is an intersecting antichain of subsets of an n-set, then

$$|\mathscr{A}| \leqslant \binom{n}{[n/2]+1}.$$

Proof Let $A \in \mathscr{A}$. If $|A| \leqslant \frac{1}{2}n$ then $\binom{n}{|A|-1} \leqslant \binom{n}{[n/2]-1} \leqslant \binom{n}{[n/2]+1}$. If $|A| > \frac{1}{2}n$, then $|A| \geqslant \left[\frac{n+2}{2}\right]$ so that $\binom{n}{|A|} \leqslant \binom{n}{[n/2]+1}$. The result now follows from Theorem 5.3.1. □

This result is a special case of a more general result due to Milner: if $\mathscr{A}$ is an antichain such that $|A_i \cap A_j| \geqslant t$ for all $A_i, A_j \in \mathscr{A}$, then $|\mathscr{A}| \leqslant \binom{n}{[\frac{1}{2}(n+t+1)]}$. The case $t=0$ is just Sperner's theorem, and the case $t=1$ is the above corollary. We prove Milner's theorem as Theorem 5.5.1.

5.4 A probability application of EKR

Suppose that $Y_1, \ldots, Y_n$ are independent variables, each taking the value unity with probability p and the value zero with probability $1-p$, where $p \geqslant \frac{1}{2}$. We use EKR to prove the following theorem.

Theorem 5.4.1 (Liggett 1977) Let $\alpha_1, \ldots, \alpha_n$ be non-negative real numbers such that $\sum_i \alpha_i = 1$. Then, with the Y_i as above,

$$P(\alpha_1 Y_1 + \ldots + \alpha_n Y_n \geqslant \tfrac{1}{2}) \geqslant p.$$

Proof Suppose first of all that no partial sum of the α_i is $\frac{1}{2}$. Let M_k denote the number of subsets A of $\{1, \ldots, n\}$ of size k such that $\sum_{i \in A} \alpha_i > \frac{1}{2}$. Then clearly $M_k + M_{n-k} = \binom{n}{k}$. Note that, for any fixed k, no two sets A contributing to M_k can be disjoint, for otherwise the sum of the α_i over all i in their union would be greater than unity. Thus by the EKR theorem we must have

$$M_k \leqslant \binom{n-1}{k-1} \quad \text{if } k \leqslant \tfrac{1}{2}n.$$

Now

$$P\left(\sum_{i=1}^{n} \alpha_i Y_i \geq \tfrac{1}{2}\right) = \sum_{k=1}^{n} M_k p^k (1-p)^{n-k}.$$

However, on writing $k = h + 1$,

$$\sum_{k=1}^{n} \binom{n-1}{k-1} p^k (1-p)^{n-k} = p \sum_{h=0}^{n-1} \binom{n-1}{h} p^h (1-p)^{n-1-h}$$
$$= p(p + 1 - p)^{n-1} = p,$$

so that we have

$$\begin{aligned}
P\left(\sum_{i=1}^{n} \alpha_i Y_i \geq \frac{1}{2}\right) - p &= \sum_{k=1}^{n} p^k (1-p)^{n-k} \left(M_k - \binom{n-1}{k-1}\right) \\
&= \sum_{k \leq n/2} p^k (1-p)^{n-k} \left(M_k - \binom{n-1}{k-1}\right) \\
&\quad + \sum_{j < n/2} p^{n-j} (1-p)^j \left(M_{n-j} - \binom{n-1}{n-j-1}\right) \\
&= \sum_{k \leq n/2} p^k (1-p)^{n-k} \left(M_k - \binom{n-1}{k-1}\right) \\
&\quad + \sum_{j < n/2} p^{n-j} (1-p)^j \left(\binom{n}{j} - M_j - \binom{n-1}{j}\right) \\
&= \sum_{k \leq n/2} p^k (1-p)^{n-k} \left(M_k - \binom{n-1}{k-1}\right) \\
&\quad + \sum_{j < n/2} p^{n-j} (1-p)^j \left(\binom{n-1}{j-1} - M_j\right) \\
&\geq \sum_{k < n/2} (p^k (1-p)^{n-k} - p^{n-k} (1-p)^k) \left(M_k - \binom{n-1}{k-1}\right).
\end{aligned}$$

We want to show that the expression on the right is ≥ 0. Since $M_k \leq \binom{n-1}{k-1}$ we therefore have to show that

$$p^k (1-p)^{n-k} \leq p^{n-k} (1-p)^k \quad \text{for all } k < \tfrac{1}{2}n,$$

i.e.

$$p^{n-2k} \geq (1-p)^{n-2k} \quad \text{for all } k < \tfrac{1}{2}n.$$

However, this is true since $p \geq \frac{1}{2}$, i.e. $p \geq 1 - p$.

Finally, if a partial sum of the α_i is equal to $\frac{1}{2}$, alter the α_i fractionally to avoid this and then apply a limiting procedure. □

5.5 Theorems of Milner and Katona

At the end of Section 5.3 we proved that if $\mathscr{A}$ is an antichain of subsets of an n-set such that $|A_i \cap A_j| \geq 1$ for all $A_i, A_j \in \mathscr{A}$, then $|\mathscr{A}| \leq \binom{n}{[\frac{1}{2}n]+1}$. This result is a special case of a more general result due to Milner, and it is our aim in this section to state and prove Milner's result.

Theorem 5.5.1 (Milner 1968) If $\mathscr{A}$ is an antichain of subsets of an n-set S such that $|A_i \cap A_j| \geq k$ for all $A_i, A_j \in \mathscr{A}$, then

$$|\mathscr{A}| \leq \left(\begin{array}{c} n \\ \left\lceil \frac{n+k+1}{2} \right\rceil \end{array} \right).$$

In the proof of this result we shall make use of a valuable result of Katona concerning the shadow of a collection of k-sets. Recall that the shadow of a collection $\mathscr{B}$ of k-subsets of an n-set S is

$$\Delta\mathscr{B} = \{A \subseteq S : |A| = k-1, A \subset B \text{ for some } B \in \mathscr{B}\}.$$

In proving Sperner's theorem we showed very simply that

$$|\Delta\mathscr{B}| \geq \frac{k}{n-k+1} |\mathscr{B}|.$$

In Chapter 7 we shall study the size of $\Delta\mathscr{B}$ in much greater depth, but at this stage we are going to study $|\Delta\mathscr{B}|$ in the special case where $\mathscr{B}$ is an intersecting family. It turns out that in this case we always have $|\Delta\mathscr{B}| \geq |\mathscr{B}|$. The proof of this fact will make use of a *shift* operator which will shift the sets away from a particular element of S.

Lemma 5.5.2 Let $\mathscr{A}$ be an intersecting collection of k-subsets of $S = \{1, \ldots, n\}$ and, for $1 < j \leq n$, define

$$S_j(A) = \begin{cases} (A - \{1\}) \cup \{j\} & \text{if } 1 \in A, j \notin A, (A - \{1\}) \cup \{j\} \notin \mathscr{A} \\ A & \text{otherwise} \end{cases}$$

for each $A \in \mathscr{A}$. Then

$$S_j(\mathscr{A}) = \{S_j(A) : A \in \mathscr{A}\}$$

is also an intersecting collection of k-subsets of S, and

$$|\Delta S_j(\mathscr{A})| \leq |\Delta\mathscr{A}|.$$

Proof We first check that $S_j(A) \cap S_j(B) \neq \varnothing$ whenever $A, B \in \mathcal{A}$. The only case which is non-trivial is when $S_j(A) = (A - \{1\}) \cup \{j\}$, $S_j(B) = B$, $1 \in B$, $j \notin B$. However, in this case $(B - \{1\}) \cup \{j\}$ and A are both in $\mathcal{A}$ and so must have an element x in common; we must have $x \neq 1$ and $x \neq j$, so $x \in S_j(A)$ and $x \in S_j(B)$.

We next check that $|\Delta\mathcal{A}| \geqslant |\Delta S_j(\mathcal{A})|$. If $X \in \Delta S_j(\mathcal{A})$ but $X \notin \Delta\mathcal{A}$ then $X \subset (A - \{1\}) \cup \{j\}$ for some A, $X \not\subset A$, so that $j \in X$, $1 \notin X$, $Y = (X - \{j\}) \cup \{1\} \in \Delta\mathcal{A}$ and $Y \notin \Delta S_j(\mathcal{A})$. We thus have

$$X \in \Delta S_j(\mathcal{A}) - \Delta\mathcal{A} \Rightarrow Y = (X - \{j\}) \cup \{1\} \in \Delta\mathcal{A} - \Delta S_j(\mathcal{A})$$

so that $|\Delta\mathcal{A} - \Delta S_j(\mathcal{A})| \geqslant |\Delta S_j(\mathcal{A}) - \Delta\mathcal{A}|$, i.e. $|\Delta\mathcal{A}| \geqslant |\Delta S_j(\mathcal{A})|$. □

Theorem 5.5.3 (Katona 1964) If $\mathcal{A} = \{(A_1, \ldots, A_m\}$ is an intersecting family of k-subsets of $S = \{1, \ldots, n\}$, then

$$|\Delta\mathcal{A}| \geqslant |\mathcal{A}|.$$

Proof If $k \geqslant \frac{1}{2}(n+1)$ the result is immediate from the symmetric chain decomposition of the set of subsets of S; so now suppose $k \leqslant \frac{1}{2}n$. First consider $k = 2$. If $m = 1$, $|\Delta\mathcal{A}| = 2 > |\mathcal{A}|$, so now suppose $m \geqslant 2$. If no member of S is in more than two members of $\mathcal{A}$, and if $A_1 = \{a, b\}$, $A_2 = \{a, c\} \in \mathcal{A}$, the only other possibility for another set in $\mathcal{A}$ is $\{b, c\}$; therefore $|\Delta\mathcal{A}| = 3 \geqslant |\mathcal{A}|$. If there is an element $a \in S$ occurring in A_1, A_2, A_3, then a must be in every set A_i (for if $a \notin A_j$, A_j needs three members to enable it to intersect each of A_1, A_2, A_3); therefore $|\Delta\mathcal{A}| = 1 + |\mathcal{A}| > |\mathcal{A}|$.

Suppose now that $k > 2$. If we repeatedly apply the shift operators S_j, $j = 2, \ldots, n$, to $\mathcal{A}$, the number of sets containing 1 must decrease and after a finite number of applications the shifts must therefore cease to make any change. We therefore eventually obtain a new collection $\mathcal{A}^0$ of k-subsets of S, with $S_j(\mathcal{A}^0) = \mathcal{A}^0$ for each $j \geqslant 2$, $|\mathcal{A}^0| = |\mathcal{A}|$, and $|\Delta\mathcal{A}^0| \leqslant |\Delta\mathcal{A}|$. The theorem will be proved for $\mathcal{A}$ if it can be proved for $\mathcal{A}^0$. Therefore we can now assume that $\mathcal{A}$ has the property that $S_j(\mathcal{A}) = \mathcal{A}$ for each $j \geqslant 2$. Thus we can suppose that, for all $A \in \mathcal{A}$ and $j \geqslant 2$,

$$1 \in A, j \notin A \Rightarrow (A - \{1\}) \cup \{j\} \in \mathcal{A}. \tag{5.8}$$

Suppose that the sets A_i are now relabelled so that $1 \in A_i$ if and only if $i \leqslant m_0$, and let $B_i = A_i - \{1\}$ for each $i \leqslant m_0$. Then we claim that the B_i are pairwise non-disjoint. Since $|A_i \cup A_h| \leqslant 2k - 1 < n$, there exists some $w \in S$, $w \notin A_i \cup A_h$. Thus on applying (5.8) with $j = w$, we find that $B_h \cup \{w\} \in \mathcal{A}$, so that by the intersection property

$$|B_i \cap B_h| = |B_i \cap (B_h \cup \{w\})| = |A_i \cap (B_h \cup \{w\})| \geqslant 1.$$

Thus the sets B_i, $i \leqslant m_0$, form an intersecting family of $(k-1)$-subsets of the $(n-1)$-set $\{2, \ldots, n\}$. If we put $\mathscr{B} = \{B_i\}$, we therefore have by induction $|\Delta\mathscr{B}| \geqslant |\mathscr{B}| = m_0$.

Finally, the collection $\mathscr{C} = \{A_i : i > m_0\}$ is also an intersecting family of subsets of $\{2, \ldots, n\}$, so by induction we have $|\Delta\mathscr{C}| \geqslant |\mathscr{C}| = m - m_0$. Adjoining the element 1 to the sets in $\Delta\mathscr{B}$ we thus have

$$|\Delta\mathscr{A}| \geqslant |\Delta\mathscr{B}| + |\Delta\mathscr{C}| \geqslant |\mathscr{B}| + |\mathscr{C}| = |\mathscr{A}|. \qquad \square$$

We now use Katona's result to prove Milner's theorem.

Proof of Theorem 5.5.1 Let $t = [\frac{1}{2}(n+k+1)]$. We are given an antichain $\mathscr{A}$ of subsets A_i of an n-set satisfying $|A_{i_1} \cap A_{i_2}| \geqslant k$, and we have to prove that $|\mathscr{A}| \leqslant \binom{n}{t}$.

If $|A_i| = t$ for all i, then the result is trivial. Next consider the case when $|A_i| \leqslant t$ for each i, but $|A_i| < t$ for at least one i. Let j be the smallest size of an A_i, $j < t$, and suppose, after relabelling if necessary, that $|A_1| = \ldots = |A_h| = j < |A_{h+1}| \leqslant \ldots \leqslant t$. The idea of the proof is now to replace $A_1, \ldots, A_h$ by sets of size $j+1$ in their shade. Now the sets in the shade of $A_1, \ldots, A_h$ are the complements of the sets in the shadow of $\mathscr{H} = \{A_1', \ldots, A_h'\}$. The condition $|A_{i_1} \cap A_{i_2}| \geqslant k$ implies that $|A_{i_1} \cup A_{i_2}| \leqslant 2j - k$, so that $|A_{i_1}' \cap A_{i_2}'| \geqslant n - 2j + k$, where $j + 1 \leqslant \frac{1}{2}(n+k+1)$, i.e. $n - 2j + k \geqslant 1$. So by Katona's theorem (Theorem 5.5.3) $|\Delta\mathscr{H}| \geqslant |\mathscr{H}| = h$. Thus we can replace $A_1, \ldots, A_h$ by h sets of size $j+1$, and clearly this replacement does not destroy the antichain or the intersection property. If we keep repeating the process we eventually obtain $|\mathscr{A}|$ sets of size t, so that $|\mathscr{A}| \leqslant \binom{n}{t}$ as required.

Finally, consider the case when some of the sets A_i have size greater than t. As in the previous case, replace all those sets of size less than t by an equal number of sets of size t. Then replace each set of size greater than t by the set of size t in the same chain in any chosen symmetric chain decomposition. (Since $t > \frac{1}{2}n$, this is possible.) The resulting collection of t-sets has $|\mathscr{A}|$ members, so again $|\mathscr{A}| \leqslant \binom{n}{t}$. $\square$

5.6 Some results related to the EKR theorem

We consider briefly a few topics which are variations on the EKR theme. If the condition $A_i \cap A_j \neq \emptyset$, i.e. $|A_i \cap A_j| \geq 1$, is replaced by $|A_i \cap A_j| \geq t$, then with each $|A_i| \leq k$, we might expect to be able to prove that $|\mathscr{A}| \leq \binom{n-t}{k-t}$. This was indeed proved for sufficiently large n by Erdös, Ko, and Rado in their 1961 paper, but it is not necessarily true for small values of n. More recently, Frankl (1976*b*) and Wilson (1984) have between them shown that the result is true for $n \geq n_0(k, t)$ where $n_0(k, t) = (t+1)(k-t+1)$.

Another possible direction in which to consider extending the EKR theorem is the following. Suppose that $\mathscr{A}$ is a collection of subsets of an n-set such that $|A| \leq \frac{1}{2}n$ for each $A \in \mathscr{A}$, and with the property that $\mathscr{A}$ does not contain $t+1$ simultaneously disjoint pairs of subsets. How large can $|\mathscr{A}|$ be? Clements (1976) proved the following result, a proof of which is outlined as Exercise 7.12.

Theorem 5.6.1 (Clements 1976) Let $\mathscr{A}$ be a collection of subsets of an n-set S, n even, where $|A| \leq \frac{1}{2}n$ for all $A \in \mathscr{A}$, and where $\mathscr{A}$ does not possess $t+1$ simultaneously disjoint pairs of subsets. Then

$$|\mathscr{A}| \leq \binom{n-1}{\frac{1}{2}n-1} + t. \qquad \square$$

A number of authors have obtained analogues of the EKR theorem for structures other than subsets of a set. Hsieh (1975) obtained an analogue for subspaces of a finite vector space, Livingston (1979) for ordered sets, Frankl and Furedi (1980) for integer sequences, Moon (1982) for Hamming schemes, Rands (1982) for t-designs, and Stanton (1980) for Chevalley groups. The more obvious problem of analogues for multisets will be discussed in Chapter 10.

Finally, we remark that equality can occur in Theorem 5.1.2 only when $A_1 \cap \ldots \cap A_m \neq \emptyset$. An upper bound for m in the case $A_1 \cap \ldots \cap A_m = \emptyset$ is given by the following theorem.

Theorem 5.6.2 (Hilton and Milner 1967) Let $\mathscr{A} = \{A_1, \ldots, A_m\}$ be an intersecting collection of k-subsets of an n-set, $k \leq \frac{1}{2}n$, such that $A_1 \cap \ldots \cap A_m = \emptyset$. Then

$$|\mathscr{A}| \leq \binom{n-1}{k-1} + 1 - \binom{n-k-1}{k-1}. \qquad \square$$

The proof uses a shifting argument similar to that in the original proof of the EKR theorem.

Exercises 5

5.1 Let $A_1, \ldots, A_m$ be subsets of an n-set S such that if $A_i \cap A_j = \emptyset$ then $A_i \cup A_j = S$. Show that $m \leq 2^{n-1} + \binom{n-1}{[(n-2)/2]}$.

5.2 Give an example of an antichain $\mathscr{A}$ for which equality holds in Corollary 5.2.4.

5.3 Prove Corollary 5.2.4 directly from EKR and Sperner, taking the cases of n even and n odd separately.

5.4 Let $1 \leq g \leq h \leq n$, $g + h \leq n$, and suppose that $A_1, \ldots, A_m$ are subsets of an n-set S such that $g \leq |A_i| \leq h$ for each i and $A_i \cap A_j \neq \emptyset$ for all i, j. Use EKR to show that $m \leq \sum_{i=g}^{h} \binom{n-1}{i-1}$. (Hilton 1974)

5.5 Let f be a non-negative real function defined on the non-negative integers. Show that if $\mathscr{A} = \{A_1, \ldots, A_m\}$ is an intersecting antichain of subsets of $\{1, \ldots, n\}$ with $|A_i| \leq \frac{1}{2}n$ for each i, then

$$\sum_{i=1}^{m} f(|A_i|) \leq \max_{0 \leq k \leq n/2} f(k) \binom{n-1}{k-1}.$$

(Greene *et al.* 1976)

5.6 Use Lemma 5.2.5 and Theorem 5.1.2 to prove Theorem 5.2.2.

5.7 Verify that if $\mathscr{A}$ satisfies the conditions of Corollary 5.2.4 then the collection obtained from $\mathscr{A}$ by choosing any one set in $\mathscr{A}$ and replacing it by its complement also satisfies the conditions. Hence obtain an alternative proof of 5.2.4 via Milner's theorem.

5.8 (i) By applying EKR to the sets $C - A_i$, show that if C is a k-set and $A_1, \ldots, A_m$ are h subsets of C with no i, j such that $A_i \cup A_j = C$, then, if $k \leq 2h$, $m \leq \frac{k-h}{k}\binom{k}{h}$.

(ii) Suppose now that $\mathscr{A}$ is a collection of subsets of an n-set containing no two sets whose union is also in $\mathscr{A}$. Suppose also that $h \leq \frac{1}{2}n$. By considering, for each k, $h \leq k \leq 2h$, and each set C of size k in $\mathscr{A}$, the number of saturated chains between ranks h and $2h$ through C and containing no smaller set in $\mathscr{A}$, show that

$$\sum_{k=h}^{2h} \frac{p_k}{\binom{n}{k}} \frac{h}{k} \leqslant 1$$

where p_k denotes the number of sets in $\mathscr{A}$ of size k. (This will be used in Chapter 12)

(Kleitman 1976*b*)

5.9 The set of all q-subsets, where $q = [\frac{1}{2}(n + k + 1)]$, is an example of a collection $\mathscr{A}$ which satisfies the conditions of Milner's theorem and which gives equality: $|\mathscr{A}| = \binom{n}{q}$. In the case when $n + k$ is odd, give another example which gives equality.

6 Ideals and a lemma of Kleitman

6.1 Kleitman's lemma

In this chapter we shall consider collections $\mathscr{D}$ of subsets of an n-set S with the property

$$A \in \mathscr{D},\ B \subset A \Rightarrow B \in \mathscr{D}.$$

We call such a $\mathscr{D}$ an *ideal* of S or a *downset* or a *simplicial complex*. The terminology 'downset' conveys the idea very clearly, and leads naturally to the corresponding idea of an upset: $\mathscr{U}$ is an *upset* or *filter* if

$$A \in \mathscr{U},\ A \subset B \Rightarrow B \in \mathscr{U}.$$

Clearly, $\mathscr{U}$ is an upset if and only if $\mathscr{U}' = \{A' : A \in \mathscr{U}\}$ is a downset.

There is a simple connection between upsets and antichains. If $\mathscr{A}$ is an antichain of subsets of S, then

$$\mathscr{A}^+ = \{X \subseteq S : X \supseteq A \text{ for some } A \in \mathscr{A}\}$$

is an upset, in which the members of $\mathscr{A}$ are the minimal members of $\mathscr{A}^+$, and similarly

$$\mathscr{A}^- = \{X \subseteq S : X \subseteq A \text{ for some } A \in \mathscr{A}\}$$

is a downset in which the members of $\mathscr{A}$ are the maximal members of $\mathscr{A}^-$. Conversely, if $\mathscr{U}$ is an upset, then the set of minimal members of $\mathscr{U}$ is an antichain.

One of the most useful results concerning ideals is due to Kleitman (1966); appropriately, it appeared in the first issue of the *Journal of Combinatorial Theory*.

Theorem 6.1.1 (Kleitman's lemma) If $\mathscr{U}$ is an upset on S and $\mathscr{D}$ is a downset on S, where $|S| = n$, then

$$|\mathscr{U}| \,.\, |\mathscr{D}| \geqslant 2^n\, |\mathscr{U} \cap \mathscr{D}|.$$

There is a natural generalization to divisors of a number (Anderson 1976). An upset of divisors of m is a set $\mathscr{U}$ of divisors of m such that if $a \in \mathscr{U}$ and $a\,|b|\,m$ then $b \in \mathscr{U}$.

Theorem 6.1.2 If $\mathscr{U}$ is an upset and $\mathscr{D}$ is a downset of divisors of $m = p_1^{k_1} \dots p_r^{k_r}$, then

$$|\mathscr{U}| \cdot |\mathscr{D}| \geq \prod_{i=1}^{r} (1 + k_i) \cdot |\mathscr{U} \cap \mathscr{D}|.$$

Since the proof of Theorem 6.1.2 is not much more difficult than that of Theorem 6.1.1, we now prove the more general result. In the next section we shall see that they are in fact consequences of a much more general result due to Ahlswede and Daykin (1978).

Proof of Theorem 6.1.2 We use induction on $n = \sum_i k_i$. If $n = 1$ or 2 the result is trivial, so we proceed to the induction step. Writing $p_r = p$ and $k_r = s$, we have $m = m'p = m''p^s$ where p does not divide m''. Partition $\mathscr{U}$ and $\mathscr{D}$ as follows:

$$\mathscr{U} = \mathscr{U}_p \cup \mathscr{U}_{\bar{p}} \qquad \mathscr{D} = \mathscr{D}_p \cup \mathscr{D}_{\bar{p}}$$

where $\mathscr{U}_p$ and $\mathscr{D}_p$ consist precisely of those members of $\mathscr{U}$ and $\mathscr{D}$ respectively which are divisible by p^s. Now if $hp^s \in \mathscr{D}$ then $h, hp, \dots, hp^{s-1}$ are all in $\mathscr{D}_{\bar{p}}$; therefore $|\mathscr{D}_{\bar{p}}| \geq s\,|\mathscr{D}_p|$. Similarly $|\mathscr{U}_{\bar{p}}| \leq s\,|\mathscr{U}_p|$. Thus

$$(s\,|\mathscr{U}_p| - |\mathscr{U}_{\bar{p}}|)(|\mathscr{D}_{\bar{p}}| - s\,|\mathscr{D}_p|) \geq 0$$

whence

$$s\,|\mathscr{U}_p| \cdot |\mathscr{D}_p| + \frac{1}{s}|\mathscr{U}_{\bar{p}}| \cdot |\mathscr{D}_{\bar{p}}| \leq |\mathscr{U}_p| \cdot |\mathscr{D}_{\bar{p}}| + |\mathscr{U}_{\bar{p}}| \cdot |\mathscr{D}_p|. \tag{6.1}$$

Using the induction hypothesis for divisors of m', we have

$$|\mathscr{U}_{\bar{p}} \cap \mathscr{D}_{\bar{p}}|\, \tau(m') \leq |\mathscr{U}_{\bar{p}}| \cdot |\mathscr{D}_{\bar{p}}| \tag{6.2}$$

where $\tau(m') = (1 + k_1) \dots (1 + k_{r-1})k_r$.

Now let $\mathscr{U}''_p$, $\mathscr{D}''_p$ denote the sets of divisors of m'' obtained by dividing each member of $\mathscr{U}_p$, $\mathscr{D}_p$ by p^s. Then, since $|\mathscr{U}_p \cap \mathscr{D}_p| = |\mathscr{U}''_p \cap \mathscr{D}''_p|$, the induction hypothesis applied to m'' gives

$$|\mathscr{U}_p \cap \mathscr{D}_p|\, \tau(m'') \leq |\mathscr{U}''_p| \cdot |\mathscr{D}''_p| = |\mathscr{U}_p| \cdot |\mathscr{D}_p|. \tag{6.3}$$

Thus

$$\begin{aligned}
|\mathcal{U} \cap \mathcal{D}|\, \tau(m) &= |\mathcal{U}_p \cap \mathcal{D}_p|\, \tau(m) + |\mathcal{U}_{\bar{p}} \cap \mathcal{D}_{\bar{p}}|\, \tau(m) \\
&= (s+1)\, |\mathcal{U}_p \cap \mathcal{D}_p|\, \tau(m'') + \left(1 + \frac{1}{s}\right) |\mathcal{U}_{\bar{p}} \cap \mathcal{D}_{\bar{p}}|\, \tau(m') \\
&\leqslant (s+1)\, |\mathcal{U}_p| \,.\, |\mathcal{D}_p| + \left(1 + \frac{1}{s}\right) |\mathcal{U}_{\bar{p}}| \,.\, |\mathcal{D}_{\bar{p}}| \\
&\qquad\qquad \text{by (6.2) and (6.3)} \\
&= |\mathcal{U}_p| \,.\, |\mathcal{D}_p| + |\mathcal{U}_{\bar{p}}| \,.\, |\mathcal{D}_{\bar{p}}| + s\, |\mathcal{U}_p| \,.\, |\mathcal{D}_p| + \frac{1}{s} |\mathcal{U}_{\bar{p}}| \,.\, |\mathcal{D}_{\bar{p}}| \\
&\leqslant |\mathcal{U}_p| \,.\, |\mathcal{D}_p| + |\mathcal{U}_{\bar{p}}| \,.\, |\mathcal{D}_{\bar{p}}| + |\mathcal{U}_p| \,.\, |\mathcal{D}_{\bar{p}}| + |\mathcal{U}_{\bar{p}}| \,.\, |\mathcal{D}_p| \\
&\qquad\qquad \text{by (6.1)} \\
&= (|\mathcal{U}_p| + |\mathcal{U}_{\bar{p}}|)(|\mathcal{D}_p| + |\mathcal{D}_{\bar{p}}|) \\
&= |\mathcal{U}| \,.\, |\mathcal{D}|.
\end{aligned}$$

□

As described in the exercises, there are other equivalent forms of Theorem 6.1.1. For example, we have the following theorem.

Theorem 6.1.3 If $\mathcal{A}$ and $\mathcal{B}$ are ideals of subsets of the n-set S, then $|\mathcal{A}| \,.\, |\mathcal{B}| \leqslant 2^n\, |\mathcal{A} \cap \mathcal{B}|$. □

6.1.1 Applications of Kleitman's lemma

We now give two simple applications of Theorem 6.1.1. The first of these is to give a very simple proof of a result which had been proved in several different ways by various authors (Daykin and Lovász 1976, Schönheim 1974*a*, Seymour 1973, Hilton 1976). The following proof is due to Anderson (1976) and Kleitman (in Greene and Kleitman 1978).

Theorem 6.1.4 If $\mathcal{A}$ is a collection of subsets of an n-set S such that if $A, B \in \mathcal{A}$ then $A \cap B \neq \varnothing$ and $A \cup B \neq S$, then $|\mathcal{A}| \leqslant 2^{n-2}$.

Proof Define $\mathcal{U}$ to be the collection of subsets of S consisting of all the sets in $\mathcal{A}$ and all their supersets, and define $\mathcal{D}$ to be the collection of subsets consisting of all sets in $\mathcal{A}$ and their subsets. Then $\mathcal{A} \subseteq \mathcal{U} \cap \mathcal{D}$, and Kleitman's lemma yields

$$|\mathcal{A}|\, 2^n \leqslant |\mathcal{U}| \,.\, |\mathcal{D}|.$$

However, $\mathscr{U}$ is a collection of intersecting subsets, so, by Theorem 1.1.1, $|\mathscr{U}| \leq 2^{n-1}$. Similarly (Exercise 1.1), $|\mathscr{D}| \leq 2^{n-1}$. Thus

$$|\mathscr{A}|\, 2^n \leq 2^{n-1} \,.\, 2^{n-1}$$

as required. □

This result is clearly best possible, since there are 2^{n-2} subsets of $\{1, \ldots, n\}$ which contain 1 but do not contain n.

Kleitman originally proved his lemma while investigating the following problem. If $\mathscr{A}$ is an intersecting family of subsets of an n-set S, then, as we recalled above, $|\mathscr{A}| \leq 2^{n-1}$. So suppose now that $\mathscr{A}_1, \ldots, \mathscr{A}_k$ are disjoint intersecting families of subsets of S; then how large can $|\cup \mathscr{A}_i|$ be?

Theorem 6.1.5 (Kleitman 1966) If $\mathscr{A}_1, \ldots, \mathscr{A}_k$ are disjoint intersecting families of subsets of an n-set S, then

$$|\cup \mathscr{A}_i| \leq 2^n - 2^{n-k}.$$

Proof The case $k = 1$ reduces to Theorem 1.1.1, so we proceed by induction on k. As explained in Theorem 1.1.1, $\mathscr{A}_k$ can be extended to a collection $\mathscr{A}_k''$ of 2^{n-1} intersecting subsets. Let $\mathscr{C} = \{X : X \notin \mathscr{A}_k''\}$; then $\mathscr{C}$ is a downset. Consider next $\mathscr{B} = \bigcup_{i=1}^{k-1} \mathscr{A}_i$. If $\mathscr{B}$ is not an upset, we can adjoin any missing superset to an appropriate $\mathscr{A}_i$ to obtain an upset $\mathscr{D}$ which contains $\mathscr{B}$ and which, by the induction hypothesis, has at most $2^n - 2^{n-k+1}$ members. Applying Kleitman's inequality to $\mathscr{C}$ and $\mathscr{D}$, we obtain

$$\left|\bigcup_{i=1}^{k} \mathscr{A}_i\right| = |\mathscr{B} \cup \mathscr{A}_k| \leq |\mathscr{B} \cap \mathscr{C}| + |\mathscr{A}_k''| \leq |\mathscr{D} \cap \mathscr{C}| + 2^{n-1}$$

$$\leq \frac{1}{2^n}|\mathscr{D}| \,.\, |\mathscr{C}| + 2^{n-1} \leq \frac{1}{2^n} \cdot (2^n - 2^{n-k+1}) \,.\, 2^{n-1} + 2^{n-1}.$$

$$= 2^{n-1} - 2^{n-k} + 2^{n-1} = 2^n - 2^{n-k}$$

□

Again, the bound obtained is best possible: if $S = \{1, \ldots, n\}$, take $\mathscr{A}_i$ to be the collection of all subsets of S containing i but none of $1, \ldots, i-1$. Then $|\mathscr{A}_i| = 2^{n-i}$ and the bound is attained.

6.2 The Ahlswede–Daykin inequality

Since Kleitman proved his inequality in 1967, a succession of generalizations of his result have been obtained. In this section we choose one fairly general theorem and deduce from it a number of inequalities, including Kleitman's, as special cases. In the exposition we shall write $\mathscr{P}(S)$ for the set of subsets of S, and, for any collections $\mathscr{A}$, $\mathscr{B}$ of subsets of S, define

$$\mathscr{A} \vee \mathscr{B} = \{E \subseteq S : E = A \cup B \text{ for some } A \in \mathscr{A} \text{ and } B \in \mathscr{B}\}$$

$$\mathscr{A} \wedge \mathscr{B} = \{E \subseteq S : E = A \cap B \text{ for some } A \in \mathscr{A} \text{ and } B \in \mathscr{B}\}.$$

Theorem 6.2.1 (Ahlswede and Daykin 1978) If α, β, φ, δ are non-negative functions defined on $\mathscr{P}(S)$ and satisfying

$$\alpha(A)\beta(B) \leqslant \varphi(A \cup B)\delta(A \cap B) \tag{6.4}$$

for all $A, B \subseteq S$, and if $\mathscr{A}$, $\mathscr{B}$ are any two collections of subsets of S, then

$$\Big(\sum_{A \in \mathscr{A}} \alpha(A)\Big)\Big(\sum_{B \in \mathscr{B}} \beta(B)\Big) \leqslant \Big(\sum_{E \in \mathscr{A} \vee \mathscr{B}} \varphi(E)\Big)\Big(\sum_{F \in \mathscr{A} \wedge \mathscr{B}} \delta(F)\Big). \tag{6.5}$$

Note If we use the abbreviation

$$f(\mathscr{A}) = \sum_{A \in \mathscr{A}} f(A)$$

we can rewrite (6.5) more compactly as

$$\alpha(\mathscr{A})\beta(\mathscr{B}) \leqslant \varphi(\mathscr{A} \vee \mathscr{B})\delta(\mathscr{A} \wedge \mathscr{B}). \tag{6.6}$$

Proof of Theorem 6.2.1 We use induction on $n = |S|$, and first consider the case $n = 1$, where we can take $S = \{1\}$. Condition (6.4) becomes

$$\begin{aligned}
\alpha(\varnothing)\beta(\varnothing) &\leqslant \varphi(\varnothing)\delta(\varnothing) \\
\alpha(\varnothing)\beta(S) &\leqslant \varphi(S)\delta(\varnothing) \\
\alpha(S)\beta(\varnothing) &\leqslant \varphi(S)\delta(\varnothing) \\
\alpha(S)\beta(S) &\leqslant \varphi(S)\delta(S).
\end{aligned} \tag{6.7}$$

If $\mathscr{A}$ or $\mathscr{B}$ has only one member then (6.5) follows immediately. For example, if $\mathscr{A} = \{\varnothing\}$ and $\mathscr{B} = \{\varnothing, S\}$ then the left-hand side of (6.5) is $\alpha(\varnothing)\{\beta(\varnothing) + \beta(S)\}$ while the right-hand side is $\{\varphi(\varnothing) + \varphi(S)\}\delta(\varnothing)$ so that (6.7) gives the required result immediately. Now suppose that $\mathscr{A} = \mathscr{B} = \{\varnothing, S\}$. In this case we have to prove that

$$\{\alpha(\varnothing) + \alpha(S)\}\{\beta(\varnothing) + \beta(S)\} \leqslant \{\varphi(\varnothing) + \varphi(S)\}\{\delta(\varnothing) + \delta(S)\}. \tag{6.8}$$

If any of $\alpha(\emptyset)$, $\beta(\emptyset)$, $\varphi(\emptyset)$, $\delta(\emptyset)$ is zero, (6.8) is immediate; therefore, after suitable scalar multiplication, we can suppose now that $\alpha(\emptyset)=\beta(\emptyset)=\varphi(\emptyset)=\delta(\emptyset)=1$. The conditions (6.7) become

$$\beta(S)\leqslant\varphi(S),\qquad \alpha(S)\leqslant\varphi(S),\qquad \alpha(S)\beta(S)\leqslant\varphi(S)\delta(S)\quad (6.7')$$

and (6.8) becomes

$$\{1+\alpha(S)\}\{1+\beta(S)\}\leqslant\{1+\varphi(S)\}\{1+\delta(S)\}. \qquad (6.8')$$

Now (6.8′) is certainly true if $\varphi(S)=0$, so we can assume now that $\varphi(S)>0$. Since $\delta(S)\geqslant\alpha(S)\beta(S)/\varphi(S)$, (6.8′) will follow if we can prove that

$$\{1+\alpha(S)\}\{1+\beta(S)\}\leqslant\{1+\varphi(S)\}\left\{1+\frac{\alpha(S)\beta(S)}{\varphi(S)}\right\}$$

i.e.

$$\alpha(S)+\beta(S)\leqslant\varphi(S)+\frac{\alpha(S)\beta(S)}{\varphi(S)}. \qquad (6.9)$$

However, (6.7′) gives $\{\varphi(S)-\alpha(S)\}\{\varphi(S)-\beta(S)\}\geqslant 0$, which yields (6.9). Therefore the proof of the case $n=1$ is complete.

We now consider the induction step. Assuming that the theorem is true for sets of size m, we consider S with $|S|=m+1$. Let $S=\{1,\ldots,m+1\}$ and define $T=\{1,\ldots,m\}$ and $W=\{m+1\}$.

In an attempt to use the induction hypothesis for m-sets, we write any $A\subseteq S$ as $A=A'\cup A''$ where $A'=A\cap T$ and $A''=A\cap W$. Define α_1, β_1, φ_1, δ_1 on $\mathcal{P}(T)$ by

$$\alpha_1(C)=\sum_{\substack{A\in\mathcal{A}\\ A'=C}}\alpha(A)\qquad \beta_1(C)=\sum_{\substack{B\in\mathcal{B}\\ B'=C}}\beta(B)$$

$$\varphi_1(C)=\sum_{\substack{E=\mathcal{A}\vee\mathcal{B}\\ E'=C}}\varphi(E)\quad \delta_1(C)=\sum_{\substack{F\in\mathcal{A}\wedge\mathcal{B}\\ F'=C}}\delta(F).$$

Then

$$\alpha(\mathcal{A})=\sum_{A\in\mathcal{A}}\alpha(A)=\sum_{C\subseteq T}\sum_{\substack{A\in\mathcal{A}\\ A'=C}}\alpha(A)=\sum_{C\subseteq T}\alpha_1(C)=\alpha_1(\mathcal{P}(T))$$

and similarly

$$\beta(\mathcal{B})=\beta_1(\mathcal{P}(T)),\quad \varphi(\mathcal{A}\vee\mathcal{B})=\varphi_1(\mathcal{P}(T)),\quad \delta(\mathcal{A}\wedge\mathcal{B})=\delta_1(\mathcal{P}(T)).$$

Suppose for the moment that we can prove that

$$\alpha_1(C)\beta_1(D)\leqslant\varphi_1(C\cup D)\delta_1(C\cap D) \qquad (6.10)$$

for all $C, D \subseteq T$. We can then apply the induction hypothesis to obtain

$$\begin{aligned}\alpha(\mathscr{A})\beta(\mathscr{B}) &= \alpha_1(\mathscr{P}(T))\beta_1(\mathscr{P}(T))\\ &\leqslant \varphi_1(\mathscr{P}(T))\delta_1(\mathscr{P}(T))\\ &= \varphi(\mathscr{A} \vee \mathscr{B})\delta(\mathscr{A} \wedge \mathscr{B})\end{aligned}$$

as required. Therefore it only remains to prove (6.10).

Choose any $C, D \subseteq T$ and let $E = C \cup D$, $F = C \cap D$. Define $\alpha_2, \beta_2, \varphi_2, \delta_2$ on $\mathscr{P}(W) = \{\varnothing, \{m+1\}\}$ by

$$\alpha_2(R) = \begin{cases}\alpha(R \cup C) & \text{if } R \cup C \in \mathscr{A}\\ 0 & \text{otherwise}\end{cases}$$

$$\beta_2(R) = \begin{cases}\beta(R \cup D) & \text{if } R \cup D \in \mathscr{B}\\ 0 & \text{otherwise}\end{cases}$$

$$\varphi_2(R) = \begin{cases}\varphi(R \cup E) & \text{if } R \cup E \in \mathscr{A} \vee \mathscr{B}\\ 0 & \text{otherwise}\end{cases}$$

$$\delta_2(R) = \begin{cases}\delta(R \cup F) & \text{if } R \cup F \in \mathscr{A} \wedge \mathscr{B}\\ 0 & \text{otherwise.}\end{cases}$$

Then it is easy to check that

$$\alpha_1(C) = \sum_{\substack{A \in \mathscr{A}\\ A' = C}} \alpha(A) = \sum_{R \subseteq W} \sum_{\substack{A \in \mathscr{A}\\ A' = C\\ A'' = R}} \alpha(A) = \sum_{R \subseteq W} \alpha_2(R) = \alpha_2(\mathscr{P}(W))$$

and that similarly $\beta_1(D) = \beta_2(\mathscr{P}(W))$, $\varphi_1(E) = \varphi_2(\mathscr{P}(W))$, $\delta_1(F) = \delta_2(\mathscr{P}(W))$. If it can be shown that

$$\alpha_2(R)\beta_2(Q) \leqslant \varphi_2(R \cup Q)\delta_2(R \cap Q) \tag{6.11}$$

for all $R, Q \subseteq W$, it will then follow, on using the induction hypothesis for $n = 1$, that

$$\begin{aligned}\alpha_1(C)\beta_1(D) &= \alpha_2(\mathscr{P}(W))\beta_2(\mathscr{P}(W))\\ &\leqslant \varphi_2(\mathscr{P}(W))\delta_2(\mathscr{P}(W))\\ &= \varphi_1(E)\delta_1(F),\end{aligned}$$

thus proving (6.10).

We complete the proof by proving (6.11). Note that the left-hand side of (6.11) is zero unless $R \cup C \in \mathscr{A}$ and $Q \cup D \in \mathscr{B}$, in which case it equals $\alpha(R \cup C)\beta(Q \cup D)$. Now $(R \cup C) \cup (Q \cup D) = (R \cup Q) \cup E$ so that $(R \cup Q) \cup E \in \mathscr{A} \vee \mathscr{B}$, and $(R \cup C) \cap (Q \cup D) = (R \cap Q) \cup F$ so that $(R \cap Q) \cup F \in \mathscr{A} \wedge \mathscr{B}$. Therefore the right-hand side of (6.11) is then $\varphi(R \cup Q \cup E)\delta((R \cap Q) \cup F)$. However, (6.4) gives $\alpha(R \cup C)\beta(Q \cup D) \leq \varphi(R \cup Q \cup E)\delta((R \cap Q) \cup F)$, so that (6.11) is true. □

Ahlswede and Daykin have pointed out that many previously proved inequalities are contained in Theorem 6.2.1 as special cases. If we take $\alpha = \beta = \varphi = \delta = 1$ we immediately obtain the following corollary.

Corollary 6.2.2 (Daykin 1977) If $\mathscr{A}$, $\mathscr{B}$ are any two collections of subsets of S, then

$$|\mathscr{A}| \,.\, |\mathscr{B}| \leq |\mathscr{A} \vee \mathscr{B}| \,.\, |\mathscr{A} \wedge \mathscr{B}|.$$

□

Corollary 6.2.3 Kleitman's lemma 6.1.1.

Proof See Exercise 6.8. □

Corollary 6.2.4 (Marica and Schönheim 1969) Let $\mathscr{A} - \mathscr{B} = \{A - B : A \in \mathscr{A}, B \in \mathscr{B}\}$. Then, for any collection $\mathscr{A}$ of subsets of S,

$$|\mathscr{A} - \mathscr{A}| \geq |\mathscr{A}|.$$

Proof Let $\mathscr{A}' = \{A' : A \in \mathscr{A}\}$. For any collections $\mathscr{A}$, $\mathscr{B}$ of subsets of S we have, by Corollary 6.2.2,

$$\begin{aligned} |\mathscr{A}| \,.\, |\mathscr{B}| &= |\mathscr{A}| \,.\, |\mathscr{B}'| \\ &\leq |\mathscr{A} \vee \mathscr{B}'| \,.\, |\mathscr{A} \wedge \mathscr{B}'| \\ &= |(\mathscr{A} \vee \mathscr{B}')'| \,.\, |\mathscr{A} \wedge \mathscr{B}'| \\ &= |\mathscr{A}' \wedge \mathscr{B}| \,.\, |\mathscr{A} \wedge \mathscr{B}'| \\ &= |\mathscr{B} - \mathscr{A}| \,.\, |\mathscr{A} - \mathscr{B}| \end{aligned}$$

Now take $\mathscr{B} = \mathscr{A}$ to obtain $|\mathscr{A} - \mathscr{A}| \geq |\mathscr{A}|$. □

Before leaving the Marica–Schönheim result we note that various generalizations of it have been obtained. For example, Ahlswede and Daykin (1979) have shown that, if $A_1, \ldots, A_m$ are distinct sets and if $B_1, \ldots, B_n$ are sets such that each A_i contains at least one B_j as a subset, there are at least m distinct differences $A_i - B_j$ (see Exercise 6.16). Also, Daykin and Lovász (1976) have shown that any non-trivial boolean function takes at least m distinct values when evaluated over m distinct sets.

For our next corollary, take $\alpha(A)=\mu(A)f(A)$, $\beta(A)=\mu(A)g(A)$, $\varphi(A)=\mu(A)$, and $\delta(A)=\mu(A)f(A)g(A)$, where f,g are *decreasing* functions (f is decreasing if $f(A)\geqslant f(B)$ whenever $A\subseteq B$) and where μ is a *log-supermodular* function, i.e.

$$\mu(A)\mu(B)\leqslant\mu(A\cup B)\mu(A\cap B)$$

for all A,B. We then have

$$\begin{aligned}\alpha(A)\beta(B)&=\mu(A)\mu(B)f(A)g(B)\\&\leqslant\mu(A\cup B)\mu(A\cap B)f(A\cap B)g(A\cap B)\\&=\varphi(A\cup B)\delta(A\cap B)\end{aligned}$$

so that, by Theorem 6.2.1,

$$\Big\{\sum_{A\in\mathscr{A}}\mu(A)f(A)\Big\}\Big\{\sum_{B\in\mathscr{B}}\mu(B)g(B)\Big)$$
$$\leqslant\Big\{\sum_{E\in\mathscr{A}\vee\mathscr{B}}\mu(E)\Big\}\Big\{\sum_{F\in\mathscr{A}\wedge\mathscr{B}}\mu(F)f(F)g(F)\Big\}. \quad (6.12)$$

We thus have the following Corollary.

Corollary 6.2.5 (The FKG inequality of Fortuin, Kasteleyn, and Ginibre (1971)) If μ is a non-negative log-supermodular function and if f and g are both decreasing (or both increasing) functions defined on $\mathscr{P}(S)$, then

$$\Big\{\sum_{A\subseteq S}\mu(A)f(A)\Big\}\Big\{\sum_{B\subseteq S}\mu(B)g(B)\Big\}\leqslant\Big\{\sum_{E\subseteq S}\mu(E)\Big\}\Big\{\sum_{F\subseteq S}\mu(F)f(F)g(F)\Big\}.$$

□

We choose here $\mathscr{A}=\mathscr{B}=\mathscr{P}(S)$ in (6.12), and we also note that if f, g are increasing then the proof easily adapts (interchange φ and δ). If, instead, we take $\mathscr{A}$, $\mathscr{B}$ in (6.12) to be downsets and f,g to be their characteristic functions,

$$f(X)=\begin{cases}1 \text{ if } X\in\mathscr{A}\\0 \text{ otherwise}\end{cases}\qquad g(X)=\begin{cases}1 \text{ if } X\in\mathscr{B}\\0 \text{ otherwise.}\end{cases}$$

Then f,g are clearly decreasing and we obtain the following corollary.

Corollary 6.2.6 Let $\mathscr{A}$, $\mathscr{B}$ be ideals of subsets of an n-set S. Then

$$\Big\{\sum_{A\in\mathscr{A}}\mu(A)\Big\}\Big\{\sum_{B\in\mathscr{B}}\mu(B)\Big\}\leqslant\Big\{\sum_{E\subseteq S}\mu(E)\Big\}\Big\{\sum_{F\in\mathscr{A}\wedge\mathscr{B}}\mu(F)\Big\}.$$

□

The special case $\mu = 1$ yields Kleitman's lemma in the form of Theorem 6.1.3 since, if $\mathscr{A}$ and $\mathscr{B}$ are ideals, $\mathscr{A} \wedge \mathscr{B} = \mathscr{A} \cap \mathscr{B}$.

We now consider some consequences of Theorem 6.2.1 for sequences. A sequence $\{a_n\}$ is *log-convex* if $a_{n-1}a_{n+1} \geq a_n^2$ for each $n \geq 1$.

Corollary 6.2.7 (Seymour and Welsh 1975) Let $\{b_n\}$ and $\{c_n\}$ be decreasing sequences of non-negative real numbers and let $\{a_n\}$ be log-convex. Then

$$\left(\sum_{k=0}^{n} a_k\right)\left(\sum_{k=0}^{n} a_k b_k c_k\right) \geq \left(\sum_{k=0}^{n} a_k b_k\right)\left(\sum_{k=0}^{n} a_k c_k\right).$$

Proof Let $d_k = a_k \Big/ \binom{n}{k}$ and let S be any n-set. For $A \subseteq S$ put $\mu(A) = d_{|A|}$, $f(A) = b_{|A|}$, $g(A) = c_{|A|}$. Then f and g are decreasing functions and so, if we can show that μ is log-supermodular, the FKG inequality will yield

$$\left(\sum_{A \subseteq S} d_{|A|} b_{|A|}\right)\left(\sum_{A \subseteq S} d_{|A|} c_{|A|}\right) \leq \left(\sum_{A \subseteq S} d_{|A|}\right)\left(\sum_{A \subseteq S} d_{|A|} b_{|A|} c_{|A|}\right)$$

i.e.

$$\left\{\sum_{k=0}^{n} \binom{n}{k} d_k b_k\right\}\left\{\sum_{k=0}^{n} \binom{n}{k} d_k c_k\right\} \leq \left\{\sum_{k=0}^{n} \binom{n}{k} d_k\right\}\left\{\sum_{k=0}^{n} \binom{n}{k} d_k b_k c_k\right\}$$

i.e.

$$\left(\sum_{k=0}^{n} a_k b_k\right)\left(\sum_{k=0}^{n} a_k c_k\right) \leq \left(\sum_{k=0}^{n} a_k\right)\left(\sum_{k=0}^{n} a_k b_k c_k\right)$$

as required. To prove that μ is log-supermodular we have to show that

$$\frac{a_{|A|}}{\binom{n}{|A|}} \cdot \frac{a_{|B|}}{\binom{n}{|B|}} \leq \frac{a_{|A \cup B|}}{\binom{n}{|A \cup B|}} \cdot \frac{a_{|A \cap B|}}{\binom{n}{|A \cap B|}}.$$

Now $|A| + |B| = |A \cup B| + |A \cap B|$, so by the log-concavity of the binomial coefficients we have (cf. Corollary 4.1.4)

$$\binom{n}{|A|}\binom{n}{|B|} \geq \binom{n}{|A \cup B|}\binom{n}{|A \cap B|}$$

However, $a_{|A|}a_{|B|} \leq a_{|A \cup B|}a_{|A \cap B|}$, so the required result follows. □

As a special case we have the following corollary.

Corollary 6.2.8 (Chebyshev) If $\{b_n\}$ and $\{c_n\}$ are decreasing sequences of non-negative real numbers, then

$$\left(\frac{1}{n}\sum_{i=1}^{n} b_i\right)\left(\frac{1}{n}\sum_{i=1}^{n} c_i\right) \leqslant \left(\frac{1}{n}\sum_{i=1}^{n} b_i c_i\right).$$

Proof Take $a_i = 1$ in Corollary 6.2.7. □

It should be clear by now what a powerful result Theorem 6.2.1 is. Further, although all the results in this section have been presented in the context of the lattice of subsets of a set, they could be presented in the more general setting of a finite distributive lattice L, with $\cup$ and $\cap$ replaced by $\vee$ and $\wedge$. We then have, for example, the following form of the FKG inequality which will be used in the next section.

Theorem 6.2.9 (FKG) If $(S, <)$ is a finite distributive lattice and if f, g are both increasing (or both decreasing) on S and μ is a non-negative function on S such that $\mu(x)\mu(y) \leqslant \mu(x \vee y)\mu(x \wedge y)$ for all $x,y \in S$, then

$$\left\{\sum_{x\in S} f(x)\mu(x)\right\}\left\{\sum_{x\in S} g(x)\mu(x)\right\} \leqslant \left\{\sum_{x\in S} \mu(x)\right\}\left\{\sum_{x\in S} \mu(x)f(x)g(x)\right\}. \qquad \square$$

We remark briefly that a partially ordered set is a *lattice* if every pair of elements x,y have a *join* $x \vee y$ (with $x \leqslant x \vee y$ and $y \leqslant x \vee y$) and a *meet* $x \wedge y$ (with $x \wedge y \leqslant x$ and $x \wedge y \leqslant y$). A lattice is *distributive* if $x \wedge (y \vee z) = (x \wedge y) \vee (x \wedge z)$ for all x, y, z.

6.3 Applications of the FKG inequality to probability theory

During the past few years a number of authors have used the FKG inequality to obtain results of a probabilistic nature. This is hardly surprising since the inequality itself was discovered in an investigation into correlation problems in statistical mechanics. We can see to begin with that Kleitman's inequality can be expressed in terms of probabilities:

$$|\mathcal{U}| \,.\, |\mathcal{D}| \geqslant 2^n \,|\mathcal{U} \cap \mathcal{D}|$$

i.e.

$$\frac{|\mathcal{U}|}{2^n} \cdot \frac{|\mathcal{D}|}{2^n} \geqslant \frac{|\mathcal{U} \cap \mathcal{D}|}{2^n}$$

i.e.

$$P(x \in \mathcal{U}) \, . \, P(x \in \mathcal{D}) \geqslant P(x \in \mathcal{U} \cap \mathcal{D})$$

i.e.

$$P(x \in \mathcal{D}) \geqslant P(x \in \mathcal{D} \mid x \in \mathcal{U})$$

where the *conditional* probability $P(A \mid B)$, the probability of A occurring given that B occurs, is defined by

$$P(A \mid B) = \frac{P(A \cap B)}{P(B)}.$$

Kleitman's lemma can therefore be interpreted as saying that the condition $x \in \mathcal{U}$ tends to pull x 'up', making it less likely for x to be in the downset $\mathcal{D}$.

In this section we attempt to give the flavour of these applications by presenting proofs of two different results, first a result originally proved by Graham, Yao, and Yao (1980) and second a proof of a conjecture known as the XYZ conjecture.

Suppose we are given two chains $a_1 < a_2 < \ldots < a_m$ and $b_1 < \ldots < b_n$, i.e. we are given two *linear orders*. Let $C = A \cup B$ where $A = \{a_i\}$ and $B = \{b_i\}$. A *linear extension* λ of C is a mapping of C onto $\{1, \ldots, m+n\}$ such that $x < y \Rightarrow \lambda(x) < \lambda(y)$. Thus a linear extension can be thought of as a ranking of the elements of $A \cup B$ which preserves the separate rankings in A and B. The set L of linear extensions of C has $\binom{m+n}{m}$ members, all of which can be considered initially to be equally likely. Now suppose that P_1, P_2, P_3 are subsets of the set of rankings of the members of C, each defined by a number of conditions of the form $a_i < b_j$. We then state the following theorem.

Theorem 6.3.1 (Graham *et al.* 1980) With P_1, P_2, P_3, L as above,

$$P(\lambda \in P_1 \mid \lambda \in L \cap P_2 \cap P_3) \geqslant P(\lambda \in P_1 \mid \lambda \in L \cap P_2).$$

This result asserts that the extra information given by P_3, that some a_i's are ranked before some b_j's, strengthens the chances of the conditions of P_1 occurring; intuitively, P_3 reinforces the view that all the a_i are in general of smaller rank than the b_i.

Before proceeding to the proof, we note that the result is not in general true if the a_i and b_i are not totally ordered.

Example 6.3.1 (Graham) Assume nothing about the orderings of a_1 and a_2, and of b_1 and b_2; therefore L is the set of all rankings of the four elements a_1, a_2, b_1, b_2. Let P_1 be defined by $a_1 < b_1$, P_2 by $a_2 < b_1$, and P_3 by $a_2 < b_2$. Then we can check that there are eight rankings satisfying P_2 and P_3, of which five satisfy P_1, so that

$$P(\lambda \in P_1 \mid \lambda \in L \cap P_2 \cap P_3) = \tfrac{5}{8}.$$

However, there are 12 rankings with $a_2 < b_1$, of which eight satisfy P_1, so that

$$P(\lambda \in P_1 \mid \lambda \in L \cap P_2) = \tfrac{2}{3} > \tfrac{5}{8}.$$

Proof of Theorem 6.3.1 (Shepp 1980) Let T denote the set of all $\binom{m+n}{m}$ m-subsets of $\{1, \ldots, m+n\}$. If $X = \{x_1 < x_2 < \ldots < x_m\}$ and $Y = \{y_1 < y_2 < \ldots < y_m\}$ are in T, write $X \leqslant Y$ if $x_i \leqslant y_i$ for each i. We then have

$$X \wedge Y = \{\min(x_1, y_1), \ldots, \min(x_m, y_m)\}$$

and

$$X \vee Y = \{\max(x_1, y_1), \ldots, \max(x_m, y_m)\};$$

it is easy to check that these are both in T. Since, for any three real numbers a,b,c,

$$\min(a, \max(b, c)) = \max(\min(a, b), \min(a, c)),$$

it follows that $X \wedge (Y \vee Z) = (X \wedge Y) \vee (X \wedge Z)$ for all $X, Y, Z \in T$. Thus $(T, <)$ is a distributive lattice.

We now exhibit a one–one correspondence between members of L and members of T. Given $X \in T$, say $X = \{x_1 < x_2 < \ldots < x_m\}$, we can construct a member $\theta(X)$ of L as follows. Place a_i in position x_i for each i, and then place b_j in the jth remaining position. For example, if $m = 4$, $n = 3$, and $X = \{1, 4, 5, 7\}$ we obtain the ranking $a_1 b_2 b_2 a_2 a_3 b_3 a_4$ since, for example, $x_2 = 4$ requires that a_2 is in the fourth position. Conversely, any linear extension gives a unique X. Next observe that if a_i is in position x_i there are $x_i - i$ 'gaps' to be filled by b_js before a_i, so that a_i will be ranked before b_j if and only if $j > x_i - i$, i.e. if $x_i \leqslant i + j - 1$. Thus each condition in the definitions of P_1, P_2, P_3 corresponds to a condition of the form $x_i \leqslant k_{i,j}$.

Thus, corresponding to P_1, P_2, and P_3 we obtain subsets $\bar{P}_i$ of T defined by a number of inequalities of the form $x_i \leq k_{i,j}$. Define functions μ, f, g on T by

$$\mu(X) = \begin{cases} 1 & \text{if } X \in \bar{P}_2 \\ 0 & \text{otherwise} \end{cases}$$

$$f(X) = \begin{cases} 1 & \text{if } X \in \bar{P}_1 \\ 0 & \text{otherwise} \end{cases}$$

$$g(X) = \begin{cases} 1 & \text{if } X \in \bar{P}_3 \\ 0 & \text{otherwise.} \end{cases}$$

Then f, g are decreasing functions, for if $X < Y$ and $f(Y) = 1$, then $Y \in \bar{P}_1$, and hence $X \in \bar{P}_1$, since $x_i \leq y_i$ and $y_i \leq k_{i,j}$ imply $x_i \leq k_{i,j}$. To show that μ satisfies $\mu(X)\mu(Y) \leq \mu(X \wedge Y)\mu(X \vee Y)$ for each X, Y, note first that it is trivial if $\mu(X) = 0$ or $\mu(Y) = 0$, so that we can now suppose that $\mu(X) = \mu(Y) = 1$, i.e. $X, Y \in \bar{P}_2$. Now each condition on X, Y is of the form $x_i \leq k$, $y_i \leq k$, but then $\min(x_i, y_i) \leq k$ and $\max(x_i, y_i) \leq k$, that so $X \wedge Y$ and $X \vee Y$ are also in $\bar{P}_2$ and so $\mu(X \wedge Y) = \mu(X \vee Y) = 1$.

The conditions of Theorem 6.2.9 have now been shown to be satisfied so we can apply that theorem to obtain

$$\frac{|T \cap \bar{P}_1 \cap \bar{P}_2|}{(m+n)!} \frac{|T \cap \bar{P}_2 \cap \bar{P}_3|}{(m+n)!} \leq \frac{|T \cap \bar{P}_2|}{(m+n)!} \cdot \frac{|T \cap \bar{P}_1 \cap \bar{P}_2 \cap \bar{P}_3|}{(m+n)!}$$

i.e.

$$P(P_1 \cap P_2 \cap L)P(P_2 \cap P_3 \cap L) \leq P(P_2 \cap L)P(P_1 \cap P_2 \cap P_3 \cap L)$$

which is the required result. □

The second application of the FKG inequality is also due to Shepp. Let $x_1, \ldots, x_n$ be random variables about which some partial information is known in the form of a set G of inequalities $x_i < x_j$. It was conjectured by Rival and Sands (see Shepp 1982) that

$$P(x_1 < x_2 \mid G) \leq P(x_1 < x_2 \mid G, x_1 < x_3)$$

since, if it is known that $x_1 < x_3$, there is all the more reason to expect that $x_1 < x_2$. This intuitively reasonable conjecture was known as the XYZ conjecture, and it became a theorem 2 years later when Shepp applied the FKG inequality to obtain a proof. It was pointed out by Shepp that an analogous conjecture

$$P(x_1 < x_2 < x_4 \mid G) \leq P(x_1 < x_2 < x_4 \mid G, x_1 < x_3 < x_4)$$

is in general false. To see this, take $n = 6$ and $G = \{x_2 < x_5 < x_6 < x_3, x_1 < x_4\}$; then

$$P(x_1 < x_2 < x_4 \mid G) = \frac{4}{15} > \frac{1}{4} = P(x_1 < x_2 < x_4 \mid G, x_1 < x_3 < x_4).$$

Theorem 6.3.2 (Shepp 1982) Let G be a set of inequalities of the form $x_i < x_j$ relating to n variables $x_1, \ldots, x_n$. Then

$$P(x_1 < x_2 \mid G) \leq P(x_1 < x_2 \mid G, x_1 < x_3)$$

or, equivalently,

$$P(x_1 < x_2, G)P(x_1 < x_3, G) \leq P(G)P(x_1 < x_2, x_1 < x_3, G).$$

Proof Let S denote the set of ordered n-tuples $\boldsymbol{x} = (x_1, \ldots, x_n)$ where each $x_i \in \{1, \ldots, N\}$. N is a large integer which will later tend to infinity. Define an order relation $\leq$ on S by

$$\boldsymbol{x} \leq \boldsymbol{y} \Leftrightarrow x_1 \geq y_1, x_i - x_1 \leq y_i - y_1 \ (i = 2, \ldots, n).$$

It is straightforward to show that $\boldsymbol{x} \wedge \boldsymbol{y}$ has ith component

$$(\boldsymbol{x} \wedge \boldsymbol{y})_i = \min(x_i - x_1, y_i - y_1) + \max(x_1, y_1)$$

and that $\boldsymbol{x} \vee \boldsymbol{y}$ is the n-tuple whose ith component is

$$(\boldsymbol{x} \vee \boldsymbol{y})_i = \max(x_i - x_1, y_i - y_1) + \min(x_1, y_1).$$

It is also easy to check that $\boldsymbol{x} \wedge \boldsymbol{y}$ and $\boldsymbol{x} \vee \boldsymbol{y}$ are both in S. We now show that $(S, <)$ is a distributive lattice. To do this we have to check that $\boldsymbol{x} \wedge (\boldsymbol{y} \vee \boldsymbol{z}) = (\boldsymbol{x} \wedge \boldsymbol{y}) \vee (\boldsymbol{x} \wedge \boldsymbol{z})$ for each $\boldsymbol{x}, \boldsymbol{y}, \boldsymbol{z}$. Put $\boldsymbol{w} = \boldsymbol{y} \vee \boldsymbol{z}$. Then

$$\begin{aligned}(\boldsymbol{x} \wedge (\boldsymbol{y} \vee \boldsymbol{z}))_i &= \min(x_i - x_1, w_i - w_1) + \max(x_1, w_1)\\ &= \min(x_i - x_1, \max(y_i - y_1, z_i - z_1))\\ &\quad + \max(x_1, \min(y_1, z_1))\end{aligned}$$

whereas

$$\begin{aligned}((\boldsymbol{x} \wedge \boldsymbol{y}) \vee (\boldsymbol{x} \wedge \boldsymbol{z}))_i &= \max((\boldsymbol{x} \wedge \boldsymbol{y})_i - (\boldsymbol{x} \wedge \boldsymbol{y})_1, (\boldsymbol{x} \wedge \boldsymbol{z})_i - (\boldsymbol{x} \wedge \boldsymbol{z})_1)\\ &\quad + \min((\boldsymbol{x} \wedge \boldsymbol{y})_1, (\boldsymbol{x} \wedge \boldsymbol{z})_1)\\ &= \max(\min(x_i - x_1, y_i - y_1), \min(x_i - x_1, z_i - z_1))\\ &\quad + \min(\max(x_1, y_1), \max(x_1, z_1)).\end{aligned}$$

However, for any three numbers a, b, c we have

$$\min(a, \max(b, c)) = \max(\min(a, b), \min(a, c)) \tag{6.13}$$

and

$$\max(a, \min(b, c)) = \min(\max(a, b), \max(a, c)), \tag{6.14}$$

so that all we need do is replace a, b, c by $x_i - x_1, y_i - y_1, z_i - z_1$ in (6.13) and replace a, b, c by x_1, y_1, z_1 in (6.14).

Thus $(S, <)$ is a distributive lattice. For an application of FKG we next introduce the characteristic functions f, g of the events $x_1 \leqslant x_2$ and $x_1 \leqslant x_3$:

$$f(\boldsymbol{x}) = \begin{cases} 1 & \text{if } x_1 \leqslant x_2 \\ 0 & \text{otherwise} \end{cases}$$

$$g(\boldsymbol{x}) = \begin{cases} 1 & \text{if } x_1 \leqslant x_3 \\ 0 & \text{otherwise.} \end{cases}$$

It is easy to check that both f and g are increasing functions. For example, if $\boldsymbol{x} < \boldsymbol{y}$ and $f(\boldsymbol{x}) = 1$, then $x_2 - x_1 \leqslant y_2 - y_1$ and $x_1 \leqslant x_2$ so that $y_1 \leqslant y_2$ and so $f(\boldsymbol{y}) = 1$. Also define

$$\mu(x) = \begin{cases} 1 & \text{if } \boldsymbol{x} \text{ satisfies the inequalities of } G \\ 0 & \text{otherwise.} \end{cases}$$

To prove that μ satisfies the condition of Theorem 6.2.9 it is sufficient to show that $\mu(\boldsymbol{x}) = \mu(\boldsymbol{y}) = 1 \Rightarrow (\boldsymbol{x} \wedge \boldsymbol{y}) = \mu(\boldsymbol{x} \vee \boldsymbol{y}) = 1$. Therefore suppose $\mu(\boldsymbol{x}) = \mu(\boldsymbol{y}) = 1$. Then $\boldsymbol{x}$ and $\boldsymbol{y}$ both satisfy the inequalities of G. If $x_i < x_j$ is one of the inequalities of G, we then have $x_i < x_j$ and $y_i < y_j$ so that

$$\begin{aligned}(\boldsymbol{x} \wedge \boldsymbol{y})_i &= \min(x_i - x_1, y_i - y_1) + \max(x_1, y_1) \\ &\leqslant \min(x_j - x_1, y_j - y_1) + \max(x_1, y_1) = (\boldsymbol{x} \wedge \boldsymbol{y})_j,\end{aligned}$$

and similarly $(\boldsymbol{x} \vee \boldsymbol{y})_i \leqslant (\boldsymbol{x} \vee \boldsymbol{y})_j$.

We can therefore apply Theorem 6.2.9 to f, g, and μ to obtain

$$\frac{\sum_{\boldsymbol{x} \in S} \mu(\boldsymbol{x})}{|S|} \cdot \frac{\sum_{\boldsymbol{x} \in S} \mu(\boldsymbol{x}) f(\boldsymbol{x}) g(\boldsymbol{x})}{|S|} \geqslant \frac{\sum_{\boldsymbol{x} \in S} f(\boldsymbol{x}) \mu(\boldsymbol{x})}{|S|} \cdot \frac{\sum_{\boldsymbol{x} \in S} g(\boldsymbol{x}) \mu(\boldsymbol{x})}{|S|},$$

i.e.

$$P(G) P(x_1 \leqslant x_2, x_1 \leqslant x_3, G) \geqslant P(x_1 \leqslant x_2, G) P(x_1 \leqslant x_3, G).$$

Up to this stage we have been considering n-tuples $\boldsymbol{x} = (x_1, \ldots, x_n)$, $x_i \in \{1, \ldots, N\}$. These are not permutations since it is possible to have $x_i = x_j$ and $i \neq j$. However, as $N \to \infty$ the probability that $x_i = x_j$ for some $i \neq j$ tends to zero, and the theorem therefore follows. We remark that we had to deal with n-tuples rather than permutations so as to obtain a distributive lattice. We note also that the complicated order relation on S was needed to give functions satisfying the conditions of the FKG theorem. □

6.4 Chvátal's conjecture

In 1974 Chvátal made an interesting conjecture concerning intersecting families in ideals. Recall that a collection $\mathscr{F}$ of subsets of an n-set S is an intersecting family if $A \cap B \neq \varnothing$ for all $A, B \in \mathscr{F}$. Among the intersecting families are those of a very simple type, called stars.

Definition 6.4.1 A collection $\mathscr{B}$ of subsets of an n-set S is a *star* if there is an element $x \in S$ such that $x \in B$ for all $B \in \mathscr{B}$. Chvátal conjectured that if we are looking for as large an intersecting family as possible in an ideal, then we can restrict our search to the stars.

Chvátal's conjecture Let $\mathscr{A}$ be an ideal of subsets of S, let $w(\mathscr{A})$ denote the maximum size of an intersecting family in $\mathscr{A}$, and let $s(\mathscr{A})$ denote the maximum size of a star in $\mathscr{A}$. Then $w(\mathscr{A}) = s(\mathscr{A})$. Note that

$$s(\mathscr{A}) = \max_{x \in S} |\{A \in \mathscr{A} : x \in A\}|.$$

It is clear from the definitions of $w(\mathscr{A})$ and $s(\mathscr{A})$ that we certainly have $w(\mathscr{A}) \geqslant s(\mathscr{A})$. Chvátal conjectures that there is equality. There are some special cases where the conjecture follows from earlier results in this book. If $\mathscr{A}$ consists of all subsets of S, then the conjecture follows from Theorem 1.1.1. If $\mathscr{A}$ consists of all the k-subsets of S and their subsets, $k \leqslant \frac{1}{2}n$, then the conjecture follows from the EKR theorem. So far the conjecture has not been proved in general, but some interesting partial results have been obtained; we shall present some of these in this section.

A lot of use has been made of the following theorem:

Theorem 6.4.1 (Berge 1976) If $\mathscr{A}$ is an ideal of subsets of an n-set S, then $\mathscr{A}$ is the disjoint union of pairs of disjoint subsets of S, together with $\varnothing$ if $|\mathscr{A}|$ is odd.

Example 6.4.1 Suppose that the ideal $\mathscr{A}$ consists of the following subsets of $\{1, \ldots, 5\}$: $\varnothing$, $\{1\}$, $\{2\}$, $\{3\}$, $\{4\}$, $\{5\}$, $\{1,2\}$, $\{1,3\}$, $\{2,3\}$, $\{1,4\}$, $\{2,4\}$, $\{1,5\}$, $\{3,5\}$, $\{4,5\}$, $\{1,2,3\}$, $\{1,2,4\}$, $\{1,3,5\}$. Then $\mathscr{A}$ is the union of the following pairs of disjoint sets, together with $\varnothing$: $\{1\}$ and $\{4,5\}$, $\{2\}$ and $\{1,3,5\}$, $\{3\}$ and $\{1,2,4\}$, $\{4\}$ and $\{1,2,3\}$, $\{5\}$ and $\{1,4\}$, $\{1,3\}$ and $\{2,4\}$, $\{1,2\}$ and $\{3,5\}$, $\{1,5\}$ and $\{2,3\}$.

It had been shown earlier by Erdös, Herzog, and Schönheim (1970) and by Marica (1971) that if $\mathscr{A}$ is an ideal, then there is a bijection $\phi : \mathscr{A} \to \mathscr{A}'$ (where $\mathscr{A}' = \{A' : A \in \mathscr{A}\}$) such that $A \subseteq \phi(A)$ for all $A \in \mathscr{A}$. Berge's result is that there is a bijection ϕ as described, with the additional property that $\phi((\phi(A))') = A'$ for all $A \in \mathscr{A}$. For if A is paired with $(\phi(A))'$ then certainly A and $(\phi(A))'$ are disjoint (since $A \subseteq \phi(A)$) and, further, $(\phi(A))'$ is paired with $(\phi((\phi(A))'))' = A$. We therefore prove the following theorem.

Theorem 6.4.2 Let $\mathscr{A}$ be an ideal. Then there exists a bijection $\phi : \mathscr{A} \to \mathscr{A}'$ such that $A \subseteq \phi(A)$ and $\phi((\phi(A))') = A'$ for all $A \in \mathscr{A}$.

Proof (Daykin, Hilton, and Miklós 1983). We use induction on $|\mathscr{A}|$. The case $|\mathscr{A}| = 1$ is trivial, and so we consider the induction step, assuming that the result holds whenever $\mathscr{A}$ is replaced by an ideal of smaller size.

Let E be a subset of S of minimum size such that $A \cup E \in \mathscr{A}'$ for some $A \in \mathscr{A}$, and let $\mathscr{D} = \{A \in \mathscr{A} : A \cup E \in \mathscr{A}'\}$. Note that $A \cap E = \varnothing$ for all $A \in \mathscr{D}$, since otherwise there is an unnecessary element in E, contradicting the minimality of E. We shall now show that $\mathscr{A} - \mathscr{D}$ is an ideal. Since $\mathscr{D} \neq \varnothing$, we shall be able to apply the induction hypothesis to $\mathscr{A} - \mathscr{D}$.

So let $X \in \mathscr{A} - \mathscr{D}$, $Y \subset X$. Then $X \in \mathscr{A}$ but $X \cup E \notin \mathscr{A}'$. Certainly $Y \in \mathscr{A}$ since $\mathscr{A}$ is an ideal, so suppose that $Y \in \mathscr{D}$. Then $Y \cup E \in \mathscr{A}'$ so that, since $\mathscr{A}'$ is an upset, we must have $X \cup E \in \mathscr{A}'$. This contradicts $X \cup E \notin \mathscr{A}'$, so the assumption that $Y \in \mathscr{D}$ must be false. So $Y \in \mathscr{A} - \mathscr{D}$ as required, and $\mathscr{A} - \mathscr{D}$ is an ideal.

Applying the induction hypothesis to $\mathscr{A} - \mathscr{D}$, we obtain a bijection $\psi : \mathscr{A} - \mathscr{D} \to (\mathscr{A} - \mathscr{D})'$ such that $A \subseteq \psi(A)$ and $\psi((\psi(A))') = A'$ for all $A \in \mathscr{A} - \mathscr{D}$. Now define ϕ on $\mathscr{A}$ by

$$\phi(A) = \begin{cases} A \cup E & \text{if } A \in \mathscr{D} \\ \psi(A) & \text{if } A \in \mathscr{A} - \mathscr{D}. \end{cases}$$

Then trivially $A \subseteq \phi(A)$ for all $A \in \mathscr{A}$. Further, if $A \in \mathscr{A} - \mathscr{D}$.

$$\phi((\phi(A))') = \phi((\psi(A))') = \psi((\psi(A))') = \mathscr{A}'$$

since $(\psi(A))' \in \mathscr{A} - \mathscr{D}$. However, if $A \in \mathscr{D}$, then

$$\phi((\phi(A))') = \phi((A \cup E)')$$

where $(A \cup E)' \in \mathscr{A}$. If we can show that $(A \cup E)' \in \mathscr{D}$ then we shall have

$$\phi((\phi(A))') = \phi((A \cup E)') = (A \cup E)' \cup E = ((A \cup E) \cap E')'$$
$$= (A \cap E')' = A'$$

since $A \cap E = \emptyset$. It therefore remains to show that if $A \in \mathscr{D}$ then $(A \cup E)' \in \mathscr{D}$. Certainly $(A \cup E)' \in \mathscr{A}$, so suppose that $(A \cup E)' \in \mathscr{A} - \mathscr{D}$. Then $(A \cup E)' \cup E \notin \mathscr{A}'$, i.e. $(A \cup E) \cap E' \notin \mathscr{A}$, i.e. $A \cap E' \notin \mathscr{A}$. However, $A \cap E' \subseteq A$ where $A \in \mathscr{A}$, so that $A \cap E' \in \mathscr{A}$ since $\mathscr{A}$ is an ideal, and we have a contradiction. Thus $(A \cup E)' \in \mathscr{D}$ and the proof is complete. □

An immediate consequence of this theorem is the following corollary.

Corollary 6.4.3 If $\mathscr{A}$ be an ideal, then $w(\mathscr{A}) \leqslant \frac{1}{2}|\mathscr{A}|$.

Proof At most one set from each disjoint pair can be in any given intersecting family. □

We note in passing that this corollary can be obtained in a different way by using Kleitman's lemma (see Exercise 6.17).

Some of the partial results concerning Chvátal's conjecture involve the maximal members of the ideal $\mathscr{A}$; these are called the bases of $\mathscr{A}$.

Definition 6.4.2 If $\mathscr{A}$ is an ideal, $A \in \mathscr{A}$ is a *base* if $A \subseteq B$, $B \in \mathscr{A} \Rightarrow A = B$.

Theorem 6.4.4 (Schönheim 1976) If the bases of the ideal $\mathscr{A}$ have a non-empty intersection then Chvátal's conjecture is true for $\mathscr{A}$, and $w(\mathscr{A}) = \frac{1}{2}|\mathscr{A}|$.

Proof Suppose that the element x is in all the bases of $\mathscr{A}$. Write $\mathscr{A} = \mathscr{B} \cup \mathscr{C}$ where x is in every set in $\mathscr{B}$ and in no set in $\mathscr{C}$. If $A \in \mathscr{B}$ then $A - \{x\} \in \mathscr{C}$, so we have $|\mathscr{B}| \leqslant |\mathscr{C}|$. However, if $D \in \mathscr{C}$ then $D \subset B$ for some base B (which contains x) so that $D \cup \{x\} \subseteq B$, and $D \cup \{x\} \in \mathscr{B}$. Thus $|\mathscr{C}| \leqslant |\mathscr{B}|$. It follows that $|\mathscr{B}| = |\mathscr{C}|$, and $|\mathscr{B}| = \frac{1}{2}|\mathscr{A}|$. Finally observe that $\mathscr{B}$ is a star, so $|\mathscr{B}| \leqslant s(\mathscr{A}) \leqslant w(\mathscr{A})$; thus $w(\mathscr{A}) \geqslant \frac{1}{2}|\mathscr{A}|$. It now follows from Corollary 6.4.3 that $w(\mathscr{A}) = \frac{1}{2}|\mathscr{A}|$ and $s(\mathscr{A}) = |\mathscr{B}| = w(\mathscr{A})$. □

Since these results were proved in the mid 1970s, Chvátal's conjecture has been proved for some further classes of ideals. For example, Stein (1983) proved the conjecture for ideals in which all but one of the bases form a *simple star*, i.e. a star in which the intersection of all the members is equal to the intersection of any two of its members. Another partial result has been obtained by Miklós. Recall from Corollary 6.4.3 that $w(\mathcal{A}) \leq [\frac{1}{2}|\mathcal{A}|]$. Miklós's result is that if $w(\mathcal{A}) = [\frac{1}{2}|\mathcal{A}|]$ then Chvátal's conjecture is true for $\mathcal{A}$. We need a preliminary lemma.

Lemma 6.4.5 Let $\mathcal{A}$ be an ideal of subset of S. If there is an $x \in S$ such that $s(x, \mathcal{A}) = |\{A \in \mathcal{A} : x \in A\}| = \frac{1}{2}|\mathcal{A}|$, then x is in every base of $\mathcal{A}$. If there is an $x \in S$ such that $s(x, \mathcal{A}) = \frac{1}{2}(|\mathcal{A}| - 1)$, then x is in all but one of the bases.

Proof Let $B_1, \ldots, B_p$ be the bases of $\mathcal{A}$. First suppose that $s(x, \mathcal{A}) = \frac{1}{2}|\mathcal{A}|$. If there is a base B_j with $x \neq B_j$, then $s(x, \mathcal{A} - \{B_j\}) = \frac{1}{2}|\mathcal{A}| > [\frac{1}{2}|\mathcal{A} - \{B_j\}|]$, contradicting Corollary 6.4.3. Therefore x is in every base. Next suppose that $s(x, \mathcal{A}) = \frac{1}{2}(|\mathcal{A}| - 1)$. Since $s(x, \mathcal{A}) \neq \frac{1}{2}|\mathcal{A}|$, it is clear from the proof of Theorem 6.4.4 that $x \notin \bigcap_i B_i$. However, if $x \notin B_i$ and $x \notin B_j$ $(i \neq j)$, then $s(x, \mathcal{A} - \{B_i, B_j\}) = \frac{1}{2}(|\mathcal{A}| - 1) > [\frac{1}{2}|\mathcal{A} - \{B_i, B_j\}|]$, contradicting Corollary 6.4.3. Thus x is in all but one of the bases. □

Theorem 6.4.6 (Miklós 1984) Let $\mathcal{A}$ be an ideal of subsets of an n-set S such that $w(A) = [\frac{1}{2}|\mathcal{A}|]$, and let $B_1, \ldots, B_p$ be the bases of $\mathcal{A}$.

(i) If $|\mathcal{A}|$ is even then $M = \bigcap_i B_i \neq \varnothing$ (so that by Theorem 6.4.4 Chvátal's conjecture is true for $\mathcal{A}$), and every intersecting family $\mathcal{B} \subseteq \mathcal{A}$ of maximum size arises in the following way: take a maximum intersecting family $\mathcal{B}_0$ of subsets of M and let $\mathcal{B} = \{A \in \mathcal{A} : B \subseteq A$ for some $B \in \mathcal{B}_0\}$.

(ii) If $|\mathcal{A}|$ is odd and $\mathcal{B} \subseteq \mathcal{A}$ is an intersecting family of maximum size then there is either a base set B_i such that $\mathcal{B} \subset \mathcal{A} - \{B_i\}$ or a set $B_{p+1} \subseteq S$, $B_{p+1} \notin \mathcal{A}$, such that $\mathcal{B} \cup \{B_{p+1}\}$ is an intersecting family and $\mathcal{A} \cup \{B_{p+1}\}$ is an ideal (i.e. either $|\mathcal{B}| = \frac{1}{2}|\mathcal{A} - \{B_i\}|$ or $|\mathcal{B} \cup \{B_{p+1}\}| = \frac{1}{2}|\mathcal{A} \cup \{B_{p+1}\}|$, in either of which cases Chvátal's conjecture follows from (i)).

Proof We consider only the case where $|\mathcal{A}|$ is even. It is clear that any family $\mathcal{B}$ as described in the statement of the theorem is an intersecting family. We now show that it is of maximum size by proving that $|\mathcal{B}| = \frac{1}{2}|\mathcal{A}|$. If $A \in \mathcal{A}$ then the symmetric difference $A + M = (A - M) \cup (M - A)$ is a disjoint union $A_1 \cup M_1$ for some

$A_1 \subseteq A, M_1 \subseteq M$. However, $A_1 \subseteq B_i$ for some i, so that $A_1 \cup M_1 \subseteq B_i$ and $A + M \in \mathscr{A}$. Also, $(A+M)+M = A$, so $\mathscr{A}$ consists of pairs $(A, A+M)$. Now either $M \cap A$ or its complement in M must be in $\mathscr{B}_0$ by maximality, so there is a $B \in \mathscr{B}_0$ such that $B \subseteq A$ or $B \subseteq M + A$; therefore at least one of each pair $(A, A+M)$ is in $\mathscr{B}$. Thus $|\mathscr{B}| \geqslant \frac{1}{2}|\mathscr{A}|$. However, $|\mathscr{B}| \leqslant \frac{1}{2}|\mathscr{A}|$ by Corollary 6.4.3, so we must have $|\mathscr{B}| = \frac{1}{2}|\mathscr{A}|$.

We next use induction to prove that $M = \bigcap_i B_i \neq \varnothing$, and assume the result holds for ideals of size less than $|\mathscr{A}|$. Imitating the proof of Theorem 6.4.2, we can choose $E \subseteq S$ of minimum size such that $A \cup E \in \mathscr{A}'$ for some $A \in \mathscr{A}$ and define $\mathscr{D} = \{A \in \mathscr{A} : A \cup E \in \mathscr{A}'\}$ to find that $\mathscr{A} - \mathscr{D}$ is an ideal. Provided that $\mathscr{A} - \mathscr{D} \neq \varnothing$, we shall apply the induction hypothesis to $\mathscr{A} - \mathscr{D}$; if $\mathscr{D} = \mathscr{A}$ then $\mathscr{A} = \mathscr{P}(S)$ and the result is trivial.

Now we are given that $\mathscr{A}$ has an interesting family $\mathscr{F}$ of size $\frac{1}{2}|\mathscr{A}|$. $\mathscr{D}$ is a union of disjoint pairs (Exercise 6.21), so $|\mathscr{D} \cap \mathscr{F}| \leqslant \frac{1}{2}|\mathscr{D}|$. If $|\mathscr{D} \cap \mathscr{F}| < \frac{1}{2}|\mathscr{D}|$ then $\mathscr{F} - \mathscr{D}$ would be an intersecting family in $\mathscr{A} - \mathscr{D}$ with $|\mathscr{F} - \mathscr{D}| > \frac{1}{2}|\mathscr{A} - \mathscr{D}|$, contradicting Corollary 6.4.3. So $|\mathscr{D} \cap \mathscr{F}| = \frac{1}{2}|\mathscr{D}|$, whence $|\mathscr{F} - \mathscr{D}| = \frac{1}{2}|\mathscr{A} - \mathscr{D}|$. We therefore have $w(\mathscr{A} - \mathscr{D}) = \frac{1}{2}|\mathscr{A} - \mathscr{D}|$, and we can apply the induction hypothesis to $\mathscr{A} - \mathscr{D}$, concluding that the intersection N of the bases of $\mathscr{A} - \mathscr{D}$ is non-empty. Since all intersecting families of maximum size in $\mathscr{A} - \mathscr{D}$ must contain N, $N \in \mathscr{F} - \mathscr{D}$. We next show that $N \cap E' \neq \varnothing$. Suppose that $N \cap E' = \varnothing$, i.e. $N \subseteq E$. Since $A \cap E = \varnothing$ for all $A \in \mathscr{D}$, $A \cap N = \varnothing$ for all $A \in \mathscr{D}$. However, since $|\mathscr{F} \cap \mathscr{D}| = \frac{1}{2}|\mathscr{D}| \neq 0$, we can choose $A \in \mathscr{F} \cap \mathscr{D}$ and for this A, $A \in \mathscr{F}$ and $N \in \mathscr{F}$, whence $A \cap N \neq \varnothing$, which is a contradiction. Therefore we must have $N \cap E' \neq \varnothing$ after all. Let $N \cap E' = M$; then M will turn out to be the intersection of all the bases of $\mathscr{A}$.

If $x \in M$, i.e. $x \in N \cap E'$, then x is in every base set of $\mathscr{A} - \mathscr{D}$; it therefore follows as in the proof of Schönheim's theorem that $s(x, \mathscr{A} - \mathscr{D}) = \frac{1}{2}|\mathscr{A} - \mathscr{D}|$. However, $\mathscr{D}$ is the union of pairs A, B with $A \cap B = \varnothing$ and $A \cup B = E'$ (Exercise 6.21), so that x occurs in half the sets of $\mathscr{D}$ and thus $s(x, \mathscr{A}) = \frac{1}{2}|\mathscr{A}|$. It therefore follows from Lemma 6.4.5 that M is in every base of $\mathscr{A}$. Since if $x \notin M$, $s(x, \mathscr{A}) < \frac{1}{2}|\mathscr{A}|$, M is the intersection of all the bases of $\mathscr{A}$; this intersection is therefore non-empty.

Since $\mathscr{F} - \mathscr{D}$ is a maximum-sized intersecting family in $\mathscr{A} - \mathscr{D}$, the induction hypothesis tells us that $\mathscr{F} - \mathscr{D}$ can be obtained by taking a maximum intersecting family of subsets of N and then taking all supersets in $\mathscr{A} - \mathscr{D}$. We can then add half the sets in $\mathscr{D}$ to obtain the intersecting family $\mathscr{F}$ of size $\frac{1}{2}|\mathscr{A}|$. If $M = N$ then $\mathscr{F}$ is as described in the theorem. Next suppose $M \neq N$, i.e. $N \cap E \neq \varnothing$. Consider any partition $M = M_1 \cup M_2$; we shall next show that M_1 or M_2 lies in $\mathscr{F} - \mathscr{D}$. Suppose that neither is in $\mathscr{F} - \mathscr{D}$. Note that

$$M_1 \cup \{M_2 \cup (N \cap E)\} = M \cup (N \cap E) = (N \cap E') \cup (N \cap E) = N$$

and

$$M_1 \cap \{M_2 \cup (N \cap E)\} = M_1 \cap (N \cap E) \subseteq (N \cap E') \cap (N \cap E) = \varnothing$$

so that M_1 and $M_2 \cup (N \cap E)$ are complementary subsets of N. Since $\mathscr{F} - \mathscr{D}$ contains a maximum intersecting family of subsets of N, it must contain one of M_1 and $M_2 \cup (N \cap E)$. Similarly it must contain one of M_2 and $M_1 \cup (N \cap E)$. Therefore if $\mathscr{F} - \mathscr{D}$ contains neither M_1 nor M_2 it must contain both $H_1 = M_1 \cup (N \cap E)$ and $H_2 = M_2 \cup (N \cap E)$. However (Exercise 6.22), there is a pair of sets in $\mathscr{D}$ such that one of them intersects M in M_1 (and hence does not intersect H_2) while the other intersects M in M_2 (and hence does not intersect H_1); therefore $\mathscr{F}$ can contain neither of the sets of this pair, contradicting the observation made earlier that half of the sets of $\mathscr{D}$ can be added to the sets of $\mathscr{F} - \mathscr{D}$ to form $\mathscr{F}$. Thus $\mathscr{F} - \mathscr{D}$ must, after all, contain M_1 or M_2, so that $\mathscr{F}$ contains a maximum intersecting family of subsets of M, say $\mathscr{B}$. Thus $\mathscr{F} \supseteq \{A \in \mathscr{A} : B \subseteq A \text{ for some } B \in \mathscr{B}\}$. However, such a collection was shown to have cardinality $\frac{1}{2}|\mathscr{A}| = |\mathscr{F}|$, so we conclude that $\mathscr{F} = \{A \in \mathscr{A} : B \subseteq A \text{ for some } B \in \mathscr{B}\}$ as required. □

Exercises 6

6.1 Prove the following two alternative forms of Kleitman's lemma.
(a) If $\mathscr{A}$, $\mathscr{B}$ are ideals of subsets of an n-set, then $|\mathscr{A}| \,.\, |\mathscr{B}| \leq 2^n |\mathscr{A} \cap \mathscr{B}|$.
(b) If $\mathscr{A}$, $\mathscr{B}$ are upsets of an n-set, then $|\mathscr{A}| \,.\, |\mathscr{B}| \leq 2^n |\mathscr{A} \cap \mathscr{B}|$.

6.2 Let $\mathscr{A}$ be an ideal of subsets of S. Show that the average size of members of $\mathscr{A}$ is $\leq \frac{1}{2}|S|$.

6.3 For $1 < k < \frac{1}{2}n$ construct k disjoint intersecting families of subsets of an n-set S whose union contains

$$g(n, k) = 2^{n-1} + \sum_{j=1}^{k-1} \binom{n-1}{j-1}$$

sets, such that none of the families can be increased without destroying disjointness. (Note that $g(n, k) < 2^n - 2^{n-k}$, so that the situation is different from the case $k=1$ where any intersecting family can be extended to one of size 2^{n-1}.)

(Kleitman 1966)

6.4 Let $\mathscr{P}$, $\mathscr{Q}$ be families of subsets of the n-set S such that

$$A \in \mathscr{P},\ B \in \mathscr{Q} \Rightarrow A \not\subseteq B \text{ and } B \not\subseteq A.$$

Show that $|\mathscr{P}|^{1/2} + |\mathscr{Q}|^{1/2} \leqslant 2^{n/2}$ as follows. Put

$$\mathscr{H} = \{X \subseteq S : X \supseteq A \text{ for some } A \in \mathscr{P} \text{ and } X \supseteq B \text{ for some } B \in \mathscr{Q}\}$$

$$\mathscr{I} = \{X \subseteq S : X \supseteq A \text{ for some } A \in \mathscr{P} \text{ but } X \not\supseteq B \text{ for all } B \in \mathscr{Q}\}$$

$$\mathscr{J} = \{X \subseteq S : X \not\supseteq A \text{ for all } A \in \mathscr{P} \text{ but } X \supseteq B \text{ for some } B \in \mathscr{Q}\}$$

$$\mathscr{K} = \{X \subseteq S : X \not\supseteq A \text{ for all } A \in \mathscr{P} \text{ and } X \not\supseteq B \text{ for all } B \in \mathscr{Q}\}.$$

Apply Kleitman's lemma to $\mathscr{U} = \mathscr{H} \cup \mathscr{I}$ and $\mathscr{D} = \mathscr{J} \cup \mathscr{K}$ to prove that $|\mathscr{I}|^{1/2} + |\mathscr{J}|^{1/2} \leqslant 2^{n-2}$, and note that $\mathscr{P} \subseteq \mathscr{I}$, $\mathscr{Q} \subseteq \mathscr{J}$.

(Seymour 1973)

6.5 Use 6.4 to prove Theorem 6.1.4.

6.6 Let $\mathscr{A}$ be a collection of subsets of an n-set S. Let

$$\mathscr{A}^+ = \{B \subseteq S : B \supseteq A \text{ for some } A \in \mathscr{A}\},\quad \mathscr{A}^- = \{B \subseteq S : B \subseteq A \text{ for some } A \in \mathscr{A}\},$$

$\mathscr{C}(\mathscr{A}) = \mathscr{A}^+ \cup \mathscr{A}^-$, $\text{mid}(\mathscr{A}) = \mathscr{A}^+ \cap \mathscr{A}^-$. Show that

(i) $\mathscr{A}^+ = (\text{mid}(\mathscr{A}))^+$

(ii) $\mathscr{C}(\mathscr{A}) = \mathscr{C}(\text{mid}(\mathscr{A}))$

(iii) $|\mathscr{A}^+| + |\mathscr{A}^-| \geqslant 2(2^n\,|\text{mid}(\mathscr{A})|)^{1/2}$

(iv) $|\mathscr{C}(\mathscr{A})| \geqslant 2(2^n\,|\text{mid}(\mathscr{A})|)^{1/2} - |\text{mid}(\mathscr{A})|)^{1/2}$

(v) if $0 \leqslant i \leqslant \frac{1}{2}n$ and $|\mathscr{A}| \geqslant 2^{n-2i}$ then $|\mathscr{C}(\mathscr{A})| \geqslant 2^{n+1-i} - 2^{n-2i}$.

(Frankl and Hilton 1977)

6.7 Verify that

(i) if $\mathscr{A}$, $\mathscr{B}$ are upsets then $\mathscr{A} \vee \mathscr{B} = \mathscr{A} \cap \mathscr{B}$,

(ii) if $\mathscr{A}$, $\mathscr{B}$ are downsets, then $\mathscr{A} \wedge \mathscr{B} = \mathscr{A} \cap \mathscr{B}$.

6.8 Deduce Kleitman's inequality from Corollary 6.2.2.

6.9 Prove FKG (6.2.9) in the special case where the lattice is $\{1 < 2 < \ldots < n\}$ by expanding the inequality

$$\sum_{i,j} (f(i) - f(j))(g(i) - g(j))\mu(i)\mu(j) \geqslant 0.$$

6.10 Suppose that μ_1, μ_2 are probability distributions (i.e. $\sum_A \mu_i(A) = 1$) such that $\mu_1(A)\mu_2(B) \leqslant \mu_1(A \cup B)\mu_2(A \cap B)$ for all $A, B \subseteq S$, and that h is an increasing function. By choosing $\alpha(A) = \mu_1(A)$, $\beta(A) = \mu_2(A)h(A)$, $\varphi(A) = \mu_1(A)h(A)$ and $\delta(A) = \mu_2(A)$, prove Holley's result that

$$\sum_A \mu_1(A)h(A) \geqslant \sum_A \mu_2(A)h(A).$$

(Holley 1974)

6.11 Show that Holley's result in 6.10 implies FKG, by taking $\mu_2 = \mu$ and $\mu_1 = g\mu / \sum_A g(A)\mu(A)$.

6.12 A *polarity* is an order-inverting bijection $\sigma : \mathcal{P}(S) \to \mathcal{P}(S)$ such that $\sigma^{-1} = \sigma$, where order-inverting means $A \subset B \Rightarrow \sigma(B) \subset \sigma(A)$. For any $\mathcal{A} \subseteq \mathcal{P}(S)$ let $\mathcal{A}^+$ denote the collection of all supersets of members of $\mathcal{A}$, and $\mathcal{A}^-$ the set of all subsets of members of $\mathcal{A}$.
(i) Show that $|(\sigma(\mathcal{A}))^+| = |\mathcal{A}^-|$.
(ii) Use Kleitman's lemma to show that $|\mathcal{A}| \leqslant |\mathcal{A}^+ \cap (\sigma(\mathcal{A}))^+|$ and $|\mathcal{A}| \leqslant |\mathcal{A}^- \cap (\sigma(\mathcal{A}))^-|$.

(Daykin, Kleitman, and West 1979)

6.13 Let $\mathcal{A}$, $\mathcal{B}$ be two complement-free collections of subsets of an n-set S such that, for all $A \in \mathcal{A}$ and $B \in \mathcal{B}$, $A \not\subseteq B$ and $B \not\subseteq A$. Prove that $|\mathcal{A}| + |\mathcal{B}| \leqslant 2^{n-1}$ as follows.
(i) Show that if $\mathcal{A} = \mathcal{A}_1 \cup \mathcal{A}_2$, $\mathcal{A}_1 \cap \mathcal{A}_2 = \varnothing$, $\mathcal{B} = \mathcal{B}_1 \cup \mathcal{B}_2$, $\mathcal{B}_1 \cap \mathcal{B}_2 = \varnothing$, $\mathcal{A}_1 = \mathcal{B}_1'$, $\mathcal{A}_2 \cap \mathcal{B}_2' = \varnothing$, $\mathcal{C} = \mathcal{A}_1 \vee \mathcal{B}_1$, $\mathcal{D} = \mathcal{A} \cup \mathcal{B}_2 \cup \mathcal{C}$, then $\mathcal{D} \cap \mathcal{D}' = \varnothing$, so that $|\mathcal{D}| \leqslant 2^{n-1}$.
(ii) Next show that $\mathcal{A} \cap \mathcal{B}_2 = \mathcal{A} \cap \mathcal{C} = \mathcal{B}_2 \cap \mathcal{C} = \varnothing$, so that $|\mathcal{A} \cup \mathcal{B}_2 \cup \mathcal{C}| = |\mathcal{A}| + |\mathcal{B}_2| + |\mathcal{C}|$.
(iii) Use Marica–Schönheim to prove that $|\mathcal{C}| \geqslant |\mathcal{A}_1| = |\mathcal{B}_1|$.
(iv) Deduce that $|\mathcal{A}| = |\mathcal{B}| = |\mathcal{D}| \leqslant 2^{n-1}$. (Hilton 1976)

6.14 Give an example of collections $\mathcal{A}$, $\mathcal{B}$ satisfying the conditions of 6.13 and with $|\mathcal{A}| + |\mathcal{B}| = 2^{n-1}$.

6.15 Use 6.13 to give another proof of Theorem 6.1.4.

6.16 Prove that if $A_1, \ldots, A_m$ are distinct subsets of S and $B_1, \ldots, B_n$ are such that each A_i contains at least one B_j, then there are at least m distinct differences $A_i - B_j$. Deduce the Marica–Schönheim result (see hints).

(Ahlswede and Daykin 1979)

6.17 Use Kleitman's lemma to prove Corollary 6.4.3.

(Schönheim, private communication)

6.18 Prove Chvátal's conjecture in the case when $\mathscr{A}$ has only two bases.

6.19 Let $\mathscr{A}$, $\mathscr{B}$ be two intersecting families of subsets of an n-set S such that $A \in \mathscr{A}$, $B \in \mathscr{B} \Rightarrow A \cup B \neq S$. Prove that $|\mathscr{A}| + |\mathscr{B}| \leq 2^{n-1}$ as follows.

(i) Extend $\mathscr{A}$, $\mathscr{B}$ to intersecting families $\mathscr{A}^*$, $\mathscr{B}^*$ of size 2^{n-1}, and let $\mathscr{D} = \{X : X \in \mathscr{A}^* \cap \mathscr{B}^*\}$, $\mathscr{E} = \mathscr{A}^* - \mathscr{D}$, $\mathscr{F} = \mathscr{B}^* - \mathscr{D}$.

(ii) Show that $E \in \mathscr{E} \Leftrightarrow E' \in \mathscr{F}$.

(iii) Apply Berge's theorem to the ideal $\mathscr{D}' = \{D' : D \in \mathscr{D}\}$, and hence show that the members of $\mathscr{D}$ can be put into pairs with the union of the two sets in any pair being S.

(iv) Then use (ii) to extend the pairing in (iii) to the whole of $\mathscr{A}^* \cup \mathscr{B}^*$. (Hilton 1978)

6.20 (i) Show that if $\mathscr{B}$ is an intersecting family of 2^{n-1} subsets of an n-set S, then $\mathscr{B}'$ is an ideal.

(ii) Use (i) and Corollary 6.4.3 to give another proof of Theorem 6.1.4. (Schönheim 1974a)

6.21 With $\mathscr{D}$ and E as in the proofs of Theorems 6.4.2 and 6.4.6, prove that if $A \in \mathscr{D}$ then $B = (A \cup E)'$ is also in $\mathscr{D}$, and $(B \cup E)' = A$. Deduce that $\mathscr{D}$ is the disjoint union of disjoint pairs A, B with $A = (B \cup E)'$ and $A \cup B = E'$.

6.22 Show that, if $M = M_1 \cup M_2$ as in the proof of Theorem 6.4.6, then one of the pairs D_1, D_2 of 6.21 is such that $D_i \cap M = M_i$.

6.23 Prove Theorem 6.1.4 by considering $|\mathscr{A} \cup (\mathscr{A} - \mathscr{A})|$ and using Corollary 6.2.4. (Daykin and Lovász 1976)

7 The Kruskal–Katona theorem

7.1 Order relations on subsets

In the original proof of Sperner's theorem, use was made of the following fact: if $\mathscr{A}$ is a collection of k-subsets of an n-set S and if $\Delta\mathscr{A}$ denotes the shadow of $\mathscr{A}$, i.e.

$$\Delta\mathscr{A} = \{B \subseteq S : |B| = k-1,\ B \subset A \text{ for some } A \in \mathscr{A}\},$$

then

$$|\Delta\mathscr{A}| \geqslant \frac{k}{n-k+1}|\mathscr{A}|.$$

This estimate for the size of $\Delta\mathscr{A}$ was obtained very easily, and was sufficient to yield the powerful normalized matching property, but it is possible to improve upon it by using deeper arguments. The aim of this chapter is to discuss such an improvement, obtained independently by Kruskal and Katona in the 1960s. Before we state their result, it will be helpful to have a look at some ways of putting an ordering on the set of k-subsets of $S = \{1, \ldots, n\}$.

7.1.1 Lexicographic order

Here we write $A <_L B$ if the *smallest* element of the symmetric difference $A + B = (A \cap B') \cup (A' \cap B)$ is in A. As usual, A' denotes the complement $S - A$ of A. This ordering is a total ordering of the k-subsets of S. For example, if the 3-subsets of $\{1, \ldots, 5\}$ are listed in lexicographic order we obtain the first column below:

1 2 3	4 5
1 2 4	3 5
1 2 5	3 4
1 3 4	2 5
1 3 5	2 4
1 4 5	2 3
2 3 4	1 5
2 3 5	1 4
2 4 5	1 3
3 4 5	1 2

This is just the 'dictionary' order, with 1 replacing a, 2 replacing b, and so on. Note also that, since $A+B=A'+B'$,

$$\begin{aligned} A <_{\mathrm{L}} B &\Leftrightarrow \text{smallest element of } A+B \text{ is in } A \\ &\Leftrightarrow \text{smallest element of } A'+B' \text{ is in } A \\ &\Leftrightarrow \text{smallest element of } A'+B' \text{ is not in } A' \\ &\Leftrightarrow \text{smallest element of } A'+B' \text{ is in } B' \\ &\Leftrightarrow B' <_{\mathrm{L}} A'. \end{aligned}$$

Thus the *complements* of the k-subsets of S, namely the $(n-k)$-subsets of S, are in *reverse* lexicographic order; this in shown in the second column above.

7.1.2 Antilexicographic order

Here we use the dictionary idea again, but consider the ordering of $1, \ldots, n$ to be reversed. Imagine therefore that the alphabetic order is reversed. We write $A <_{\mathrm{A}} B$ if the *largest* element of $A+B$ is in A. Since j is playing the role of $n+1-j$, we obtain the antilexicographic order from the lexicographic order by noting that $A <_{\mathrm{A}} B \Leftrightarrow \bar{A} <_{\mathrm{L}} \bar{B}$ where, if $A=\{a_1, \ldots, a_k\}$, $\bar{A}$ denotes $\{n+1-a_1, \ldots, n+1-a_k\}$. The antilexicographic ordering of the 3-subsets of $\{1, \ldots, 5\}$ can therefore be read off from the first column above:

5 4 3		3 4 5
5 4 2		2 4 5
5 4 1		1 4 5
5 3 2		2 3 5
5 3 1	i.e.	1 3 5
5 2 1		1 2 5
4 3 2		2 3 4
4 3 1		1 3 4
4 2 1		1 2 4
3 2 1		1 2 3.

7.1.3 Squashed ordering

We write $A <_{\mathrm{S}} B$ if the *largest* element of $A+B$ is in B, so that the squashed order is clearly the reverse of the antilexicographic order. Under the squashed order all subsets containing the element $m \leqslant n$ come after all the subsets consisting only of elements less than m. The

3-subsets of $\{1, \ldots, 5\}$ in squashed order are

1 2 3
1 2 4
1 3 4
2 3 4
1 2 5
1 3 5
2 3 5
1 4 5
2 4 5
3 4 5.

Although the squashed order appears less natural than the lexicographic order, it turns out that it plays a particularly important role in the investigation of the size of the shadow of a collection of k-sets. Roughly speaking, if we want to minimize $|\Delta\mathscr{A}|$ over all choices of $\mathscr{A}$ of a given size, we want to squash the $|\mathscr{A}|$ sets so that they are made up from as few elements as possible, and this corresponds to choosing the first $|\mathscr{A}|$ sets in the squashed ordering. This is essentially the kernel of the Kruskal–Katona result.

In our investigations it will sometimes be convenient to represent *subsets* by *sequences* of 0s and 1s. For example, if $n = 5$, the subset $\{1, 2, 4\}$ of $\{1, \ldots, 5\}$ could be represented by the sequence 11010. The ith digit is 1 if the element i is in the set, and is 0 otherwise. However, for reasons which will become clearer shortly, it is convenient to represent subsets by *reverse* sequences; for example $\{1, 2, 4\}$ would be represented by the sequence 01011. Here the ith digit *from the right* is 1 precisely when i is in the set. This representation will be particularly useful when we come to study multisets; for example, we will be able to represent the divisor $2 . 3^2 . 7$ of $2^2 . 3^2 . 5 . 7$ by the reverse sequence 1021, the sequence of powers in the factorization $7^1 . 5^0 . 3^2 . 2^1$.

The reason for representing by reverse sequences is the following. We have $A <_S B \Leftrightarrow$ largest element in one but not the other is in $B \Leftrightarrow$ first place in representing reverse sequences where they differ has 0 for A and 1 for $B \Leftrightarrow$ sequence for A is before the sequence for B in the *lexicographic* ordering of the sequences, where $(x_1, \ldots, x_n) <_L (y_1, \ldots, y_n) \Leftrightarrow x_i < y_i$ for the first i with $x_i \neq y_i$. Thus *squashed* ordering of *sets* corresponds to *lexicographic* ordering of the representing reverse *sequences*. For example, the 3-subsets of $1, \ldots, 5$

listed in their squashed order above are represented by the following sequences which are in lexicographic order:

0 0 1 1 1
0 1 0 1 1
0 1 1 0 1
0 1 1 1 0
1 0 0 1 1
1 0 1 0 1
1 0 1 1 0
1 1 0 0 1
1 1 0 1 0
1 1 1 0 0.

7.2 The l-binomial representation of a number

In order to shed further light on the squashed ordering of sets, we now prove the following theorem.

Theorem 7.2.1 Given positive integers m and l, there exists a unique representation of m in the form

$$m = \binom{a_l}{l} + \binom{a_{l-1}}{l-1} + \ldots + \binom{a_t}{t} \tag{7.1}$$

where $a_l > a_{l-1} > \ldots > a_t \geqslant t \geqslant 1$. (This representation of m is called the *l-binomial representation of m*.)

Proof We can easily find such a representation of m. First choose a_l as the largest integer for which $\binom{a_l}{l} \leqslant m$. Then choose a_{l-1} as the largest integer for which $\binom{a_{l-1}}{l-1} \leqslant m - \binom{a_l}{l}$. If $a_{l-1} \geqslant a_l$ then we would have $m \geqslant \binom{a_l}{l} + \binom{a_l}{l-1} = \binom{1+a_l}{l}$, contradicting the maximality of a_l. Therefore $a_{l-1} < a_l$. Continuing this process we eventually reach a stage where the choice of a_t for some $t \geqslant 2$ actually gives equality,

$$\binom{a_t}{t} = m - \binom{a_l}{l} - \binom{a_{l-1}}{l-1} - \ldots - \binom{a_{t+1}}{t+1},$$

or we get right down to choosing a_1 as the integer such that

$$\binom{a_1}{1} \leqslant m - \binom{a_l}{l} - \ldots - \binom{a_2}{2} < \binom{a_1+1}{1}$$

in which case we have

$$0 \leqslant m - \binom{a_l}{l} - \ldots - \binom{a_1}{1} < 1,$$

so that

$$m = \binom{a_l}{l} + \ldots + \binom{a_1}{1}.$$

We now show that such a representation of m is unique. If $l=1$ there is clearly only one representation, namely $m = \binom{m}{1}$, so we use induction and suppose that the representation of any m is unique if $l = k-1$. Suppose then that we have two different representations

$$m = \binom{a_k}{k} + \ldots + \binom{a_t}{t} = \binom{b_k}{k} + \ldots + \binom{b_r}{r}$$

for some $t \geqslant 1$, $r \geqslant 1$. If $a_k = b_k$ we then obtain two representations for $m - \binom{a_k}{k}$ with $l = k-1$; by the induction hypothesis these representations are identical, so that $a_i = b_i$ for each i. Next suppose that $a_k \neq b_k$. Without loss of generality we can assume that $a_k < b_k$. We then obtain

$$\begin{aligned} m &= \binom{a_k}{k} + \binom{a_{k-1}}{k-1} + \ldots + \binom{a_t}{t} \\ &\leqslant \binom{a_k}{k} + \binom{a_k - 1}{k-1} + \ldots + \binom{a_k - (k-t)}{t} + \ldots + \binom{a_k - k + 1}{1} \\ &= \binom{a_k + 1}{k} - 1 \end{aligned}$$

on using the identity

$$\binom{n}{k} + \binom{n-1}{k-1} + \ldots + \binom{n-k}{0} = \binom{n+1}{k}. \qquad (7.2)$$

We thus obtain $m < \binom{a_k+1}{k} \leqslant \binom{b_k}{k} \leqslant m$, which is a contradiction. Thus, by induction, the l-binomial representation of m is unique. □

Example 7.2.1 Take $l=4$ and consider $m=26$. The largest x for which $\binom{x}{4} \leqslant 26$ is $x=6$. We have

$$\begin{aligned} 26 &= \binom{6}{4} + 11 \\ &= \binom{6}{4} + \binom{5}{3} + 1 \\ &= \binom{6}{4} + \binom{5}{3} + \binom{2}{2}. \end{aligned}$$

What exactly is going on here? The representation is closely related to squashed ordering. Suppose that (7.1) holds. In the squashed ordering of l-sets the first $\binom{a_l}{l}$ sets are precisely the l-subsets of $\{1, \ldots, a_l\}$. The next $\binom{a_{l-1}}{l-1}$ are those obtained by adjoining a_l+1 to the $(l-1)$-subsets of $\{1, \ldots, a_{l-1}\}$. The next $\binom{a_{l-2}}{l-2}$ are those obtained by adjoining $\{a_{l-1}+1, a_l+1\}$ to the $(l-2)$-subsets of $\{1, \ldots, a_{l-2}\}$ and so on until the final $\binom{a_t}{t}$ are the union of $\{a_{t+1}+1, \ldots, a_l+1\}$ with the t-subsets of $\{1, \ldots, a_t\}$. Thus the mth l-set in the squashed order is

$$\{a_l+1, a_{l-1}+1, \ldots, a_{t+1}+1; a_t, a_t-1, \ldots, a_t-t+1\}. \quad (7.3)$$

Example 7.2.2 Take $l=3$.

$$9 = \binom{4}{3} + \binom{3}{2} + \binom{2}{1}, \text{ so the 9th set is } \{5, 4, 2\}.$$

$$5 = \binom{4}{3} + \binom{2}{2}, \text{ so the 5th set is } \{5, 2, 1\}.$$

$$4 = \binom{4}{3}, \text{ so the 4th set is } \{4, 3, 2\}.$$

Example 7.2.3 In which place in the squashed order is (a) $\{2, 3, 5\}$ among the 3-sets and (b) $\{2, 3, 5, 7\}$ among the 4-sets?

Solution

(a) The set $\{2, 3, 5\}$ is rewritten as $\{5, 3, 2\}$ and compared with

(7.3). We see that $a_3+1=5$ and $a_2=3$, so the set is the mth set where $m=\binom{4}{3}+\binom{3}{2}=7$.

(b) Write the set as $\{7,5,3,2\}$ and compare with (7.3) to obtain $a_4+1=7$, $a_3+1=5$, $a_2=3$. Thus the set is the mth set where $m=\binom{6}{4}+\binom{4}{3}+\binom{3}{2}=22$.

The reversal of the ordering of the elements so as to compare with (7.3) suggests that we could work with the reverse representing sequences instead.

Example 7.2.4 In which place in the squashed ordering of 4-sets is the set $\{3,4,5,6\}$?

Solution This set is represented by the sequence 111100, so we have to find where 111100 occurs in the lexicographic ordering of sequences of length 6 with four 1s. However, 111100 is clearly the last sequence in the lexicographic ordering, so it is in the place number $\binom{6}{4}=15$.

Clearly, in general, if the sequence consists of r 1s followed by s 0s, then its place in the lexicographic ordering of those sequences with r 1s is $\binom{r+s}{r}$.

Example 7.2.5 In what place in the squashed ordering of 4-sets is the set $\{2,3,5,7\}$?

Solution We have to place 1010110 in the lexicographic ordering of sequences with four 1s. It is preceded by all such sequences which begin with 0; there are $\binom{6}{4}$ of them. The other sequences preceding it begin 10 . . . . Therefore we now consider the position of 10110 among those sequences with three 1s. It is preceded by all the $\binom{4}{3}$ sequences beginning with 0. The other sequences before 10110 begin with 10 . . . , so we have to place 110 in the ordering of sequences with two 1s. By the remark after the previous example, its position is

$\binom{3}{2}$. Thus the required position number is

$$\binom{6}{4}+\binom{4}{3}+\binom{3}{2}=22.$$

In general, we peel off 1s from the left until a sequence consisting of a string of 1s followed by a string of 0s is obtained.

7.3 The Kruskal–Katona theorem

Consider the first m l-sets in the squashed order, as described before Example 7.2.2, when m is given by (7.1). Note that the $(l-1)$-sets contained in them are

all $\binom{a_l}{l-1}$ $(l-1)$-subsets of $\{1, \ldots, a_l\}$

all $\binom{a_{l-1}}{l-2}$ $(l-2)$-subsets of $\{1, \ldots, a_{l-1}\}$, unioned with $\{1+a_l\}$

⋮

all $\binom{a_t}{t-1}$ $(t-1)$-subsets of $\{1, \ldots, a_t\}$, unioned with

$$\{1+a_l,\ 1+a_{l-1}, \ldots, 1+a_{t+1}\}.$$

Thus there are

$$\binom{a_l}{l-1}+\binom{a_{l-1}}{l-2}+\ldots+\binom{a_t}{t-1}$$

such $(l-1)$-sets, and they are precisely the first $\binom{a_l}{l-1}+\ldots+\binom{a_t}{t-1}$ $(l-1)$-sets in the squashed order. (Care has to be taken if $t=1$, for then the representation is not as in Theorem 7.2.1 $(t-1=0)$, but on rewriting $\binom{a_t}{0}$ as $\binom{a_{t+1}}{0}$ it can be combined with $\binom{a_{t+1}}{1}$ to give $\binom{a_{t+1}+1}{1}$ and the process continued until the representation is as in Theorem 7.2.1.) We are therefore in a position to understand what the Kruskal–Katona theorem stated below essentially says: to minimize the shadow of a collection of m l-sets, where m is given, take the first m l-sets in the squashed order.

Theorem 7.3.1 (Kruskal 1963, Katona 1966*a*) Let $\mathscr{A}=\{A_1, \ldots, A_m\}$ be a collection of l-subsets of S, and suppose that the

l-binomial representation of m is

$$m = \binom{a_l}{l} + \binom{a_{l-1}}{l-1} + \ldots + \binom{a_t}{t}$$

where $a_l > a_{l-1} > \ldots > a_t \geqslant t \geqslant 1$. Then

$$|\Delta \mathscr{A}| \geqslant \binom{a_l}{l-1} + \binom{a_{l-1}}{l-2} + \ldots + \binom{a_t}{t-1}.$$

Note that this result has nothing really to do with $n = |S|$; all that matters is how many sets are in $\mathscr{A}$ (for recent work on the case where $\bigcup_i A_i = S$, see Mörs (1985)).

Various proofs of Theorem 7.3.1 have been given. The original proofs of Kruskal and Katona are quite complicated, but since they were published a number of authors have given shorter proofs; for example, we mention Daykin (1974*a*), Hilton (1979) and Frankl (1984). Hilton's proof is by induction, the main step being to show that, for any $i \in \{1, \ldots, n\}$ and any positive integers b, c with $b + c = a$, the 'earliest' possible collection $\mathscr{A}$ of a k-sets with minimum shadow and containing b sets without i and c sets with i is given by

$$\mathscr{A} = F_k(b)^{\bar{i}} \cup \{\{i\} \cup F_{k-1}(c)^{\bar{i}}\}$$

where $F_l(x)^{\bar{i}}$ denotes the first x l-sets in the squashed order which do not contain i. Later we shall give a proof involving the idea of compression in the more general context of multisets (the generalized Macaulay theorem of Clements and Lindström). However, the proof we now present is due to Frankl and depends on a shifting process; we begin by trying to replace an element of a set by 1 if 1 is not already present. This shifting idea goes back to the original proof of the Erdös–Ko–Rado theorem, where the shifting is away from n; recall that a shifting away from 1 was used in Section 5.5.

Lemma 7.3.2 Let $\mathscr{A}$ be a collection of l-subsets of $S = \{1, \ldots, n\}$, and, for $1 < j \leqslant n$, define

$$S_j(A) = \begin{cases} (A - \{j\}) \cup \{1\} & \text{if } j \in A,\ 1 \notin A,\ (A - \{j\}) \cup \{1\} \notin \mathscr{A} \\ A & \text{otherwise.} \end{cases}$$

Also define

$$S_j(\mathscr{A}) = \{S_j(A) \colon A \in \mathscr{A}\}.$$

Then

$$\Delta(S_j(\mathscr{A})) \subseteq S_j(\Delta \mathscr{A}).$$

Note It follows immediately that $|\Delta\mathcal{A}| \geqslant |\Delta(S_j(\mathcal{A}))|$.

Proof of Lemma 7.3.2 We have to show that if $A \in \mathcal{A}$ then $\Delta S_j(A) \subseteq S_j(\Delta\mathcal{A})$. First suppose that $A = S_j(A)$. If $B \in \Delta S_j(A)$ then $B \in \Delta A$ and so $A = B \cup \{i\}$ for some $i \leqslant n$. We show that $B = S_j(B)$ where (this is important!) S_j is the shift operator for $\Delta\mathcal{A}$, not $\mathcal{A}$; it will then follow that $B \in S_j(\Delta\mathcal{A})$. If $i = j$ then $j \notin B$ so that $B = S_j(B)$. If $i = 1$ and $j \in B$ then $(B - \{j\}) \cup \{1\} = A - \{j\} \in \Delta\mathcal{A}$, so $S_j(B) = B$. If $i = 1$ and $j \notin B$ then $S_j(B) = B$. If $i \neq 1$ or j, $B = S_j(B)$ unless possibly when $j \in B$ and $1 \notin B$. But then $j \in A$ and $1 \notin A$ and, since $S_j(A) = A$, we must have $(A - \{j\}) \cup \{1\} \in \mathcal{A}$. Thus $(B - \{j\}) \cup \{1\} \in \Delta\mathcal{A}$, and so $S_j(B) = B$.

Next suppose that $A \neq S_j(A)$. Then $j \in A$, $1 \notin A$, $S_j(A) = (A - \{j\}) \cup \{1\}$. If $B \in \Delta S_j(A)$ and $1 \notin B$, then $B = A - \{j\}$ and so $S_j(B) = B$. If $B \in \Delta S_j(A)$ and $1 \in B$, then $B^* = (B - \{1\}) \cup \{j\} \subset A$ and so $B^* \in \Delta\mathcal{A}$. If $B \notin \Delta\mathcal{A}$, then $S_j(B^*) = B$ so that $B \in S_j(\Delta\mathcal{A})$. If, finally, $B \in \Delta\mathcal{A}$, note that $j \notin B$ so that $B = S_j(B)$, whence $B \in S_j(\Delta\mathcal{A})$. □

Lemma 7.3.3 If $S_j(\mathcal{A}) = \mathcal{A}$ for all $j \geqslant 2$, and if

$$\mathcal{A}_0 = \{A \in \mathcal{A}: 1 \notin A\}, \qquad \mathcal{A}_\uparrow = \{A - \{1\} : 1 \in A \in \mathcal{A}\},$$

then

$$|\mathcal{A}_\uparrow| \geqslant |\Delta\mathcal{A}_0|.$$

Proof If $E \in \Delta\mathcal{A}_0$ then $E \cup \{i\} \in \mathcal{A}_0$ for some $i > 1$ so that, on applying S_j with $j = i$, $E \cup \{1\} \in \mathcal{A}$. Thus $E \in \Delta\mathcal{A}_0 \Rightarrow E \in \mathcal{A}_\uparrow$. □

Proof of Theorem 7.3.1 (Frankl 1984) If we repeatedly apply the shift operators S_j, $j = 2, \ldots, n$, to $\mathcal{A}$, the number of sets containing 1 increases, so that after a finite number of applications the shifts must therefore cease to make any change. We have then obtained a new collection $\mathcal{A}^*$ of the same size as $\mathcal{A}$, with $S_j(\mathcal{A}^*) = \mathcal{A}^*$ for each $j \geqslant 2$, and with $|\Delta\mathcal{A}| \geqslant |\Delta\mathcal{A}^*|$. The theorem will therefore be proved for $\mathcal{A}$ if we can prove it for $\mathcal{A}^*$. Therefore we now suppose that $\mathcal{A}$ has the property that $S_j(\mathcal{A}) = \mathcal{A}$ for each $j \geqslant 2$.

Now the theorem is true for $l = 1$ and any $n = |S|$, so we now assume that the theorem is true whenever $n < p$ and whenever $l < k$ and $n = p$, and consider the case of a collection $\mathcal{A}$ of k-subsets of a p-set T. We have

$$|\mathcal{A}| = \binom{a_k}{k} + \ldots + \binom{a_s}{s}$$

where $a_k > \ldots > a_s \geqslant s \geqslant 1$.

Suppose, if possible, that

$$|\mathscr{A}_\uparrow| < \binom{a_k - 1}{k-1} + \ldots + \binom{a_s - 1}{s-1}.$$

Then, since $|\mathscr{A}_0| = |\mathscr{A}| - |\mathscr{A}_\uparrow|$,

$$|\mathscr{A}_0| > \left\{\binom{a_k}{k} - \binom{a_k - 1}{k-1}\right\} + \ldots + \left\{\binom{a_s}{s} - \binom{a_s - 1}{s-1}\right\}$$
$$= \binom{a_k - 1}{k} + \ldots + \binom{a_s - 1}{s}$$

so that, by the induction hypothesis,

$$|\Delta\mathscr{A}_0| \geqslant \binom{a_k - 1}{k-1} + \ldots + \binom{a_s - 1}{s-1}.$$

Thus $|\mathscr{A}_\uparrow| \geqslant |\Delta\mathscr{A}_0| > |\mathscr{A}_\uparrow|$, which is a contradiction. Therefore we must have

$$|\mathscr{A}_\uparrow| \geqslant \binom{a_k - 1}{k-1} + \ldots + \binom{a_s - 1}{s-1};$$

but then, by the induction hypothesis,

$$|\Delta\mathscr{A}_\uparrow| \geqslant \binom{a_k - 1}{k-2} + \ldots + \binom{a_s - 1}{s-2}$$

so that

$$|\Delta\mathscr{A}| \geqslant |\mathscr{A}_\uparrow| + |\Delta\mathscr{A}_\uparrow| \geqslant \binom{a_k}{k-1} + \ldots + \binom{a_s}{s-1}$$

as required. □

Corollary 7.3.4 Let $\mathscr{A}$ be a collection of k-sets and, for $r < k$, let

$$\Delta^{(r)}\mathscr{A} = \{B : |B| = r,\ B \subset A \text{ for some } A \in \mathscr{A}\}.$$

Then if

$$|\mathscr{A}| = \binom{a_k}{k} + \ldots + \binom{a_t}{t},$$

$a_k > \ldots > a_t \geqslant t \geqslant 1$,

$$|\Delta^{(r)}\mathscr{A}| \geqslant \binom{a_k}{r} + \ldots + \binom{a_t}{t-k+r}.$$ □

The Kruskal–Katona theorem is a very powerful result, but, owing to the complicated nature of its statement, it is sometimes not very suitable for applications. Instead, the following variation due to Lovász is often more convenient. In it we extend the definition of the binomial coefficient $\binom{n}{k}$ to $\binom{x}{k}$, where x is a positive real number, by

$$\binom{x}{k} = \frac{x(x-1)\ldots(x-k+1)}{k!}.$$

These numbers satisfy the usual recurrence relation $\binom{x}{k} + \binom{x}{k-1} = \binom{x+1}{k}$; further, if $\binom{n-1}{k} < m < \binom{n}{k}$, we can find a real number x such that $m = \binom{x}{k}$.

Theorem 7.3.5 (Lovász 1979) If $\mathscr{A}$ is a collection of k-sets and $|\mathscr{A}| = \binom{x}{k}$ $(x \geqslant k)$, then $|\Delta\mathscr{A}| \geqslant \binom{x}{k-1}$.

Proof (Frankl 1984) As in the proof of Kruskal–Katona, we can suppose that $S_j(\mathscr{A}) = \mathscr{A}$ for each $j \geqslant 2$. Again, we use double induction on k and $m = |\mathscr{A}|$. If $|\mathscr{A}_\uparrow| < \binom{x-1}{k-1}$ then $|\mathscr{A}_0| = |\mathscr{A}| - |\mathscr{A}_\uparrow| > \binom{x}{k} - \binom{x-1}{k-1} = \binom{x-1}{k}$, so that, by the induction hypothesis, $|\Delta\mathscr{A}_0| \geqslant \binom{x-1}{k-1}$. However, Lemma 7.3.3 then gives $|\mathscr{A}_\uparrow| \geqslant \binom{x-1}{k-1}$ which is a contradiction. Therefore we must have $|\mathscr{A}_\uparrow| \geqslant \binom{x-1}{k-1}$, but then, by the induction hypothesis, $|\Delta\mathscr{A}_\uparrow| \geqslant \binom{x-1}{k-2}$ so that

$$|\Delta\mathscr{A}| \geqslant |\mathscr{A}_\uparrow| + |\Delta\mathscr{A}_\uparrow| \geqslant \binom{x-1}{k-1} + \binom{x-1}{k-2} = \binom{x}{k-1}$$

as required. □

This Lovász version of Theorem 7.3.1 leads to a simple proof of the following result from which Lemma 5.2.5 can be deduced.

Corollary 7.3.6 If $\mathscr{A}$ is a collection of k-subsets of S, and if $|\mathscr{A}| \leq \binom{u}{k}$ then

$$|\Delta\mathscr{A}| \geq \frac{\binom{u}{k-1}}{\binom{u}{k}}|\mathscr{A}|.$$

Proof Let $|\mathscr{A}| = \binom{x}{k}$, $x \leq u$. Then, by Theorem 7.3.5,

$$|\Delta\mathscr{A}| \geq \binom{x}{k-1} = \frac{\binom{x}{k-1}}{\binom{x}{k}}|\mathscr{A}| \geq \frac{\binom{u}{k-1}}{\binom{u}{k}}|\mathscr{A}|$$

since

$$\frac{\binom{x}{k-1}}{\binom{x}{k}} = \frac{k}{x-k+1}$$

is a decreasing function of x. □

7.4 Some easy consequences of Kruskal–Katona

7.4.1 The Erdös–Ko–Rado theorem

As our first example of the power of the Kruskal–Katona theorem we give the elegant deduction of the EKR theorem due to Daykin (1974*b*) and Clements (1976) independently.

Theorem 7.4.1 (EKR) Let $\mathscr{A} = \{A_1, \ldots, A_m\}$ be a collection of k-subsets of an n-set S, $k \leq \frac{1}{2}n$, such that $A_i \cap A_j \neq \varnothing$ for each pair i, j. Then $m \leq \binom{n-1}{k-1}$.

Proof (Daykin 1974*b*, Clements 1976) Suppose that $m > \binom{n-1}{k-1}$. Then $|\mathscr{A}'| = m > \binom{n-1}{n-k}$ where, as usual, $\mathscr{A}' = \{S - A : A \in \mathscr{A}\}$. The condition $A_i \cap A_j \neq \varnothing$ is equivalent to $A_i \nsubseteq A_j'$, so the sets in $\Delta^{(k)}\mathscr{A}'$

must be distinct from the sets in $\mathscr{A}$; thus

$$|\mathscr{A}| + |\Delta^{(k)}\mathscr{A}'| \leqslant \binom{n}{k}. \tag{7.4}$$

But $|\Delta^{(k)}\mathscr{A}'| \geqslant \binom{n-1}{k}$ by Theorem 7.3.1, so that

$$|\mathscr{A}| + |\Delta^{(k)}\mathscr{A}'| > \binom{n-1}{k-1} + \binom{n-1}{k} = \binom{n}{k},$$

contradicting (7.4). □

7.4.2 Subadditivity of $F_k(n)$

Let $F_k(n)$ denote the first n k-sets in the squashed order. The following result will be used frequently in later chapters.

Theorem 7.4.2 (Clements 1974) For any positive integers n_1, n_2,

$$|\Delta F_k(n_1 + n_2)| \leqslant |\Delta F_k(n_1)| + |\Delta F_k(n_2)|.$$

Proof We consider subsets of $\{1, 2, 3, \ldots\}$. Let h denote the largest element occurring in the sets in $F_k(n_1)$, and let $\mathscr{B}$ denote the collection of the first n_2 k-sets with least element greater than h. Then $\Delta F_k(n_1)$ and $\Delta\mathscr{B}$ are clearly disjoint, so that $|\Delta(F_k(n_1) \cup \mathscr{B})| = |\Delta F_k(n_1)| + |\Delta\mathscr{B}|$. However, $\mathscr{B}$ is isomorphic to $F_k(n_2)$. It therefore follows from Kruskal–Katona that

$$\begin{aligned}|\Delta F_k(n_1 + n_2)| &\leqslant |\Delta(F_k(n_1) \cup \mathscr{B})| = |\Delta F_k(n_1)| + |\Delta\mathscr{B}| \\ &= |\Delta F_k(n_1)| + |\Delta F_k(n_2)|.\end{aligned}$$

□

7.4.3 Extensions of EKR

Suppose we are given a collection $\mathscr{A} = \{A_1, \ldots, A_m\}$ of k-subsets of an n-set S, and an integer l. Define

$$I_l(\mathscr{A}) = \{B \subseteq S : |B| = l,\ B \cap A \neq \varnothing \text{ for all } A \in \mathscr{A}\}.$$

Thus $I_l(\mathscr{A})$ consists of all l-subsets of S which intersect every member of $\mathscr{A}$. What can be said about the size of $I_l(\mathscr{A})$? Hilton (1977) observed that, since the condition $B \cap A \neq \varnothing$ is equivalent to $B \nsubseteq A'$, it follows that if $l < n - k$ then

$$|I_l(\mathscr{A})| + |\Delta^{(l)}\mathscr{A}'| \leqslant \binom{n}{l}$$

where, as usual, $\mathscr{A}' = \{A' : A \in \mathscr{A}\}$. Thus, if

$$m = \binom{a_{n-k}}{n-k} + \ldots + \binom{a_p}{p}$$

is the $(n-k)$-binomial representation of m,

$$|I_l(\mathscr{A})| \leqslant \binom{n}{l} - \binom{a_{n-k}}{l} - \ldots - \binom{a_p}{p+l-(n-k)}.$$

We can therefore prove the following theorem.

Theorem 7.4.3 (Kleitman 1968) Let $\mathscr{A} = \{A_1, \ldots, A_m\}$ be a collection of k-subsets of $\{1, \ldots, n\}$, let $l + k \leqslant n$, and let $B_1, \ldots, B_u$ be all the l-subsets of $\{1, \ldots, n\}$ such that $A_i \cap B_j \neq \varnothing$ for all i, j. Then either $u \leqslant \binom{n-1}{l-1}$ or $m < \binom{n-1}{k-1}$.

Proof (Hilton 1977) Suppose that $m \geqslant \binom{n-1}{k-1}$. Then $m \geqslant \binom{n-1}{n-k}$ so that, by the above remarks,

$$u = |I_l(\mathscr{A})| \leqslant \binom{n}{l} - \binom{n-1}{l} = \binom{n-1}{l-1}.$$ □

This result was conjectured by Hilton and Milner (1967) in the special case when $k = l$.

Corollary 7.4.4 If $\mathscr{A}$, $\mathscr{B}$ are two collections of k-subsets of an n-set, $k \leqslant \frac{1}{2}n$, such that $A \cap B \neq \varnothing$ whenever $A \in \mathscr{A}$, $B \in \mathscr{B}$, then $|\mathscr{A}| \leqslant \binom{n-1}{k-1}$ or $|\mathscr{B}| \leqslant \binom{n-1}{k-1}$. □

The EKR theorem arises when $\mathscr{A}$ and $\mathscr{B}$ coincide.

7.5 Compression

Let $\mathscr{A} = \{A_1, \ldots, A_m\}$ be a collection of k-sets. Then the *compression* $C\mathscr{A}$ of $\mathscr{A}$ is defined to be the collection of sets consisting of the *first* $|\mathscr{A}|$ k-sets in the squashed order. The Kruskal–Katona theorem asserts that $|\Delta\mathscr{A}| \geqslant |\Delta C\mathscr{A}|$. In fact we have seen that the shadow of the first so many k-sets consists of the first so many

$(k-1)$-sets, so that the Kruskal–Katona theorem actually asserts the stronger result

$$\Delta C\mathcal{A} \subseteq C\Delta\mathcal{A}.$$

More generally, if $\mathcal{A}$ contains sets of different sizes, let $\mathcal{A} = \mathcal{A}_0 \cup \mathcal{A}_1 \cup \ldots \cup \mathcal{A}_n$ where $\mathcal{A}_i = \{A \in \mathcal{A} : |A| = i\}$. We then define the *compression* $C\mathcal{A}$ of $\mathcal{A}$ to be $C\mathcal{A} = \bigcup_i C\mathcal{A}_i$; thus compression is performed at each rank level. We also say that $\mathcal{A}$ is *compressed* if $C\mathcal{A} = \mathcal{A}$. We have therefore seen in Section 7.2 that the following result holds.

Theorem 7.5.1 If $\mathcal{A}$ is compressed, so is $\Delta\mathcal{A}$. □

Compression will play an important role when we extend the Kruskal–Katona theorem to multisets. It will also be important when we study antichains in greater detail in the next chapter. However, for the moment we content ourselves with the following simple observation.

Theorem 7.5.2 If $\mathcal{A}$ is an ideal then so is $C\mathcal{A}$.

Proof Let $\mathcal{A} = \bigcup_{i=0}^{k} \mathcal{A}_i$, where $\mathcal{A}_i = \{A \in \mathcal{A} : |A| = i\}$. Then $\Delta\mathcal{A}_i \subseteq \mathcal{A}_{i-1}$ for each i so that, on writing $\Delta\mathcal{A} = \bigcup_i \Delta\mathcal{A}_i$, we have

$$\Delta C\mathcal{A} = \Delta\Big(\bigcup_i C\mathcal{A}_i\Big) = \bigcup_i \Delta(C\mathcal{A}_i)$$

$$\subseteq \bigcup_i C\Delta\mathcal{A}_i \quad \text{by Kruskal–Katona}$$

$$\subseteq \bigcup_i C\mathcal{A}_{i-1} \subseteq C\mathcal{A}.$$

□

Exercises 7

7.1 (a) Find the 4-binomial representations of 23, 27 and 37.
(b) How many 3-element sets are there in $\Delta F_4(23)$?

7.2 What is the 25th set in the squashed ordering of 3-sets?

7.3 In what position in the squashed ordering of 5-sets is (a) $\{1, 3, 5, 6, 8\}$ and (b) $\{3, 4, 5, 7, 8\}$?

7.4 Show the following:

(i) if $A <_S B$ then $B' <_S A'$;

(ii) if $\mathscr{A}$ consists of the first m $(\frac{1}{2}n)$-subsets of an n-set S, where n is even, then $\mathscr{A}'$ consists of the last m $(\frac{1}{2}n)$-subsets;

(iii) if $m \geqslant \frac{1}{2}\binom{n}{n/2}$ then $|\mathscr{A} \cup \mathscr{A}'| = \binom{n}{n/2}$.

7.5 Show that if A is a set of size k then its successor in the squashed ordering of k-sets is

$$B = \{1, \ldots, b\} \cup \{a+1\} \cup (A - \{1, \ldots, a\})$$

where a is the least integer such that $a \in A$ and $a+1 \notin A$, and $b = |A \cap \{1, \ldots, a\}| - 1$. Hence verify that the successor of $\{2, 3, 4\}$ is $\{1, 2, 5\}$.

7.6 Let $m = \binom{a_k}{k} + \ldots + \binom{a_t}{t}$ be the k-binomial representation of m, let s be the smallest index for which $a_s > s$, and let $M = \binom{a_k}{k+1} + \ldots + \binom{a_s}{s+1}$. Show the following:

(a) $|\Delta F_{k+1}(M)| \leqslant m$ and $|\Delta F_{k+1}(M+1)| > m$;

(b) $|\Delta F_{k+1}(M+m)| = m + |\Delta F_k(m)|$.

7.7 By considering $\mathscr{A} = F_k(r)^{\hat{1}} \cup (\{1\} \cup F_{k-1}(s)^{\hat{1}})$, prove that

$$|\Delta F_k(r+s)| \leqslant |\Delta F_{k-1}(s)| + \max(s, |\Delta F_k(r)|)$$

(Hilton 1979)

7.8 Show that if $m = r + s$, where the k-binomial representation of m is as in 7.6, then $s \leqslant |\Delta F_k(r)| \Leftrightarrow$

$$s \leqslant \begin{cases} \binom{a_k - 1}{k-1} + \ldots + \binom{a_t - 1}{t-1} & \text{when } a_t > t \\ \binom{a_k - 1}{k-1} + \ldots + \binom{a_t - 1}{t-1} - 1 & \text{when } a_t = t. \end{cases}$$

(Hilton 1979)

7.9 (a) Show the following:

(i) if $v < \binom{n}{k-1}$ then

$$\left|\Delta F_k\left(\binom{n}{k} + v\right)\right| = \binom{n}{k-1} + |\Delta F_{k-1}(v)|;$$

(ii) if $0 \leqslant m_1 \leqslant m_2 \leqslant \binom{n}{k} \leqslant m_1 + m_2$ then

$$|\Delta F_k(m_1)| + |\Delta F_k(m_2)| \geqslant \binom{n}{k-1} + \left|\Delta F_k\left(m_1 + m_2 - \binom{n}{k}\right)\right|.$$

(b) Let $\mathscr{A}$ be a collection of k-subsets of $S = S_1 \cup S_2$ where $S_1 \cap S_2 = \varnothing$, $|S_1| \leqslant |S_2| = n$, $\binom{n}{k} \leqslant |\mathscr{A}| \leqslant \binom{n}{k} + \binom{|S_2|}{k}$, and where, for each $A \in \mathscr{A}$, $A \subseteq S_1$ or $A \subseteq S_2$. Use (a) to show that

$$|\Delta \mathscr{A}| \geqslant \binom{n}{k-1} + \left|\Delta F_k\left(|\mathscr{A}| - \binom{n}{k}\right)\right|.$$

(Katona 1973)

7.10 Deduce the EKR theorem from Theorem 7.4.3.

7.11 Prove that $|\Delta F_k(m+1)| \leqslant |\Delta F_k(m)| + (k-1)$.

7.12 (i) Show that, if n is even and $\frac{1}{2}\binom{n}{n/2} < m \leqslant \binom{n}{n/2}$, then there exists an antichain $\mathscr{A}_0$ of m $(\frac{1}{2}n)$-subsets of $S = \{1, \ldots, n\}$ such that the number of pairs $A, B \in \mathscr{A}_0$ with $A \cap B = \varnothing$ is $m - \frac{1}{2}\binom{n}{n/2}$.

(ii) With m as in (i) show that every antichain $\mathscr{A}$ of m subsets of S each of size $\leqslant \frac{1}{2}n$ must have at least $m - \frac{1}{2}\binom{n}{n/2}$ simultaneously disjoint pairs of members.

(iii) Deduce Theorem 5.6.1. (Clements 1976)

7.13 Prove the following version of the Kruskal–Katona theorem. Let $\mathscr{A}$ be a collection of k-subsets of an n-set S. Then $|\nabla \mathscr{A}| \geqslant |\nabla \mathscr{B}|$ where $\mathscr{B}$ consists of the *last* $|\mathscr{A}|$ k-subsets of S in the squashed ordering.

7.14 Deduce Lemma 5.2.5 from Corollary 7.3.6.

7.15 In investigating all selections of 12 numbers from $\{1, \ldots, 20\}$, the selections $1 \leqslant a_1 < a_2 < \ldots < a_{12} \leqslant 20$ are considered in the lexicographic order of sequences $a_1 \ldots a_{12}$. Show that

$$1, 2, 4, 5, 8, 10, 11, 12, 13, 15, 16, 18$$

is the 29 937th sequence in the lexicographic ordering.

(Lehmer 1964)

7.16 Show that if $\mathscr{A}$ consists of the first m $(n-k)$-subsets of an n-set S, $k<\frac{1}{2}n$, $m=\binom{n-1}{k-1}-\binom{n-k-1}{k-1}+1$, then $|\Delta^{(k)}\mathscr{A}|=\binom{n-1}{k}$.

8 Antichains

8.1 Squashed antichains

The study of antichains has been very central to the subject matter of most of the previous chapters. Now that we have met the Kruskal–Katona theorem we are better equiped to study antichains and their structure at greater depth. Much of our time will be spent studying the *parameters* of an antichain.

Definition 8.1.1 If $\mathscr{A}$ is an antichain of subsets of an n-set S, the *parameters* p_i are the numbers defined by $p_i = |\mathscr{A}_i|$ where $\mathscr{A}_i = \{A \in \mathscr{A} : |A| = i\}$.

Many of the results obtained so far are really theorems about these parameters; we summarize these results before going on to obtain some new ones.

Sperner's fundamental theorem, in its LYM form, asserts that

$$\sum_{i=0}^{n} \frac{p_i}{\binom{n}{i}} \leqslant 1.$$

The EKR theorem is concerned with intersecting antichains, and, in its LYM form of Theorem 5.2.2, asserts that if $p_i = 0$ for all $i > k$, where $k \leqslant \frac{1}{2}n$, then

$$\sum_{i \leqslant k} \frac{p_i}{\binom{n-1}{i-1}} \leqslant 1.$$

If $\mathscr{A}$ is an intersecting antichain, but there is no restriction on the size of the sets in $\mathscr{A}$, then, by Theorem 5.3.1,

$$\sum_{i \leqslant n/2} \frac{p_i}{\binom{n}{i-1}} + \sum_{i > n/2} \frac{p_i}{\binom{n}{i}} \leqslant 1.$$

Also, Bollobás's result (Corollary 5.2.3) asserts that if the antichain $\mathcal{A}$ satisfies $A \in \mathcal{A} \Rightarrow A' \in \mathcal{A}$ then

$$\sum_{i \leq n/2} \frac{p_i}{\binom{n-1}{i-1}} + \sum_{i > n/2} \frac{p_i}{\binom{n-1}{i}} \leq 2.$$

All these results put necessary conditions on the parameters which must be satisfied if an antichain of a particular type with these parameters is to exist. This raises an obvious question: is there a necessary and sufficient condition on numbers p_i for the existence of an antichain with these numbers as its parameters? In answering this question (in the affirmative) it is necessary first of all to discuss a special family of antichains which are, in a certain sense, compressed.

Suppose that the non-negative numbers $p_0, \ldots, p_n$ are given. We attempt to construct an antichain $\mathcal{A}^*$ with parameters $p_0, \ldots, p_n$ in the following way. Let g be the largest value of i for which $p_i \neq 0$. Define $\mathcal{A}_g^*$ to consist of the *first* p_g sets of size g in the squashed order. This definition is satisfactory, provided that

$$p_g \leq \binom{n}{g}. \tag{8.1}$$

In the shadow of these p_g g-sets there will be $n_{g-1} = |\Delta F_g(p_g)|$ sets of size $g-1$, namely $F_{g-1}(n_{g-1})$; none of these sets can be in $\mathcal{A}^*$. Therefore define $\mathcal{A}_{g-1}^*$ to consist of the *next* p_{g-1} sets of size $g-1$ in the squashed order. This definition is satisfactory provided that there are p_{g-1} sets available to be chosen, i.e. provided that

$$p_{g-1} + n_{g-1} \leq \binom{n}{g-1}. \tag{8.2}$$

We then have $p_{g-1} + n_{g-1}$ sets of size $g-1$ which are in $\mathcal{A}_{g-1}^*$ or are contained in sets in $\mathcal{A}_g^*$, namely the sets of $F_{g-1}(p_{g-1} + n_{g-1})$. In the shadow of these $p_{g-1} + n_{g-1}$ sets there will be $n_{g-2} = |\Delta F_{g-1}(p_{g-1} + n_{g-1})|$ sets of size $g-2$, namely $F_{g-2}(n_{g-2})$, none of which can be in $\mathcal{A}^*$. Therefore define $\mathcal{A}_{g-2}^*$ to consist of the *next* p_{g-2} sets of size $g-2$ in the squashed order. This definition is satisfactory provided that there are p_{g-2} sets of size $g-2$ available to be chosen, i.e. provided that

$$p_{g-2} + n_{g-2} \leq \binom{n}{g-2}. \tag{8.3}$$

Continuing in this way we construct an antichain $\mathcal{A}^* = \mathcal{A}_g^* \cup \mathcal{A}_{g-1}^* \cup \ldots \cup \mathcal{A}_0^*$ with parameters $p_i = |\mathcal{A}_i^*|$, such that, for each i, the sets in $\mathcal{A}_i^*$, together with the i-sets contained in larger members of $\mathcal{A}^*$, constitute an initial segment of the sets of size i in the squashed order.

The construction works provided that the inequalities (8.1), (8.2), (8.3), etc. all hold. We have therefore proved the following theorem.

Theorem 8.1.1 (Clements 1973, Daykin, Godfrey, and Hilton 1974) Let $p_0, \ldots, p_n$ be non-negative integers, and let k and g be the smallest and largest values of i for which $p_i \neq 0$. Define integers $n_k, \ldots, n_g$ by

$$\begin{aligned} n_g &= 0 \\ n_{g-1} &= |\Delta F_g(p_g)| \\ n_{g-2} &= |\Delta F_{g-1}(p_{g-1} + n_{g-1})| \\ &\vdots \\ n_k &= |\Delta F_{k+1}(p_{k+1} + n_{k+1})|. \end{aligned}$$

Then, provided that

$$p_i + n_i \leqslant \binom{n}{i} \tag{8.4}$$

for each $i = k, \ldots, g$, there exists an antichain $\mathscr{A}^*$ with parameters p_i such that, for each i, the sets in $\mathscr{A}^*$ of size i, together with the sets of size i contained in larger members of $\mathscr{A}^*$, constitute an initial segment of the sets of size i in the squashed order. □

We call $\mathscr{A}^*$ the *squashed antichain* with parameters $p_0, \ldots, p_n$.

Example 8.1.1 Construct the squashed antichain on $\{1, \ldots, 6\}$ with parameters $p_0 = p_1 = p_5 = p_6 = 0$, $p_2 = 2$, $p_3 = 3$, $p_4 = 1$.

Solution Take the first 4-set, namely $\{1, 2, 3, 4\}$. Its 3-subsets are $\{1, 2, 3\}$, $\{1, 2, 4\}$, $\{1, 3, 4\}$, $\{2, 3, 4\}$, the first four 3-sets in the squashed order. Therefore take the next three 3-sets, namely $\{1, 2, 5\}$, $\{1, 3, 5\}$, $\{2, 3, 5\}$. The 2-sets contained in the sets so far chosen are $\{1, 2\}$, $\{1, 3\}$, $\{2, 3\}$, $\{1, 4\}$, $\{2, 4\}$, $\{3, 4\}$, $\{1, 5\}$, $\{2, 5\}$, $\{3, 5\}$; therefore take the next two 2-sets, namely $\{4, 5\}$ and $\{1, 6\}$. Thus the squashed antichain consists of the following sets: $\{4, 5\}$, $\{1, 6\}$, $\{1, 2, 5\}$, $\{1, 3, 5\}$, $\{2, 3, 5\}$, $\{1, 2, 3, 4\}$.

The importance of squashed antichains arises from the following result which says essentially that the conditions (8.4) are automatically satisfied if *any* antichain with parameters p_i exists.

Theorem 8.1.2 (Clements 1973, Daykin *et al.* 1974) Let $\mathscr{A}$ be any antichain on $\{1, \ldots, n\}$; suppose that its parameters are $p_0, \ldots, p_n$, and that k and g are the smallest and largest values of i for which $p_i \neq 0$. Then the conditions (8.4) are satisfied by the p_i.

Proof. Let $\mathcal{A} = \mathcal{A}_k \cup \ldots \cup \mathcal{A}_g$ where $\mathcal{A}_i = \{A \in \mathcal{A} : |A| = i\}$. Since $|\mathcal{A}_g| = p_g$, $p_g \leqslant \binom{n}{g}$, so that (8.4) holds for $i = g$. Now there are $|\Delta(\mathcal{A}_g)|$ sets of size $g-1$ in the shadow of $\mathcal{A}_g$, and none of these can be in $\mathcal{A}_{g-1}$, so that there are at least p_{g-1} other sets of size $g-1$, namely the sets in $\mathcal{A}_{g-1}$. Thus $p_{g-1} \leqslant \binom{n}{g-1} - |\Delta(\mathcal{A}_g)|$, i.e. $p_{g-1} + |\Delta(\mathcal{A}_g)| \leqslant \binom{n}{g-1}$. However, by the Kruskal–Katona theorem, $|\Delta(\mathcal{A}_g)| \geqslant |\Delta F_g(p_g)| = n_{g-1}$, so that $p_{g-1} + n_{g-1} \leqslant \binom{n}{g-1}$, i.e. (8.4) holds for $i = g-1$.

Next, consider the $p_{g-1} + |\Delta(\mathcal{A}_g)|$ sets of size $g-1$ in $\mathcal{A}$ or contained in larger sets of $\mathcal{A}$. There are $\geqslant p_{g-1} + n_{g-1}$ such sets, so by the Kruskal–Katona theorem they contain at least $|\Delta F_{g-1}(p_{g-1} + n_{g-1})| = n_{g-2}$ sets in their shadow. Since none of these can be in $\mathcal{A}_{g-2}$, there must be at least p_{g-2} other sets of size $g-2$, so that we must have $p_{g-2} + n_{g-2} \leqslant \binom{n}{g-2}$, and (8.4) is satisfied for $i = g-2$. Continuing in this way we find that (8.4) holds for each i. □

Corollary 8.1.3 If $\mathcal{A}$ is an antichain on $\{1, \ldots, n\}$ with parameters $p_0, \ldots, p_n$, then there exists a squashed antichain $S\mathcal{A}$ with the same parameters. □

Note that we have, in the process of studying squashed antichains, obtained the necessary and sufficient conditions we were looking for.

Theorem 8.1.4 The non-negative integers $p_0, \ldots, p_n$ are the parameters of an antichain on $\{1, \ldots, n\}$ if and only if the conditions (8.4) hold.

Proof. Theorem 8.1.2 proves the 'only if' part; Theorem 8.1.1 proves the 'if' part. □

This result can be presented in a slightly different form. First of all note that if one of the conditions (8.4) fails to hold, say for $i = j$, then all the subsequent ones, for $i < j$, will also fail to hold. Therefore we can replace (8.4) by the one condition $p_k + n_k \leqslant \binom{n}{k}$. Further, if we rewrite $|\Delta F_j(m)|$ as $\delta_j(m)$, we obtain the following corollary.

Corollary 8.1.5 Let $p_0, \ldots, p_n$ be non-negative integers and let k be the smallest value of i for which $p_i \neq 0$. Then there exists an antichain on $\{1, \ldots, n\}$ with parameters p_i if and only if

$$p_k + \delta_{k+1}(p_{k+1} + \delta_{k+2}(p_{k+2} + \delta_{k+3}(\ldots))) \leq \binom{n}{k}. \qquad \square$$

We finish this section by pointing out one important property of squashed antichains. It is implicit in the proof of Theorem 8.1.2 that the number of sets of size i contained in sets of $\mathcal{A}$ of size greater than i is at least n_i, while the corresponding number of sets for the squashed antichain is precisely n_i. We therefore have the following theorem.

Theorem 8.1.6 (Clements 1973) If $\mathcal{A}$ is an antichain on $\{1, \ldots, n\}$ with parameters $p_k, p_{k+1}, \ldots, p_g$, where $p_k \neq 0$, $p_g \neq 0$, let $\Delta^{(k-1)}(\mathcal{A})$ denote the set of all $(k-1)$-subsets of $\{1, \ldots, n\}$ contained in at least one member of $\mathcal{A}$. Then the minimum value of $|\Delta^{(k-1)}(\mathcal{A})|$, taken over all antichains $\mathcal{A}$ with these given parameters, is $n_{k-1} = |\Delta F_k(n_k)|$, attained when $\mathcal{A}$ is squashed. $\square$

8.2 Using squashed antichains

We begin this section by showing how improvements on the LYM inequality can be obtained by an elegant use of the existence of squashed antichains. If $\mathcal{A}$ is an antichain with parameters $p_0, \ldots, p_n$, define

$$\text{top}(\mathcal{A}) = \sum_{i>n/2} \frac{p_i}{\binom{n-1}{i}}$$

$$\text{bot}(\mathcal{A}) = \sum_{i<n/2} \frac{p_i}{\binom{n-1}{i-1}}$$

$$\text{mid}(\mathcal{A}) = \begin{cases} 0 & \text{if } n \text{ is odd} \\ \dfrac{1}{2}\dfrac{p_m}{\binom{n-1}{m}} & \text{if } n = 2m \end{cases}$$

and let $\tau(\mathcal{A}) = \text{top}(\mathcal{A}) + \text{mid}(\mathcal{A})$, $\xi(\mathcal{A}) = \text{bot}(\mathcal{A}) + \text{mid}(\mathcal{A})$.

Theorem 8.2.1 (Daykin 1984) If $\mathcal{A}$ is an antichain with parameters $p_0, \ldots, p_n$, then either $\tau(\mathcal{A}) \leq 1$ or $\xi(\mathcal{A}) \leq 1$.

Proof. Because of Corollary 8.1.3 we can assume that $\mathscr{A}$ is squashed. First suppose that n is odd, $n = 2m + 1$, and let A be the last member of $\mathscr{A}$ of size $m + 1$ in the squashed order. If $n \in A$ then, for each $i \leqslant m$, there is a subset B of A of size i which contains n, and so all the members of $\mathscr{A}$ of size i coming later in the squashed order than B must contain n. If we delete n from each of them we obtain an antichain of subsets of $\{1, \ldots, n-1\}$, and the LYM inequality yields

$$\sum_{i \leqslant m} \frac{p_i}{\binom{n-1}{i-1}} \leqslant 1, \quad \text{i.e.} \quad \xi(\mathscr{A}) \leqslant 1.$$

If, however, $n \notin A$, then no set in $\mathscr{A}$ of size $\geqslant m + 1$ can contain n, so these sets form an antichain on $\{1, \ldots, n-1\}$ and the LYM inequality yields $\tau(\mathscr{A}) \leqslant 1$. Finally, if $n = 2m$, take A to be the 'middle' set of size m in $\mathscr{A}$, and argue as above. □

Corollary 8.2.2 (Greene and Hilton 1979) If $p_i = p_{n-i}$ for each i then $\tau(\mathscr{A}) + \xi(\mathscr{A}) \leqslant 2$.

Proof $\tau(\mathscr{A}) = \xi(\mathscr{A})$, so both are $\leqslant 1$. □

This generalizes Bollobás's result (Corollary 5.2.3), since $p_i = p_{n-i}$ holds there. Note also that we can deduce EKR from Corollary 8.2.2 (see Exercise 8.2).

We now present a theorem the truth of which was first conjectured by Kleitman and Milner (1973); the proof will be another nice application of the existence of squashed antichains.

Theorem 8.2.3 (Daykin *et al.* 1974) If there exists an antichain with parameters $p_0, \ldots, p_n$, then there exists another antichain with parameters $q_0, \ldots, q_n$ where

$$q_i = \begin{cases} 0 & \text{if } i < \tfrac{1}{2}n \\ p_i & \text{if } i = \tfrac{1}{2}n \\ p_i + p_{n-i} & \text{if } i > \tfrac{1}{2}n. \end{cases}$$

Proof Choose the largest h such that $p_h + p_{n-h} > 0$. If n is even and $h = \frac{1}{2}n$, then there is nothing to prove. If n is odd and $h = \frac{1}{2}(n+1)$ then, by Sperners's theorem, $p_{h-1} + p_h \leqslant \binom{n}{h}$, so we can choose $p_{h-1} + p_h$ sets of size h as the required antichain.

We now consider the case $h > \frac{1}{2}(n+1)$. We shall suppose that the theorem is true for all antichains $\mathscr{A}$ for which $\frac{1}{2}n \leqslant h < k$, and attempt to prove the theorem for any antichain for which $h = k$. We therefore suppose that we have an antichain $\mathscr{A}$ with parameters $p_0, \ldots, p_n$, with $p_i = 0$ for $i < n-k$ and for $i > k$, but with $p_k + p_{n-k} > 0$. We shall suppose first, for technical reasons, that $k \geqslant \frac{1}{2}n + 3$. The way the proof goes is to derive from $\mathscr{A}$ a succession of seven antichains, the last one having the required parameters q_i.

Consider first all sets in $\mathscr{A}$ of size k. Their shadow consists of at least $|\Delta F_k(p_k)|$ subsets of size $k-1$, so we can choose $|\Delta F_k(p_k)|$ of these subsets to replace the p_k k-subsets, giving us a new antichain $\mathscr{A}_1$. Next take complements to obtain an antichain $\mathscr{A}_2$, and then obtain $\mathscr{A}_3$ from $\mathscr{A}_2$ in exactly the same way as $\mathscr{A}_1$ was obtained from $\mathscr{A}$. The antichain $\mathscr{A}_3$ has parameters r_i as follows:

$$r_i = 0 \quad \text{if } i \geqslant k$$

$$r_{k-1} = p_{n-k+1} + |\Delta F_k(p_{n-k})|$$

$$r_i = p_{n-i} \quad \text{if } n-k+1 < i < k-1$$

$$r_{n-k+1} = p_{k-1} + |\Delta F_k(p_k)|$$

$$r_i = 0 \quad \text{if } i \leqslant n-k.$$

Observe now that $\mathscr{A}_3$ is a valid subject for an application of the induction hypothesis. We can therefore deduce the existence of an antichain $\mathscr{A}_4$ with parameters t_i given by

$$t_i = 0 \quad \text{if } i \geqslant k$$
$$t_{k-1} = p_{k-1} + p_{n-k+1} + |\Delta F_k(p_k)| + |\Delta F_k(p_{n-k})|$$
$$t_i = p_i + p_{n-i} \quad \text{if } \tfrac{1}{2}n < i < k-1,$$
$$t_{n/2} = p_{n/2} \quad \text{if } n \text{ is even.}$$

Now recall the subadditivity result of Theorem 7.4.2, by which

$$|\Delta F_k(p_k)| + |\Delta F_k(p_{n-k})| \geqslant |\Delta F_k(p_k + p_{n-k})|.$$

We can therefore omit $|\Delta F_k(p_k)| + |\Delta F_k(p_{n-k})| - |\Delta F_k(p_k + p_{n-k})|$ of the $(k-1)$-sets in $\mathscr{A}_4$ to obtain an antichain $\mathscr{A}_5$ with the same parameters as $\mathscr{A}_4$ except that t_{k-1} is replaced by

$$t'_{k-1} = p_{k-1} + p_{n-k+1} + |\Delta F_k(p_k + p_{n-k})|.$$

Now use Theorem 8.1.1 to obtain a squashed antichain $\mathcal{A}_6$ with the same parameters as $\mathcal{A}_5$. The t'_{k-1} $(k-1)$-sets in $\mathcal{A}_6$ are precisely the first t'_{k-1} $(k-1)$-sets in the squashed ordering, so if we replace the first $|\Delta F_k(p_k + p_{n-k})|$ of these sets by the first $p_{n-k} + p_k$ k-sets, we obtain an antichain whose parameters are the q_i required by the theorem.

The condition $k \geqslant \frac{1}{2}n + 3$ is needed for the inequality $\frac{1}{2}n < i < k - 1$ to make sense. If $k < \frac{1}{2}n + 3$ the argment above can easily be adapted. □

We next use the existence of squashed antichains to obtain results about antichains which are *complement free*. We say that $\mathcal{A}$ is complement free if $A \in \mathcal{A} \Rightarrow A' \notin \mathcal{A}$. For such antichains we can obtain a slight improvement on the LYM inequality and then deduce an improvement on Sperner's bound for the size of such an antichain.

Theorem 8.2.4 (Clements and Gronau 1981) If $\mathcal{A}$ is a complement free antichain on $S = \{1, \ldots, n\}$ with parameters $p_0, \ldots, p_n$, then

$$\sum_{\substack{i=0 \\ i \neq n/2}}^{n} \frac{p_i}{\binom{n}{i}} + \frac{p_{n/2}}{\binom{n}{\frac{1}{2}n - 1}} \leqslant 1.$$

Proof If n is odd, this is just the LYM inequality, so we now assume that n is even. We can also suppose that $p_{n/2} \neq 0$, for otherwise we just have the LYM inequality again.

By Theorem 8.2.3, there is an antichain $\mathcal{A}_1$ with parameters $q_0, \ldots, q_n$ where $q_i = 0$ if $i < \frac{1}{2}n$, $q_{n/2} = p_{n/2}$, $q_i = p_i + p_{n-i}$ if $i > \frac{1}{2}n$. Now let $\mathcal{A}_2 = \mathcal{A}'_1 = \{A : A' \in \mathcal{A}_1\}$. Then $\mathcal{A}_2$ is an antichain with parameters $r_i = q_{n-i}$, and, by Theorem 8.1.3, we can find a squashed antichain $\mathcal{A}_3$ with the same parameters r_i as $\mathcal{A}_2$. Note that the largest size of a set in $\mathcal{A}_3$ is $\frac{1}{2}n$.

We now (and only now) use the fact that $\mathcal{A}$ is complement free. At most half of the $\frac{1}{2}n$-subsets of S can be in $\mathcal{A}$, so we must have $p_{n/2} \leqslant \frac{1}{2}\binom{n}{n/2}$. In $\mathcal{A}_3$ the $p_{n/2}$ sets of size $\frac{1}{2}n$ are the *first* $p_{n/2}$ sets of size $\frac{1}{2}n$ in the squashed order so that, since they all occur in the first half of the ordering, none of them can contain n. They are therefore subsets of $\{1, \ldots, n-1\}$ so that, by the normalized matching property,

$$|\Delta F_{n/2}(p_{n/2})| \geq \frac{\binom{n-1}{\frac{1}{2}n-1}}{\binom{n-1}{\frac{1}{2}n}} p_{n/2} = p_{n/2}.$$

We can therefore now construct another antichain $\mathscr{A}_4$ by replacing the $p_{n/2}$ sets of size $\frac{1}{2}n$ in $\mathscr{A}_3$ by $p_{n/2}$ of the $|\Delta F_{n/2}(p_{n/2})|$ sets of size $\frac{1}{2}n-1$ in their shadow. The antichain $\mathscr{A}_4$ has parameters t_i where $t_i = 0$ for $i \geq \frac{1}{2}n$, $t_{n/2-1} = p_{n/2-1} + p_{n/2+1} + p_{n/2}$, $t_i = p_i + p_{n-i}$ for $i < \frac{1}{2}n - 1$. The ordinary LYM inequality for $\mathscr{A}_4$ yields

$$\sum_{i \leq n/2-2} \frac{p_i + p_{n-i}}{\binom{n}{i}} + \frac{p_{n/2-1} + p_{n/2+1} + p_{n/2}}{\binom{n}{\frac{1}{2}n-1}} \leq 1,$$

which is the required result. □

Note that the proof of Theorem 8.2.4 uses the fact that $\mathscr{A}$ is complement free only to deduce that $\mathscr{A}$ contains at most half of the subsets of S of size $\frac{1}{2}n$. The LYM-type inequality of the theorem therefore holds for any antichain with this weaker condition.

As a corollary to the theorem, we immediately obtain the following.

Corollary 8.2.5 (Purdy 1977) If $\mathscr{A}$ is a complement-free antichain on an n-set S where n is even, then $|\mathscr{A}| \leq \binom{n}{\frac{1}{2}n-1}$. □

8.3 Parameters of intersecting antichains

Some of the results of the previous sections and chapters seem to suggest that there is a close connection between antichains on $\{1, \ldots, n-1\}$ and intersecting antichains on $\{1, \ldots, n\}$. For example, we can compare Sperner's theorem with the EKR result of Theorem 5.1.2, or the LYM inequality with Theorem 5.2.2. In this section we show that a direct connection does exist, and that there is a result for intersecting antichains corresponding to Corollary 8.1.5.

Consider for a moment the EKR result of Theorem 5.1.1. The bound given for the size of an intersecting antichain of sets of size $\leq k$ is $\binom{n-1}{k-1}$, and we can attain this bound by taking all the k-subsets of

$\{1, \ldots, n\}$ which contain one chosen element. Therefore the k-sets in the extreme case are just $(k-1)$-sets in a $(n-1)$-set, with the nth element adjoined to each. We are led to ask the following question: in general, given an intersecting antichain on $\{1, \ldots, n\}$, can we in some natural way correspond to it an antichain on $\{1, \ldots, n-1\}$ in such a way that new light is shed on the theorems that we have been proving?

Now if we are given an intersecting family $\mathscr{A}$ of k-subsets of $\{1, \ldots, n\}$, the squashed family corresponding to it does not have the property that all its members contain a particular element; therefore squashing down the order does not appear to be a useful idea. However, if we squash in the opposite direction, replacing the sets of $\mathscr{A}$ by the *last* $|\mathscr{A}|$ k-subsets of S in the squashed order, then the resulting sets are likely to contain the element n, for the later part of the squashed ordering consists precisely of the k-subsets which contain n. This suggests that we should imitate the approach to squashed antichains of Section 8.1, but this time squash to the end of the ordering.

We have seen (for example in Exercise 7.2) that if $A < B$ in the squashed ordering, then $B' < A'$. This suggests that we proceed as follows. given an antichain $\mathscr{A}$ on $\{1, \ldots, n\}$ with parameters p_i, let $\mathscr{A}' = \{A : A' \in \mathscr{A}\}$. Then $\mathscr{A}'$ is an antichain with parameters $q_i = p_{n-i}$. Form the squashed antichain with parameters q_i, and then take complements of the sets of this squashed antichain. The resulting antichain has parameters p_i and is called the *antilexicographic antichain* with parameters p_i (recall that the antilexicographic order is the reverse of the squashed order). Since the complements of its members have been squashed to the beginning of the order, its members have been squashed to the end of the order as was required. An alternative way of describing this construction is the following. Let r be the smallest value of i for which $p_i \neq 0$. Replace the p_r sets of $\mathscr{A}$ of size r by the last p_r r-subsets of $\{1, \ldots, n\}$ in the squashed order, i.e. by $L_r(p_r)$. Then replace the p_{r+1} sets of size $r+1$ in $\mathscr{A}$ by the last p_{r+1} subsets of size $r+1$ which do not contain any of the sets in $L_r(p_r)$ and so on.

In what follows we shall write S$\mathscr{A}$ and AL$\mathscr{A}$ for the squashed and antilexicographic antichains corresponding to $\mathscr{A}$ (i.e. with the same parameters as $\mathscr{A}$).

Theorem 8.3.1 (Greene and Hilton 1979) Let $\mathscr{A}$ be an intersecting antichain on $\{1, \ldots, n\}$. Then the antilexicographic antichain AL$\mathscr{A}$ with the same parameters as $\mathscr{A}$ is also intersecting.

Before proceeding to the proof, we state and prove the following lemma.

Lemma 8.3.2 $\mathscr{A}$ is an intersecting antichain if and only if, for all i,

(i) $\Delta^{(i)}\mathscr{A}' \cap \mathscr{A}_i = \varnothing$.

(ii) $\mathscr{B}_i = \Delta^{(i)}\mathscr{A}' \cup \mathscr{A}_1 \cup \ldots \cup \mathscr{A}_i$ is an antichain.

(Here, as usual, $\mathscr{A}_i = \{A \in \mathscr{A} : |A| = i\}$ and $\Delta^{(i)}\mathscr{A}' = \{X : |X| = i, X \subseteq A'$ for some $A' \in \mathscr{A}'\}$.)

Proof Suppose that $\mathscr{A}$ is an intersecting antichain. Since $A \cap B \neq \varnothing$ for all $A, B \in \mathscr{A}$, $A \nsubseteq B'$ whenever $A \in \mathscr{A}$, $B' \in \mathscr{A}'$. Thus (i) holds for each i. Suppose (ii) does not hold. Then we can find $A \in \mathscr{A}_j$, $B \in \Delta^{(i)}\mathscr{A}'$, $j < i$, $A \subset B$. However, $B \subseteq D'$ for some $D' \in \mathscr{A}'$, so $A \subset D'$, i.e. $A \cap D = \varnothing$ for some $A, D \in \mathscr{A}$ which is a contradiction. Therefore (ii) holds.

Conversely, suppose (i) and (ii) hold for all i. Taking $i = n$ in (ii) immediately gives that $\mathscr{A}$ is an antichain. Suppose it is not intersecting. Then $A \cap B = \varnothing$ for some $A, B \in \mathscr{A}$, and we can suppose that $A \in \mathscr{A}_i$, $B \in \mathscr{A}_j$, $i + j \leq n$. Then $A \subseteq B'$ where $B' \in \mathscr{A}'_{n-j}$, so that $A \in \mathscr{A}_i \cap \Delta^{(i)}\mathscr{A}'$, contradicting (i). Therefore $\mathscr{A}$ is an intersecting antichain. □

Proof of Theorem 8.3.1 By Lemma 8.3.2, $\mathscr{B}_i = \Delta^{(i)}\mathscr{A}' \cup \mathscr{A}_1 \cup \ldots \cup \mathscr{A}_i$ is an antichain for each i. Now if we construct $\mathrm{AL}(\mathscr{A}_1 \cup \ldots \cup \mathscr{A}_i)$ there will be at least $|\Delta^{(i)}\mathscr{A}'|$ sets of size i at the beginning of the squashed order not containing any set in $\mathrm{AL}(\mathscr{A}_1 \cup \ldots \cup \mathscr{A}_i)$, for otherwise $\mathrm{AL}\mathscr{B}_i$ would not be constructible. Thus

$$S(\Delta^{(i)}\mathscr{A}') \cup \mathrm{AL}(\mathscr{A}_1 \cup \ldots \cup \mathscr{A}_i)$$

is an antichain. However, the squashed antichain $\mathrm{S}\mathscr{A}'$ has a shadow at level i no greater than the shadow of $\mathscr{A}'$ at that level, so that $\Delta^{(i)}(\mathrm{S}\mathscr{A}') \subseteq \mathrm{S}(\Delta^{(i)}\mathscr{A}')$; thus

$$\Delta^{(i)}(\mathrm{S}\mathscr{A}') \cup \mathrm{AL}(\mathscr{A}_1 \cup \ldots \cup \mathscr{A}_i) \tag{8.5}$$

is an antichain.

If we denote $\mathrm{AL}\mathscr{A}$ by $\mathscr{G}$, then $\mathscr{G}' = \mathrm{S}\mathscr{A}'$ and (8.5) becomes

$$\Delta^{(i)}\mathscr{G}' \cup \mathscr{G}_1 \cup \ldots \cup \mathscr{G}_i.$$

Thus $\mathscr{G}$ satisfies condition (ii) of the lemma. However, $\mathrm{S}(\Delta^{(i)}\mathscr{A}') \cap \mathrm{AL}(\mathscr{A}_1 \cup \ldots \cup \mathscr{A}_i) = \varnothing$ so that $\Delta^{(i)}(\mathrm{S}\mathscr{A}') \cap \mathrm{AL}(\mathscr{A}_1 \cup \ldots \cup \mathscr{A}_i) = \varnothing$. Thus $\Delta^{(i)}\mathscr{G}' \cap \mathscr{G}_i = \varnothing$, i.e. $\mathscr{G}$ also satisfies condition (i). □

We now show that if the intersecting antichain $\mathscr{A}$ is such that $|A| \leqslant \frac{1}{2}n$ for each $A \in \mathscr{A}$, then AL$\mathscr{A}$ is intersecting for the very simple reason that all the sets in AL$\mathscr{A}$ contain n.

Theorem 8.3.3 (Greene *et al.* 1976) Let $\mathscr{A}$ be an intersecting antichain of subsets of $\{1, \ldots, n\}$ such that $|A| \leqslant \frac{1}{2}n$ for each $A \in \mathscr{A}$. Then every set in AL$\mathscr{A}$ contains n.

Proof If all the sets in $\mathscr{A}$ have size k for some $k \leqslant \frac{1}{2}n$, then $|\mathscr{A}| \leqslant \binom{n-1}{k-1}$ by EKR, so that AL$\mathscr{A}$ consists of the last $|\mathscr{A}| \leqslant \binom{n-1}{k-1}$ k-sets in the squashed order. However, the last $\binom{n-1}{k-1}$ k-sets are precisely those which contain n.

If $\mathscr{A}$ contains sets of more than one size, let $k \leqslant \frac{1}{2}n$ be the largest size of sets in $\mathscr{A}$. Let $\nabla^{(k)}(\mathscr{A})$ denote the collection of all those k-subsets of $\{1, \ldots, n\}$ which are in $\mathscr{A}$ or contain a member of $\mathscr{A}$; this is just the shade of $\mathscr{A}$ at level k. Then $\nabla^{(k)}(\mathscr{A})$ is an intersecting family (since $\mathscr{A}$ is), so we must have $|\nabla^{(k)}(\mathscr{A})| \leqslant \binom{n-1}{k-1}$ by the EKR theorem. Now just as shadows are minimized by taking squashed antichains, shades are minimized by taking antilexicographic antichains, so that $|\nabla^{(k)}(\text{AL}\mathscr{A})| \leqslant \nabla^{(k)}(\mathscr{A})| \leqslant \binom{n-1}{k-1}$. Thus $\nabla^{(k)}(\text{AL}\mathscr{A})$ consists of some or all of the last $\binom{n-1}{k-1}$ k-subsets of $\{1, \ldots, n\}$. Since these all contain n, no set in AL$\mathscr{A}$ is a subset of a k-set not containing n; hence every set in AL$\mathscr{A}$ contains n. □

It follows that, if we are given an intersecting antichain on $\{1, \ldots, n\}$ with parameters p_i where $p_i = 0$ for all $i > \frac{1}{2}n$, there exists another intersecting antichain with the same parameters in which every set contains n. We can remove n from each of its sets to obtain an antichain on $\{1, \ldots, n-1\}$ with parameters q_i where $q_i = p_{i+1}$. We therefore have the following theorem.

Theorem 8.3.4 (Greene *et al.* 1976) Let $p_0, \ldots, p_n$ be non-negative integers with $p_0 = 0$ and $p_i = 0$ whenever $i > \frac{1}{2}n$. Then there exists an intersecting antichain on $\{1, \ldots, n\}$ with parameters p_i if and only if there exists an antichain on $\{1, \ldots, n-1\}$ with parameters $q_i = p_{i+1}$, $0 \leqslant i < n$. □

This is the correspondence we were looking for.

Exercises 8

8.1 Find the squashed antichain $S\mathscr{A}$ and the antilexicographic antichain $AL\mathscr{A}$ corresponding to $\mathscr{A}$ if
(a) $\mathscr{A}$ consists of all the k-subsets of $\{1, \ldots, n\}$
(b) $\mathscr{A}$ consists of the subsets $\{1,3,4\}$, $\{2,3\}$, $\{2,4\}$, $\{1,2,5\}$ of $\{1, \ldots, 5\}$.

8.2 Deduce EKR from Corollary 8.2.2 by considering $\mathscr{A} \cup \mathscr{A}'$.

8.3 Use Corollary 8.1.5 to prove Theorem 7.4.3.

8.4 State a theorem similar to Corollary 8.1.5 for intersecting antichains.

8.5 Give another proof of Theorem 5.2.2 using Theorem 8.3.4.

8.6 Deduce Purdy's theorem in the form $|\mathscr{A}| \leqslant \binom{n}{[n/2]+1}$ for any complement-free antichain $\mathscr{A}$ of subsets of an n-set (n even or odd) from Theorem 5.2.3 as follows. Let $\mathscr{A} = \{A_1, \ldots, A_r\}$ be a complement-free antichain on $\{1, \ldots, n\}$. Define a collection $\mathscr{B}$ of subsets of $\{1, \ldots, n+1\}$ as follows:
(a) if $A_i \subset A_j'$ put $A_i \cup \{n+1\}$, A_i', $A_j \cup \{n+1\}$, A_j' in $\mathscr{B}$;
(b) if $A_i \supset A_j'$ put $A_i' \cup \{n+1\}$, A_i, $A_j' \cup \{n+1\}$, A_j in $\mathscr{B}$;
(c) if, for some i, $A_i \not\subset A_j'$ and $A_k' \not\subset A_i$ for all j, k, put $A_i \cup \{n+1\}$, A_i' in $\mathscr{B}$.
Then show that (i) $A \in \mathscr{B} \Rightarrow \{1, \ldots, n+1\} - A \in \mathscr{B}$, (ii) $\mathscr{B}$ is an antichain, and (iii) $|\mathscr{B}| = 2|\mathscr{A}|$. Finally, apply 5.2.3.

(Greene and Hilton 1979)

8.7 Show that, if all the parameters p_i of an antichain are zero except for p_k and p_{k+1}, the following strengthening of the LYM inequality holds:

$$\frac{p_k}{\binom{n}{k}} + \frac{\delta_{k+1}(p_{k+1})}{\binom{n}{k}} \leqslant 1.$$

(Greene and Kleitman 1978)

8.8 Does there exist an antichain on $\{1, \ldots, 8\}$ with parameters p_i if $(p_0, \ldots, p_8)$ is equal to (i) $(0,0,0,21,35,0,0,0,0)$, (ii) $(0,0,9,8,30,8,0,0,0)$, and (iii) $(0,0,0,8,30,8,0,0,0)$?

8.9 Suppose that we have equality in Theorem 8.2.4, where $n = 2m$.
(i) Show that the only parameters p_i of $\mathscr{A}$ which need not be zero are p_{m-1}, p_m, p_{m+1}.
(ii) Show that $p_m = 0$ or $\binom{2m-1}{m}$.

(iii) Show that if $|S| = 2m$ and $\mathscr{B}$ is a non-empty collection of $(m-1)$-subsets of S such that $A_1 \in \mathscr{B}$; $|A_2| = m-1$, $A_1 \cap A_2 = \varnothing \Rightarrow A_2 \in \mathscr{B}$, then $\mathscr{B} = (m-1)S$. Hence show that if $p_m = 0$ then p_{m-1} or p_{m+1} is also zero. (Note: $(m-1)S$ denotes the set of all $(m-1)$-subsets of S.) (For further results, see Clements and Gronau (1981).)

9 The generalized Macaulay theorem for multisets

9.1 The theorem of Clements and Lindström

In this chapter we shall consider the extension of the Kruskal–Katona theorem to multisets. Recall that the Kruskal–Katona result essentially states that, for given m, the number of $(k-1)$-sets in the shadow $\Delta\mathscr{A}$ of a collection $\mathscr{A}$ of m k-sets is minimized by taking $\mathscr{A}$ to consist of the *first* m k-sets in the squashed order. In 1969 Clements and Lindström obtained a similar result for multisets, and it is the aim of this chapter to present their work and to look at one or two applications.

We saw in Chapter 7 that the squashed ordering of sets is equivalent to the lexicographic ordering of the reversed $(0, 1)$-sequences which represent the sets. We can identify the subsets with the corresponding sequences, which are in many ways easier to work with. In studying multisets we shall therefore take up this idea and work with n-tuples instead of subsets. Thus let us consider the multiset $S(k_1, \ldots, k_n)$ to consist of all n-tuples or vectors $\boldsymbol{x} = (x_1, \ldots, x_n)$ such that each x_i is an integer satisfying $0 \leq x_i \leq k_i$. We shall define the *rank* $|\boldsymbol{x}|$ of $\boldsymbol{x}$ to be $|\boldsymbol{x}| = x_1 + \ldots + x_n$. The vectors in $S(k_1, \ldots, k_n)$ of a given rank can be ordered lexicographically: if $\boldsymbol{a} = (a_1, \ldots a_n)$ and $\boldsymbol{b} = (b_1, \ldots, b_n)$ write $\boldsymbol{a} < \boldsymbol{b}$ if $a_1 < b_1$ or if $a_1 = b_1, \ldots, a_{i-1} = b_{i-1}$, $a_i < b_i$ for some $i > 1$. As an example, consider $S(2, 3, 4)$. The vectors of rank 3 in lexicographic order are

0 0 3
0 1 2
0 2 1
0 3 0
1 0 2
1 1 1
1 2 0
2 0 1
2 1 0.

We shall call a vector of rank k a *k-vector* rather than a k-subset. If $\mathcal{A}$ is a collection of m k-vectors of $S(k_1, \ldots, k_n)$ then the shadow $\Delta\mathcal{A}$ of $\mathcal{A}$ is given by

$$\Delta\mathcal{A} = \{x = (x_1, \ldots, x_n) : |x| = k-1; (x_1, \ldots, x_{i-1}, x_i + 1, x_{i+1}, \ldots, x_n) \in \mathcal{A} \text{ for some } i \leqslant n\}.$$

The natural extension of the Kruskal–Katona theorem would be that, for fixed $k_1, \ldots, k_n$, k and m, the size of $\Delta\mathcal{A}$ is minimized by taking $\mathcal{A}$ to consist of the first m k-vectors in the lexicographic ordering. This is in fact what Clements and Lindström proved, under the condition that $k_1 \leqslant k_2 \leqslant \ldots \leqslant k_n$. Their result is known as the generalized Macaulay theorem, as it is related to an earlier result due to Macaulay (1927).

One of the key ideas in this chapter will be that of compression. We have already met this idea briefly in Chapter 7. If $\mathcal{A}$ is a collection of m k-vectors of $S(k_1, \ldots, k_n)$ the *compression* $C\mathcal{A}$ of $\mathcal{A}$ is the collection consisting of the first m k-vectors of $S(k_1, \ldots, k_n)$ in the lexicographic order. More generally, if $\mathcal{A} = \bigcup_i \mathcal{A}_i$, where $\mathcal{A}_i$ is the collection of i-vectors in $\mathcal{A}$, we define the compression $C\mathcal{A}$ of $\mathcal{A}$ to be $C\mathcal{A} = \bigcup_i C\mathcal{A}_i$; thus compression is carried out at each rank level. Further, we say that $\mathcal{A}$ is compressed if $C\mathcal{A} = \mathcal{A}$. We can now state the main result of this chapter.

Theorem 9.1.1 (Clements and Lindström 1969) Let $k_1 \leqslant k_2 \leqslant \ldots \leqslant k_n$ and let $\mathcal{A}$ be a collection of k-vectors of the multiset $S(k_1, \ldots, k_n)$. Then

$$\Delta(C\mathcal{A}) \subseteq C(\Delta\mathcal{A}). \tag{9.1}$$

Note that this immediately implies that

$$|\Delta(C\mathcal{A})| \leqslant |\Delta\mathcal{A}|$$

since $|C\mathcal{B}| = |\mathcal{B}|$ for all $\mathcal{B}$.

The proof of Theorem 9.1.1 will require a number of lemmas. The following lemma has already been proved for $k_1 = \ldots = k_n = 1$.

Lemma 9.1.2 If $\mathcal{A}$ is a collection of k-vectors of $S(k_1, \ldots, k_n)$ and if $\mathcal{A}$ is compressed, then $\Delta\mathcal{A}$ is also compressed.

Proof Suppose that $\boldsymbol{a} = (a_1, \ldots, a_n)$ and $\boldsymbol{b} = (b_1, \ldots, b_n)$ have rank $k-1$, that $\boldsymbol{a} < \boldsymbol{b}$ and that $\boldsymbol{b} \in \Delta\mathcal{A}$. We have to show that

$\boldsymbol{a} \in \Delta\mathcal{A}$. Since $\boldsymbol{b} \in \Delta\mathcal{A}$, there exists $j \leq n$ such that $(b_1, \ldots, b_j + 1, \ldots, b_n)$ is in $\mathcal{A}$. Let i be the first suffix for which $a_i < b_i$. We consider two cases. First suppose that $j \leq i$. Then

$$(a_1, \ldots, a_j + 1, \ldots, a_n) < (b_1, \ldots, b_j + 1, \ldots, b_n) \in \mathcal{A}.$$

Since $\mathcal{A}$ is compressed, $(a_1, \ldots, a_j + 1, \ldots, a_n) \in \mathcal{A}$, so that $\boldsymbol{a} \in \Delta\mathcal{A}$.

Suppose next that $j > i$. If $a_r < k_r$ for some $r > i$ then $(a_1, \ldots, a_r + 1, \ldots, a_n) < (b_1, \ldots, b_j + 1, \ldots, b_n) \in \mathcal{A}$ so that $(a_1, \ldots, a_r + 1, \ldots, a_n) \in \mathcal{A}$ and $\boldsymbol{a} \in \Delta\mathcal{A}$. If $a_i < b_i - 1$, then $(a_1, \ldots, a_i + 1, \ldots, a_n) < (b_1, \ldots, b_i, \ldots, b_j + 1, \ldots, b_n) \in \mathcal{A}$ so that again $\boldsymbol{a} \in \Delta\mathcal{A}$. Finally, if $i < j$ and $\boldsymbol{a} = (b_1, \ldots, b_{i-1}, b_i - 1, \; k_{i+1}, \ldots, k_n)$ then, since $|\boldsymbol{a}| = |\boldsymbol{b}|$, we must have $\boldsymbol{b} = (b_1, \ldots, b_{i-1}, \; b_i, \ldots, k_t - 1, \ldots, k_n)$ for some $t \leq n$. But then j must be t, and so $(b_1, \ldots, b_i, k_{i+1}, \ldots, k_n) \in \mathcal{A}$; thus $\boldsymbol{a} \in \Delta\mathcal{A}$. □

We now introduce the idea of *i-compression*. Let $\mathcal{A}$ be a collection of k-vectors of the multiset $S(k_1, \ldots, k_n)$; denote by $\mathcal{A}_{i:d}$ the collection of those members of $\mathcal{A}$ whose ith component is d, and let $C\mathcal{A}_{i:d}$ denote the *first* $|\mathcal{A}_{i:d}|$ k-vectors with ith component d. (Thus we are compressing in the restricted field of only those vectors whose ith component is d; this involves a reduction in the number of effective components by 1 and will therefore enable an induction argument to proceed.) We say that $\mathcal{A}$ is *i-compressed* if $C\mathcal{A}_{i:d} = \mathcal{A}_{i:d}$ for each $d = 0, \ldots, k_i$.

Let us illustrate these ideas with an example. Consider 2-vectors in $S(1, 1, 1, 1)$ and take

$$\mathcal{A} = \{(0, 0, 1, 1), (0, 1, 0, 1), (1, 0, 0, 1)\}.$$

We show that $\mathcal{A}$ is 1-compressed. First consider $d = 0$. The 2-vectors in $S(1, 1, 1, 1)$ with first component 0 are, in order, $(0, 0, 1, 1)$, $(0, 1, 0, 1)$, and $(0, 1, 1, 0)$, and the two vectors in $\mathcal{A}_{1:0}$ are precisely the first two of these. Next consider $d = 1$. The vectors with first component 1 are, in order, $(1, 0, 0, 1)$, $(1, 0, 1, 0)$, and $(1, 1, 0, 0)$, and the only vector in $\mathcal{A}_{1:1}$ is the first of these. Thus $\mathcal{A}$ is 1-compressed. It can easily be checked by similar arguments that $\mathcal{A}$ is i-compressed for each of $i = 1, \ldots, 4$. Note, however, that $\mathcal{A}$ is *not* compressed since $(0, 1, 1, 0) < (1, 0, 0, 1)$ and $(0, 1, 1, 0) \notin \mathcal{A}$. Therefore being i-compressed for all i is not the same as being compressed.

Starting with any collection $\mathcal{A}$ of k-vectors of $S(k_1, \ldots, k_n)$, we now define a sequence of collections $\mathscr{A}^1, \mathscr{A}^2, \ldots$ as follows. Define

$$\mathscr{A}^1 = \mathscr{A}$$

$$\mathscr{A}^2 = \text{union of all } C\mathscr{A}^1_{1:d} \quad (d = 0, \ldots, k_1)$$

$$\vdots$$

$$\mathscr{A}^{n+1} = \text{union of all } C\mathscr{A}^n_{n:d} \quad (d = 0, \ldots, k_n)$$

$$\mathscr{A}^{n+2} = \text{union of all } C\mathscr{A}^{n+1}_{1:d} \quad (d = 0, \ldots, k_1)$$

and so on cyclically:

$$\mathscr{A}^{m+1} = \text{union of all } C\mathscr{A}^m_{r:d} \quad (d = 0, \ldots, k_r)$$

where $r \equiv m \pmod n$, $1 \leqslant r \leqslant n$. If $\mathscr{A}^j \neq \mathscr{A}^{j+1}$, at least one member of $\mathscr{A}^j$ is being replaced by an earlier vector in the lexicographic order. Eventually no more replacements will be possible, so that we have the following lemma.

Lemma 9.1.3 There exists a positive integer p such that $\mathscr{A}^p$ is i-compressed for all $i = 1, \ldots, n$. □

The next lemma gives us some more practice in dealing with i-compression. We suppose that $k_1 \leqslant \ldots \leqslant k_n$.

Lemma 9.1.4 Let $n \geqslant 3$, $\boldsymbol{a} = (a_1, \ldots, a_n)$, $\boldsymbol{b} = (b_1, \ldots, b_n)$, $|\boldsymbol{a}| = |\boldsymbol{b}| = k$, $\boldsymbol{a} < \boldsymbol{b}$ and $b_n = 0$ or $a_n = k_n$. Then, if $\boldsymbol{b} \in \mathscr{A}$ where $\mathscr{A}$ is i-compressed for $i = 1$, 2, and n, it follows that $\boldsymbol{a} \in \mathscr{A}$.

Proof We shall find an increasing sequence of k-vectors from $\boldsymbol{a}$ to $\boldsymbol{b}$ such that any two consecutive members of the sequence agree in the first, second or nth component. It will then follow that all the members of the sequence, and $\boldsymbol{a}$ in particular, are in $\mathscr{A}$. We deal first of all with the case when $a_n = k_n$.

If $a_1 = b_1$ then the sequence $\boldsymbol{a} < \boldsymbol{b}$ suffices, so now suppose that $a_1 < b_1$. The first subcase which we consider is when $a_i > 0$ for some i, $2 \leqslant i \leqslant n-1$. In this case we have

$$\boldsymbol{a} = (a_1, a_2, \ldots, a_n) < (a_1 + 1, a_2'', \ldots, a_{n-1}'', a_n)$$

where $a_2'', \ldots, a_{n-1}''$ are chosen so that $a_2'' + \ldots + a_{n-1}'' = a_2 + \ldots + a_{n-1} - 1$, and so that $(a_2'', \ldots, a_{n-1}'')$ is as early as possible in the lexicographic order. If $a_1 + 1 = b_1$ we have $\boldsymbol{a} < (b_1, a_2'', \ldots, a_{n-1}'', a_n) \leqslant \boldsymbol{b}$ as required. If $a_1 + 1 < b_1$ and $a_i'' > 0$ for some i, $2 \leqslant i \leqslant n-1$, repeat the process. Either $b_1 - a_1$ applications of the process will give a sequence as required, or at some stage we shall enter the second subcase where we have $\boldsymbol{a} < \ldots < \boldsymbol{a}''$, $a_1'' < b_1$, $a_2'' = \ldots = a_{n-1}'' = 0$. But then we have $\boldsymbol{a}'' = (a_1'', a_2'', \ldots, a_{n-1}'', k_n) < (b_1, a_2'', \ldots, a_{n-1}'', k_n - b_1 + a_1'') \leq \boldsymbol{b}$ as required. Note that the middle vector here is in $S(k_1, \ldots, k_n)$ since $0 \leq (k_1 - b_1) + a_1'' \leq k_n - b_1 + a_1'' \leq k_n$.

We next consider the case when $b_n = 0$. The above argument can be applied to the vectors $\boldsymbol{b}' = (k_1 - b_1, \ldots, k_n - b_n)$, $\boldsymbol{a}' = (k_1 - a_1, \ldots, k_n - a_n)$. The complements of the vectors in the sequence thus constructed from $\boldsymbol{b}'$ to $\boldsymbol{a}'$ form a sequence of vectors from $\boldsymbol{a}$ to $\boldsymbol{b}$ as required. □

Before proceeding to the proof of the Clements–Lindström theorem, we need one further result concerning i-compression.

Lemma 9.1.5 Suppose that Theorem 9.1.1 is true in $n-1$ dimensions and that $\mathscr{B}$ is a collection of $(k-1)$-vectors of $S(k_1, \ldots, k_n)$, $k_1 \leq \ldots \leq k_n$, such that $\Delta\mathscr{A} \subseteq \mathscr{B}$. Then $\Delta(\mathscr{A}^j) \subseteq \mathscr{B}^j$ for all $j \geq 1$.

Proof Since the lemma is true for $j = 1$, we use induction on j and consider the induction step from j to $j+1$. We let S_k denote the set of all k-vectors of $S(k_1, \ldots, k_n)$.

Suppose then that

$$\Delta(\mathscr{A}^j) \subseteq \mathscr{B}^j. \tag{9.2}$$

Our aim is to prove that $\Delta(\mathscr{A}^{j+1}) \subseteq \mathscr{B}^{j+1}$. In proceeding from j to $j+1$ we shall use the fact that

$$\mathscr{A}^{j+1} = \bigcup_d C((\mathscr{A}^j)_{i:d}), \quad i \equiv j \bmod n, \tag{9.3}$$

so we must study $\Delta C(\mathscr{A}^j)_{i:d}$.

If $d > 0$, the members of $\Delta((\mathscr{A}^j)_{i:d})$ are of two types: there are those whose ith component is d and those whose ith component is $d-1$. First consider those with ith component d. They constitute $\Delta((\mathscr{A}^j)_{i:d}) \cap (S_{k-1})_{i:d}$ and, by (9.2), we have $\Delta((\mathscr{A}^j)_{i:d}) \cap (S_{k-1})_{i:d} \subseteq (\mathscr{B}^j)_{i:d}$. Now $(\mathscr{A}^j)_{i:d}$ has only $n-1$ effective components so we can apply Theorem 9.1.1 to it to obtain

$$\begin{aligned}\Delta(C((\mathscr{A}^j)_{i:d})) \cap (S_{k-1})_{i:d} &\subseteq C(\Delta((\mathscr{A}^j)_{i:d})) \cap (S_{k-1})_{i:d} \\ &\subseteq C((\mathscr{B}^j)_{i:d}).\end{aligned} \tag{9.4}$$

Next consider those members of $\Delta((\mathscr{A}^j)_{i:d})$ whose ith component is $d-1$. Since different members of $\mathscr{A}^j$ with ith component d clearly give rise to different members of $\Delta(\mathscr{A}^j)$ with ith component $d-1$,

(9.2) gives

$$|(\mathscr{A}^j)_{i:d}| = |(\Delta(\mathscr{A}^j))_{i:d-1}| \leqslant |(\mathscr{B}^j)_{i:d-1}|$$

for $d \geqslant 1$, so that

$$|C((\mathscr{A}^j)_{i:d})| \leqslant |C((\mathscr{B}^j)_{i:d-1})|. \tag{9.5}$$

In (9.4) we dealt with those members of $\Delta(C((\mathscr{A}^j)_{i:d}))$ with ith component d. The other members constitute $\Delta(C((\mathscr{A}^j)_{i:d})) \cap (S_{k-1})_{i:d-1}$ which, by Lemma 9.1.2, consists of the first $|C((\mathscr{A}^j)_{i:d})|$ members of $(S_{k-1})_{i:d-1}$. Since $C((\mathscr{B}^j)_{i:d-1})$ consists of the first $|C((\mathscr{B}^j)_{i:d-1})|$ members of $(S_{k-1})_{i:d-1}$, (9.5) yields

$$\Delta(C((\mathscr{A}^j)_{i:d})) \cap (S_{k-1})_{i:d-1} \subseteq C((\mathscr{B}^j)_{i:d-1}). \tag{9.6}$$

From (9.4) and (9.6) we now obtain

$$\Delta(C((\mathscr{A}^j)_{i:d})) \subseteq C((\mathscr{B}^j)_{i:d}) \cup C((\mathscr{B}^j)_{i:d-1})$$

for each $d \geqslant 1$. Since, by (9.2), we also have

$$\Delta(C((\mathscr{A}^j)_{i:0}) \subseteq C(\Delta((\mathscr{A}^j)_{i:0}) \subseteq C((\mathscr{B}^j)_{i:0})$$

we finally obtain, from (9.3),

$$\Delta(\mathscr{A}^{j+1}) = \bigcup_d \Delta C((\mathscr{A}^j)_{i:d}) \subseteq \bigcup_d C((\mathscr{B}^j)_{i:d}) = \mathscr{B}^{j+1}. \qquad \square$$

Proof of Theorem 9.1.1 We consider a collection of k-vectors of the multiset $S(k_1, \ldots, k_n)$ where $k_1 \leqslant \ldots \leqslant k_n$. The theorem is easily seen to be true when $n = 2$ (see Exercise 9.2) so we assume that the theorem is true for $n-1$ dimensions and consider the induction step from $n-1$ to n.

By Lemma 9.1.3 there is a positive integer p such that $\mathscr{V} = \mathscr{A}^p$ is i-compressed for all $i = 1, \ldots, n$. Let $\mathscr{W} = (\Delta\mathscr{A})^p$. If we take $\mathscr{B} = \Delta\mathscr{A}$ in Lemma 9.1.5 we get $\Delta\mathscr{V} \subseteq \mathscr{W}$. The proof now proceeds by showing that $\mathscr{V}$ can be altered to $C\mathscr{A}$ and $\mathscr{W}$ to a subset of $C\Delta\mathscr{A}$ in such a way that $\Delta C\mathscr{A} \subseteq C\Delta\mathscr{A}$ is obtained.

First we dispose of the case when $\mathscr{V} = S_k$, i.e. when every k-vector is in $\mathscr{V}$. Then $|\mathscr{V}| = |\mathscr{A}^p| = |\mathscr{A}| = |C\mathscr{A}|$ so that $C\mathscr{A} = S_k$. Also, $S_{k-1} = \Delta S_k = \Delta\mathscr{V} \subseteq \mathscr{W}$, so we must have $\mathscr{W} = S_{k-1}$. Since $|\mathscr{W}| = |\Delta\mathscr{A}| = |C\Delta\mathscr{A}|$ it follows that $C\Delta\mathscr{A} = S_{k-1}$, and hence $\Delta C\mathscr{A} = \Delta S_k = S_{k-1} = C\Delta\mathscr{A}$ as required.

We now can assume that $\mathscr{V} \neq S_k$. Let $\boldsymbol{a}$ be the first vector of S_k which is not in $\mathscr{V}$, and let $\boldsymbol{b}$ be the last vector of $\mathscr{V}$. If $\boldsymbol{b} < \boldsymbol{a}$ then $\mathscr{V} = C\mathscr{A}$ and $\Delta C\mathscr{A} \subseteq \mathscr{W}$ where $|\mathscr{W}| = |\Delta\mathscr{A}|$. Next assume that $\boldsymbol{b} > \boldsymbol{a}$. If $b_n = 0$ then Lemma 9.1.4 applied to $\mathscr{V}$ would give $\boldsymbol{a} \in \mathscr{V}$, which is a contradiction, and so we must have $b_n > 0$. Define

$$b^* = (b_1, \ldots, b_{n-1}, b_n - 1)$$

and also define

$$a^* = (a_1, \ldots, a_{n-1}, a_n - 1) \quad \text{if } a_n > 0.$$

Since $\Delta\mathcal{V} \subseteq \mathcal{W}$, we have $b^* \in \mathcal{W}$. Now all the k-vectors in the shade of b^*, other than b, must come after b in the ordering; therefore the vector b^* in the shadow of $\mathcal{V}$ arises from b but from no other vector in $\mathcal{V}$. Now alter $\mathcal{V}$ and $\mathcal{W}$ as follows. Define

$$\mathcal{V}^* = (\mathcal{V} - \{b\}) \cup \{a\}$$

$$\mathcal{W}^* = \begin{cases} (\mathcal{W} - \{b^*\}) \cup \{a^*\} & \text{if } a_n > 0 \\ \mathcal{W} & \text{otherwise.} \end{cases}$$

We show that $\Delta\mathcal{V}^* \subseteq \mathcal{W}^*$. Since $b^* \notin \Delta(\mathcal{V} - \{b\})$, it suffices to prove that $\Delta\{a\} \subseteq \mathcal{W}^*$. Certainly, if $a_n > 0$ then $a^* \in \Delta\{a\}$ and $a^* \in \mathcal{W}^*$. To deal with the other members of $\Delta\{a\}$, first note that $a_n < k_n$, since otherwise Lemma 9.1.4 would give $a \in \mathcal{V}$. If $a_i > 0$ for some $i \leqslant n-1$ then $(a_1, \ldots, a_i - 1, \ldots, a_n)$ is also in $\Delta\mathcal{V}$ since $(a_1, \ldots, a_i - 1, \ldots, a_n + 1)$ precedes a in the ordering and a is the first vector of S_k not in $\mathcal{V}$. However, $\Delta\mathcal{V} \subseteq \mathcal{W}$, so we have shown that $\Delta\{a\} \subseteq \mathcal{W}$. To show that $\Delta\{a\} \subseteq \mathcal{W}^*$ we must show that $b^* \notin \Delta\{a\}$. Therefore suppose that $b^* = (a_1, \ldots, a_i - 1, \ldots, a_n)$. Then $b^* = (a_1, \ldots, a_i - 1, \ldots, a_n + 1)$, contradicting the fact that $a < b$.

Having started with $\mathcal{V}$ and $\mathcal{W}$, $\Delta\mathcal{V} \subseteq \mathcal{W}$, we have shown how to change them to $\mathcal{V}^*$ and $\mathcal{W}^*$ with $\Delta\mathcal{V}^* \subseteq \mathcal{W}^*$. Now $\mathcal{V}^*$ is i-compressed for all i just as $\mathcal{V}$ was, so we can repeat the process, each time replacing the last vector in $\mathcal{V}$ by an earlier vector. After a finite number of steps we shall compress $\mathcal{V}$ to $C\mathcal{A}$ (since $|\mathcal{V}| = |C\mathcal{A}|$) and $\mathcal{W}$ to a set $\mathcal{U}$ satisfying $\Delta C\mathcal{A} \subseteq \mathcal{U}$. Now $C\Delta\mathcal{A}$ consists of the first $|\Delta\mathcal{A}|$ members of S_{k-1} while, by Lemma 9.1.2, $\Delta C\mathcal{A}$ consists of the first $|\Delta C\mathcal{A}|$ members of S_{k-1}. However, $|\Delta C\mathcal{A}| \leqslant |\mathcal{U}| = |\mathcal{W}| = |\Delta\mathcal{A}|$, so that $\Delta C\mathcal{A} \subseteq C\Delta\mathcal{A}$ as required. □

9.2 Some corollaries

The generalized Macaulay theorem plays a role similar to that of the Kruskal–Katona theorem. Some of its applications to antichains, leading to results similar to those of Chapter 8, will be discussed in the next chapter, but meanwhile we give a few of the simpler consequences of the theorem in this section, following this with two nice applications due to Clements in the closing sections of the chapter.

If $\mathscr{A}$ is a collection of k-vectors of $S(k_1, \ldots, k_n)$, define $\Delta^r(\mathscr{A})$ inductively by $\Delta^1(\mathscr{A}) = \Delta\mathscr{A}$ and $\Delta^{r+1}(\mathscr{A}) = \Delta(\Delta^r(\mathscr{A}))$ for each $r \geq 1$. Then $\Delta^r(\mathscr{A})$ is the shadow of $\mathscr{A}$ at level $k-r$.

Corollary 9.2.1 For each positive integer r and each collection $\mathscr{A}$ of k-vectors in $S(k_1, \ldots, k_n)$,

$$\Delta^r(C\mathscr{A}) \subseteq C\Delta^r(\mathscr{A}).$$

Proof The case $r=1$ is just Theorem 9.1.1. Assume that it is true for r and proceed by induction. $\Delta^{r+1}(C\mathscr{A}) = \Delta(\Delta^r(C\mathscr{A})) \subseteq \Delta(C\Delta^r(\mathscr{A})) \subseteq C\Delta^{r+1}(\mathscr{A})$. □

The next corollary concerns *ideals*. If $\mathscr{A} = \mathscr{A}_1 \cup \mathscr{A}_2 \cup \ldots$ is a collection of members of $S(k_1, \ldots, k_n)$, where $\mathscr{A}_i$ denotes the set of elements of $\mathscr{A}$ of rank i, we call $\mathscr{A}$ an ideal if $\Delta\mathscr{A}_i \subseteq \mathscr{A}_{i-1}$ for each i.

Corollary 9.2.2 If $\mathscr{A}$ is an ideal of subsets of a multiset then so is $C\mathscr{A}$.

Proof The proof is precisely that of Theorem 7.5.2. □

The next corollary re-expresses Theorem 9.1.1 in terms of the shade $\nabla\mathscr{A}$ of $\mathscr{A}$.

Corollary 9.2.3 Under the conditions of Theorem 9.1.1,

$$\nabla(L\mathscr{A}) \subseteq L(\nabla\mathscr{A})$$

where $L\mathscr{A}$ denotes the *last* $|\mathscr{A}|$ k-vectors of $S(k_1, \ldots, k_n)$.

Proof Consider the mapping which sends $(x_1, \ldots, x_n)$ to $(k_1 - x_1, \ldots, k_n - x_n)$. If $\mathscr{A}$ is mapped onto $\mathscr{A}'$, then $C\mathscr{A}'$ is mapped onto $L\mathscr{A}$ and $\Delta\mathscr{A}'$ is mapped onto $\nabla\mathscr{A}$. Thus $\Delta(C\mathscr{A}') \subseteq C(\Delta\mathscr{A}')$ becomes $\nabla(L\mathscr{A}) \subseteq L(\nabla\mathscr{A})$. □

Our next result extends the *subadditivity property* of sets (Theorem 7.4.2) to multisets.

Corollary 9.2.4 (Clements 1977b) Let $F_k(r)$ denote the first r k-vectors of $S(k_1, \ldots, k_n)$. Then, if $a + b \leq |(k)S|$, the number of

k-vectors in $S(k_1, \ldots, k_n)$,

$$|\Delta F_k(a+b)| \leq |\Delta F_k(a)| + |\Delta F_k(b)|.$$

Proof (Kleitman: see Clements and Gronau (1981)) We introduce the multiset $S'(1, 1, k_1, \ldots, k_n)$ and let $F'_k(r)$ denote the first r k-vectors of S'. Let

$$\mathscr{A} = \{(0, 0, \boldsymbol{x}) : \boldsymbol{x} \text{ is a } (k+1)\text{-vector of } S\}$$
$$\mathscr{B} = \{(0, 1, \boldsymbol{x}) : \boldsymbol{x} \in F_k(a)\}$$
$$\mathscr{C} = \{(1, 0, \boldsymbol{x}) : \boldsymbol{x} \in F_k(b)\}.$$

Let $(k)S$ denote the set of k-vectors of S. The first $|(k+1)S|$ $(k+1)$-vectors of S' are those in $\mathscr{A}$; the next $a+b$ are the first $a+b$ $(k+1)$-vectors of the form $(0, 1, \boldsymbol{y})$ where $\boldsymbol{y}$ is a k-vector in S. Thus

$$|\Delta F'_{k+1}(a+b+|(k+1)S|)| = |\Delta(k+1)S| + |\Delta F_k(a+b)|. \quad (9.7)$$

However, by Theorem 9.1.1, we also have

$$\begin{aligned} |\Delta F'_{k+1}(a+b+|(k+1)S|)| &\leq \Delta(\mathscr{A} \cup \mathscr{B} \cup \mathscr{C}) \\ &= |\Delta(k+1)S| + |\Delta F_k(a)| + |\Delta F_k(b)|. \quad (9.8) \end{aligned}$$

Comparing (9.7) with (9.8) we finally obtain

$$|\Delta F_k(a+b)| \leq |\Delta F_k(a)| + |\Delta F_k(b)|$$

as required. □

The final corollary will be needed in the next section. As usual, we write $|\boldsymbol{x}|$ for the rank of $\boldsymbol{x} = (x_1, \ldots, x_n)$; thus $|\boldsymbol{x}| = x_1 + \ldots + x_n$. If $\mathscr{A}$ is any collection of vectors we define the *weight* $w(\mathscr{A})$ of $\mathscr{A}$ by

$$w(\mathscr{A}) = \sum_{\boldsymbol{x} \in \mathscr{A}} |\boldsymbol{x}|.$$

So far we have considered the lexicographic ordering of the vectors in a collection $\mathscr{A}$ only when all the vectors in $\mathscr{A}$ have the same rank. We can, however, order lexicographically any collection of vectors of $S(k_1, \ldots, k_n)$, irrespective of their ranks. If $\boldsymbol{x} = (x_1, \ldots, x_n)$ and $\boldsymbol{y} = (y_1, \ldots, y_n)$ we write $\boldsymbol{x} < \boldsymbol{y}$ if $x_1 < y_1$ or if $x_1 = y_1, \ldots, x_{i-1} = y_{i-1}$, $x_i < y_i$ for some $i > 1$. When we consider ideals we have two comparisons between vectors: their ordering as vectors and their possible inclusion relation as subsets of the multiset. If $\boldsymbol{x}$ is a subset of $\boldsymbol{y}$ in the multiset, i.e. if $x_i \leq y_i$ for each i, then certainly $\boldsymbol{x} \leq \boldsymbol{y}$ in the lexicographic ordering, but conversely if $\boldsymbol{x} \leq \boldsymbol{y}$ it is not necessarily true that $\boldsymbol{x}$ is a subset of $\boldsymbol{y}$.

Corollary 9.2.5 (Clements and Lindström 1969) If $\mathcal{A}$ is an ideal in $S(k_1, \ldots, k_n)$ and $|\mathcal{A}| = m$, then $w(\mathcal{A}) \leq w(F_m)$ where F_m denotes the first m vectors of $S(k_1, \ldots, k_n)$ in the lexicographic order.

Proof By Corollary 9.2.2, $C\mathcal{A}$ is an ideal, and clearly $w(C\mathcal{A}) = w(\mathcal{A})$, so we might as well assume that $\mathcal{A}$ is compressed. If $\mathcal{A} = F_m$ there is nothing to prove, so suppose $\mathcal{A} \neq F_m$. Let $\boldsymbol{a} = (a_1, \ldots, a_n)$ be the first vector which is not in $\mathcal{A}$ and let $\boldsymbol{b} = (b_1, \ldots, b_n)$ be the last vector in $\mathcal{A}$. Then $\boldsymbol{a} < \boldsymbol{b}$. Suppose for the moment that we can prove that $|\boldsymbol{a}| > |\boldsymbol{b}|$. We can then delete $\boldsymbol{b}$ from $\mathcal{A}$ and replace it by $\boldsymbol{a}$ to obtain an ideal $\mathcal{A}_1$ with $w(\mathcal{A}_1) > w(\mathcal{A})$ and with $\mathcal{A}_1$ compressed. If $\mathcal{A}_1 \neq F_m$ we can repeat this replacement process, and after a finite number of applications we obtain $\mathcal{A}_j = F_m$, and $w(F_m) = w(\mathcal{A}_j) > w(\mathcal{A})$.

We therefore have to show that $|\boldsymbol{a}| > |\boldsymbol{b}|$. First note that if $|\boldsymbol{a}| = |\boldsymbol{b}|$ then it follows from the facts that $\boldsymbol{a} < \boldsymbol{b}$, $\boldsymbol{b} \in \mathcal{A}$, and $\mathcal{A}$ is compressed (at each rank level) that $\boldsymbol{a} \in \mathcal{A}$, which is a contradiction. Therefore we now suppose that $|\boldsymbol{a}| < |\boldsymbol{b}|$ and hope for another contradiction in this case. Since $\boldsymbol{a} < \boldsymbol{b}$ we have, for some i, $a_i < b_i$ and $a_j = b_j$ for all $j < i$. If we had $a_{i+1} = \ldots = a_n = 0$ we would then have $\boldsymbol{a}$ contained in $\boldsymbol{b}$ as subsets, so that, since $\boldsymbol{b} \in \mathcal{A}$ where $\mathcal{A}$ is an ideal, we would have $\boldsymbol{a} \in \mathcal{A}$, which is a contradiction. Therefore we can now suppose that $a_j > 0$ for some $j > i$. Put $d = |\boldsymbol{b}| - |\boldsymbol{a}| > 0$, and let $d' = d - \min(d, b_i - a_i - 1)$; then we have $d' \leq b_{i+1} + \ldots + b_n$ (if $d' = d - b_i + a_i + 1 > 0$ then $b_{i+1} + \ldots + b_n = (a_{i+1} + \ldots + a_n) + d - b_i + a_i \geq 1 + d - b_i + a_i = d'$). Now consider the vector $\boldsymbol{c} = (c_1, \ldots, c_n)$ where $c_1 = a_1, \ldots, c_{i-1} = a_{i-1}$, $c_i = b_i - d + d'$ and where $c_{i+1}, \ldots, c_n$ are obtained from $b_{i+1}, \ldots, b_n$ by subtracting a total of d' in such a way that $c_{i+1}, \ldots, c_n$ are non-negative integers; this is possible since $d' \leq b_{i+1} + \ldots + b_n$. We have $c_i \leq b_i$ for each i, so that, since $\boldsymbol{b} \in \mathcal{A}$ and $\mathcal{A}$ is an ideal, $\boldsymbol{c} \in \mathcal{A}$. However, $|\boldsymbol{a}| = |\boldsymbol{c}|$ and $\boldsymbol{a} < \boldsymbol{c}$ since $a_i < c_i$; therefore the facts that $\boldsymbol{c} \in \mathcal{A}$ and $\mathcal{A}$ is compressed now force the conclusion that $\boldsymbol{a} \in \mathcal{A}$ which is a contradiction. □

9.3 A minimization problem in coding theory

In this section we use Corollary 9.2.5 to solve a problem which arises in coding theory. Suppose that we are to send a message which is in the form of a string of non-negative integers, the permitted integers being $0, 1, \ldots, m-1$ where $m = (k_1 + 1) \ldots (k_n + 1)$. Each of these m integers is to be encoded as a code vector $(x_1, \ldots, x_n)$ with

$0 \leq x_i \leq k_i$ for each i, i.e. as a vector in $S(k_1, \ldots, k_n)$. Now during the transmission of the code vectors some errors may possibly arise; one of the x_i may be received falsely with the result that the received vector will be interpreted as representing an integer different from that intended. Let us assume that at most one error will be made in the transmission of any code vector. Then each received vector will be either the one that was sent or one of its *neighbours*, where we say that two vectors in $S(k_1, \ldots, k_n)$ are neighbours if they agree in all but one component. Now there are many ways of choosing which vector should represent which integer, and this gives rise to the following question: how do we choose the vectors to represent the integers so as to minimize the likely error involved on receiving a message? If we let $\boldsymbol{x}$ represent the integer $g(\boldsymbol{x})$ where, as $\boldsymbol{x}$ varies over the set $S = S(k_1, \ldots, k_n)$, $g(\boldsymbol{x})$ varies over $\{0, 1, \ldots, m-1\}$, a reasonable definition of the error associated with the assignment g would be

$$\varepsilon(g) = \sum_{\boldsymbol{x}<\boldsymbol{y}} f(\boldsymbol{x}, \boldsymbol{y})\, |g(\boldsymbol{x}) - g(\boldsymbol{y})|$$

where $f(\boldsymbol{x}, \boldsymbol{y})$ is 1 if $\boldsymbol{x}$ and $\boldsymbol{y}$ are neighbours and 0 otherwise. The problem now becomes: which assignment g minimizes $\varepsilon(g)$?

For any integer l define $h_l(a, b)$ by

$$h_l(a, b) = \begin{cases} 1 & \text{if } a \leq l \text{ and } b > l \\ 0 & \text{otherwise.} \end{cases}$$

Then

$$\begin{aligned} \sum_{l=0}^{m-1} \sum_{g(\boldsymbol{x}) \leq l < g(\boldsymbol{y})} f(\boldsymbol{x}, \boldsymbol{y}) &= \sum_{l=0}^{m-1} \sum_{\boldsymbol{x}, \boldsymbol{y} \in S} h_l(g(\boldsymbol{x}), g(\boldsymbol{y})) f(\boldsymbol{x}, \boldsymbol{y}) \\ &= \sum_{\substack{\boldsymbol{x}, \boldsymbol{y} \in S \\ \boldsymbol{x} \neq \boldsymbol{y}}} f(\boldsymbol{x}, \boldsymbol{y}) \sum_{l=0}^{m-1} h_l(g(\boldsymbol{x}), g(\boldsymbol{y})) \\ &= \sum{}^{*} f(\boldsymbol{x}, \boldsymbol{y}) \sum_{l=0}^{m-1} (h_l(g(\boldsymbol{x}), g(\boldsymbol{y})) + h_l(g(\boldsymbol{y}), g(\boldsymbol{x}))) \end{aligned} \tag{9.9}$$

where $\sum^*$ denotes the sum over unordered pairs $(\boldsymbol{x}, \boldsymbol{y})$ with $\boldsymbol{x} \neq \boldsymbol{y}$. Now if $g(\boldsymbol{x}) < g(\boldsymbol{y})$ then $\sum_{l=0}^{m-1} h_l(g(\boldsymbol{x}), g(\boldsymbol{y}))$ is the number of integers l such that $g(\boldsymbol{x}) \leq l < g(\boldsymbol{y})$, and so is $g(\boldsymbol{y}) - g(\boldsymbol{x})$, whereas $\sum_{l=0}^{m-1} h_l(g(\boldsymbol{y}), g(\boldsymbol{x})) = 0$. A similar result holds if $g(\boldsymbol{y}) < g(\boldsymbol{x})$, so the expression (9.9) becomes

$$\sum{}^{*} f(\boldsymbol{x}, \boldsymbol{y})\, |g(\boldsymbol{y}) - g(\boldsymbol{x})| = \varepsilon(g).$$

Thus

$$\begin{aligned}
\varepsilon(g) &= \sum_{l=0}^{m-1} \sum_{g(x)\leq l<g(y)} f(x, y) \\
&= \sum_{l=0}^{m-1} \left\{ \sum_{\substack{g(x)\leq l\\ y\in S}} f(x, y) - \sum_{\substack{g(x)\leq l\\ g(y)\leq l}} f(x, y) \right\} \\
&= \sum_{l=0}^{m-1} \sum_{g(x)\leq l} (k_1 + \ldots + k_n) - \sum_{l=0}^{m-1} \sum_{\substack{g(x)\leq l\\ g(y)\leq l}} f(x, y) \\
&= \sum_{l=0}^{m-1} (l+1)(k_1 + \ldots + k_n) - 2\sum_{l=0}^{m-1} e(l)
\end{aligned}$$

where $e(l)$ is the number of neighbouring pairs of vectors involving only those vectors representing $0, \ldots, l$. Thus to minimize $\varepsilon(g)$ we have to maximize $\sum_{l=0}^{m-1} e(l)$. Ideally the maximum of this sum will occur when each $e(l)$ is individually maximized, but that might be asking for too much. Surprisingly, we find that we *can* maximize each $e(l)$ simultaneously!

If T is any subset of $S = S(k_1, \ldots, k_n)$, let $E(T)$ denote the number of unordered pairs of neighbours in T. If each $k_i = 1$ it is natural to think of a pair of neighbours, differing in only one component, as an *edge* of an n-dimensional hypercube. We also use the word edge in the more general multiset situation to denote a pair of neighbouring vectors. Lindsey (1964) showed that, for given l, the way to choose a subset T of S consisting of l vectors so as to maximize $E(T)$ is to choose T to consist of the first l vectors of S in the lexicographic order. Thus $e(l)$ is maximized by representing each integer j by the $(j+1)$th vector of S in the lexicographic order.

Before formally stating and proving Lindsey's result, we point out a connection between $E(T)$ and $w(T)$, where w is the weight function of Corollary 9.2.5.

Lemma 9.3.1 If T is an ideal of vectors of $S(k_1, \ldots, k_n)$ then $E(T) = w(T)$.

Proof For fixed $y = (y_1, \ldots, y_n) \in T$, the number of edges (x, y) with $x < y$ is $y_1 + \ldots + y_n = |y|$. Thus

$$E(T) = \sum_{y\in T} |y| = w(T).$$

Theorem 9.3.2 (Lindsey 1964) Let F_l denote the first l vectors of $S = S(k_1, \ldots, k_n)$ in the lexicographic order, and let T be any set of l vectors of S. Then $E(T) \leq E(F_l)$.

Proof (Clements 1971) We shall show that the theorem follows from Corollary 9.2.5. Suppose that T is a set of l vectors such that $E(T)$ is maximum; we shall show that T can be replaced by F_l without decreasing the number of edges. Since the theorem is trivially true when $n=1$, we proceed by induction and consider the induction step from $n-1$ to n. We recall the notation: $T_{1:i}$ is the set of all those members of T whose first component is i.

Thus let T be a set of l vectors of S such that $E(T)$ is maximum. Let $a_i = |T_{1:i}|$. We can, without loss of generality, suppose that $a_0 \geqslant a_1 \geqslant \ldots \geqslant a_{k_1}$ since we could if necessary replace T by a set for which this is true and without altering $E(T)$; this can be achieved by permuting $0, 1, \ldots, k_1$ appropriately in the first component. If $i \leqslant j$ we shall say that an edge $(\boldsymbol{x}, \boldsymbol{y})$ is an (i, j)-edge if $\boldsymbol{x} \in T_{1:i}$ and $\boldsymbol{y} \in T_{1:j}$. Let $N_{i,j}(T)$ denote the number of (i, j)-edges of T. Then

$$E(T) = \sum_{i=0}^{k_1} \sum_{j=i}^{k_1} N_{i,j}(T). \tag{9.10}$$

Suppose now that T is replaced by the set T' consisting of the first a_i members of $S_{1:i}$ for each i; in other words, T is 1-compressed into T'. Then by the induction hypothesis $N_{i,i}(T') \geqslant N_{i,i}(T)$ for each i. Also, if $i<j$ we have $N_{i,j}(T) \leqslant a_j$, whereas $N_{i,j}(T') = a_j$, since the edges contributing to $N_{i,j}(T')$ are the pairs of vertices in T' coinciding in all other components and there are $\min(a_i, a_j) = a_j$ such pairs. Thus $N_{i,j}(T) \leqslant N_{i,j}(T')$ for all $i \leqslant j$, so that (9.10) gives $E(T) \leqslant E(T')$.

Finally, T' is an ideal (see Exercise 9.5), so by Lemma 9.3.1 $E(T) \leqslant E(T') = w(T')$. Thus, by Corollary 9.2.5, $E(T) \leqslant w(T') \leqslant w(F_l) = E(F_l)$.

Corollary 9.3.3 The error $\varepsilon(g)$ is minimized by choosing the assignment g given by $g(\boldsymbol{x}) = j$ where $\boldsymbol{x}$ is the $(j+1)$th vector of $S(k_1, \ldots, k_n)$ in lexicographic order. □

9.4 Uniqueness of maximum-sized antichains in multisets

We now return to the problem of finding the largest antichains in a multiset. We saw earlier, using the symmetric chain decompositions, that if $\mathscr{A}$ is an antichain of subsets of $S(k_1, \ldots, k_n)$ then $|\mathscr{A}| \leqslant N_\lambda$ where $\lambda = [\frac{1}{2}\Sigma_i k_i]$ and where N_i is the number of subsets of rank i. The special case of each $k_i = 1$ is of course Sperner's theorem, and we recall that in that case the antichains $\mathscr{A}$ with $|\mathscr{A}| = N_\lambda$ have been identified: if n is even there is a unique maximum-sized antichain, namely the set of all $[n/2]$-subsets, whereas if n is odd there are precisely two, the set of all $\frac{1}{2}(n-1)$-subsets and the set of all $\frac{1}{2}(n+1)$-subsets. At first sight it might have seemed possible to obtain an antichain of size $\binom{n}{\frac{1}{2}(n-1)}$ consisting of a mixture of sets of sizes $\frac{1}{2}(n-1)$ and $\frac{1}{2}(n+1)$, but we saw that such a mixture cannot give a large enough antichain.

Now consider the corresponding problem for multisets. Let $k_1 \leq k_2 \leq \ldots \leq k_n$ be given and let $K = k_1 + \ldots + k_n$. Is it true that if K is even the only maximum-sized antichain consists of all subsets of rank $\frac{1}{2}K$? Similarly, if K is odd, is it true that there are only two antichains of maximum size, namely all subsets of size $\frac{1}{2}(K-1)$ and all subsets of size $\frac{1}{2}(K+1)$? If $k_n > \frac{1}{2}K$ the answer is generally no. For example, Clements (1984*b*) has pointed out that in the multiset $S(1, 2, 4)$ there is a maximum-sized antichain which is a mixture of sets of ranks 3 and 4:

$$\{(0,0,4), (1,0,3), (0,1,2), (1,1,1), (1,2,0), (0,2,1)\}.$$

However, if $k_n \leq \frac{1}{2}K$ we do obtain the result we hoped for.

Theorem 9.4.1 (Clements 1984*b*) Let $\mathscr{A}$ be a maximum-sized antichain of subsets of the multiset $S(k_1, \ldots, k_n)$, where $k_1 \leq k_2 \leq \ldots \leq k_n \leq \frac{1}{2}K$, $K = k_1 + \ldots + k_n$. Then

(i) if K is even, $\mathscr{A}$ is the set of all subsets of rank $\frac{1}{2}K$;

(ii) if K is odd, $\mathscr{A}$ is either the set of all subsets of rank $\frac{1}{2}(K-1)$ or the set of all subsets of rank $\frac{1}{2}(K+1)$.

Proof Let $\lambda = [\frac{1}{2}K]$. Since $2k_n \leq K$ we know, by Theorem 4.1.1, that

$$N_0 < N_1 < \ldots < N_\lambda; \qquad N_\lambda > \ldots > N_K \quad (K \text{ even})$$

and

$$N_0 < N_1 < \ldots < N_\lambda = N_{\lambda+1}; \qquad N_{\lambda+1} > \ldots > N_K \quad (K \text{ odd}).$$

Let $\mathscr{A}$ have parameters p_i; thus p_i is the number of members of $\mathscr{A}$ of rank i. By the LYM inequality we have

$$\sum_i \frac{p_i}{N_i} \leq 1$$

and it follows from this that if $p_i > 0$ for any i with $N_i < N_\lambda$ then

$$|\mathscr{A}| = \sum_i p_i = \sum_i \frac{p_i}{N_i} N_i < N_\lambda \sum_i \frac{p_i}{N_i} \leq N_\lambda,$$

i.e. $|\mathscr{A}| < N_\lambda$. Thus $\mathscr{A}$ can contain only sets of rank $\frac{1}{2}K$ if K is even and of ranks $\frac{1}{2}(K-1)$ or $\frac{1}{2}(K+1)$ if K is odd. The theorem will therefore follow if we can show that $\mathscr{A}$ cannot consist of a mixture of sets of ranks $\frac{1}{2}(K-1)$ and $\frac{1}{2}(K+1)$. Now if it were possible to have an antichain $\mathscr{A}=\mathscr{B}\cup\mathscr{C}$ where $\mathscr{B}$ consists of sets of rank $\frac{1}{2}(K-1)$ and $\mathscr{C}$ consists of sets of rank $\frac{1}{2}(K+1)$ and where $|\mathscr{B}|+|\mathscr{C}|=N_\lambda$, then $\nabla\mathscr{B}\cap\mathscr{C}=\varnothing$ since $\mathscr{A}$ is an antichain, so that $|\nabla\mathscr{B}|\leqslant|\mathscr{B}|$. But, by the normalized matching property $|\nabla\mathscr{B}|\geqslant|\mathscr{B}|$, and so we must have $|\nabla\mathscr{B}|=|\mathscr{B}|$. We can therefore complete the proof of the theorem by excluding this possibility as follows.

Lemma 9.4.2 Let $\mathscr{B}$ be a collection of subsets of $S(k_1,\ldots,k_n)$, each of rank λ, where $k_1\leqslant k_2\leqslant\ldots\leqslant k_n\leqslant\frac{1}{2}(K-1)=\lambda$, K odd. Then if $\mathscr{B}$ does not contain all of the λ-subsets, $|\nabla\mathscr{B}|>|\mathscr{B}|$.

Proof By Corollary 9.2.3, $|\nabla\mathscr{B}|\geqslant|\nabla(L\mathscr{B})|$, so we might as well assume that $\mathscr{B}$ consists of the last $|\mathscr{B}|$ λ-subsets of $S(k_1,\ldots,k_n)$. We consider two cases.

(i) Suppose first of all that all members of $\mathscr{B}$ are of the form $\boldsymbol{x}=(x_1,\ldots,x_n)$ where $x_1\geqslant 1$. (We are reverting to thinking of subsets as vectors.) For $\boldsymbol{x}=(x_1,\ldots,x_n)\in\mathscr{B}$, write $\boldsymbol{x}^c=(k_1-x_1,\ldots,k_n-x_n)$ and put $\mathscr{B}^c=\{\boldsymbol{x}^c:\boldsymbol{x}\in\mathscr{B}\}$. Then $\mathscr{B}^c$ can be considered as a collection of $(K-\lambda)$-vectors of $S(k_1-1,k_2,\ldots,k_n)$. This multiset has rank $K^c=K-1$ where K^c is even, $k_n\leqslant\frac{1}{2}K^c$, and $K-\lambda>\frac{1}{2}K^c$; therefore, by Theorem 4.1.1, $N^c_{K-\lambda-1}>N^c_{K-\lambda}$ and the normalized matching property gives

$$|\nabla\mathscr{B}|=|\Delta\mathscr{B}^c|\geqslant\frac{N^c_{K-\lambda-1}}{N^c_{K-\lambda}}|\mathscr{B}^c|>|\mathscr{B}^c|=|\mathscr{B}|.$$

(ii) Suppose now that not all members of $\mathscr{B}$ have first component $\geqslant 1$. Let $\mathscr{B}^0$ denote the set of members of $\mathscr{B}$ with first component 0. Form the set $\mathscr{D}$ by taking all those λ-vectors of $S(k_1,\ldots,k_n)$ which have first component 0 but which are not in $\mathscr{B}$ and deleting the first component; thus $\mathscr{D}$ is a set of λ-vectors in $S(k_2,\ldots,k_n)$. Similarly form $\mathscr{E}$ by taking all those $(\lambda+1)$-vectors of $S(k_1,\ldots,k_n)$ with first component 0 which are not in $\nabla\mathscr{B}$ and deleting the first component; thus $\mathscr{E}$ is a set of $(\lambda+1)$-vectors in $S(k_2,\ldots,k_n)$. We then have $\Delta\mathscr{E}\subseteq\mathscr{D}$. If we now let N'_i denote the number of i-vectors in $S(k_2,\ldots,k_n)$ we then have

$$|\mathscr{B}|=|\mathscr{B}^0|+N'_{\lambda-1}+\ldots+N'_{\lambda-k_1}$$

and

$$|\nabla\mathscr{B}|=|(\nabla\mathscr{B})^0|+N'_\lambda+\ldots+N'_{\lambda+1-k_1}$$

so that

$$\begin{aligned}|\nabla\mathscr{B}|-|\mathscr{B}| &= (N'_\lambda-|\mathscr{B}^0|)-(N'_{\lambda-k_1}-|(\nabla\mathscr{B})^0|)\\ &= (N'_\lambda-|\mathscr{B}^0|)-(N'_{\lambda+1}-|(\nabla\mathscr{B})^0|)\\ &= |\mathscr{D}|-|\mathscr{E}|\\ &\geqslant |\Delta\mathscr{E}|-|\mathscr{E}|.\end{aligned}$$

If $\mathscr{E}=\varnothing$ we have $|\nabla\mathscr{B}|-|\mathscr{B}|\geqslant|\mathscr{D}|>0$ as required; therefore suppose now that $\mathscr{E}\neq\varnothing$. Since $\lambda\geqslant\max(k_n, \frac{1}{2}(K-k_1))$ we have $N'_\lambda>N'_{\lambda+1}$ by Theorem 4.1.1, so that

$$|\Delta\mathscr{E}|\geqslant\frac{N'_\lambda}{N'_{\lambda+1}}|\mathscr{E}|>|\mathscr{E}|$$

whence $|\nabla\mathscr{B}|-|\mathscr{B}|\geqslant|\Delta\mathscr{E}|-|\mathscr{E}|>0$. Thus the lemma, and hence the theorem, is proved. □

Exercises 9

9.1 Verify Theorem 9.1.1 for the multiset $S(1, 2, 3, 4)$ and $k=4$ in the case when $\mathscr{A}=\{(1, 0, 2, 1),\ (0, 1, 1, 2),\ (2, 0, 2, 0)\}$ by listing all the members of $C\mathscr{A}$, $\Delta\mathscr{A}$, $\Delta C\mathscr{A}$ and $C\Delta\mathscr{A}$.

9.2 Prove the case $n=2$ of Theorem 9.1.1 as follows. Consider the multiset $S(k_1, k_2)$ where $k_1\leqslant k_2$ and let $\mathscr{A}$ be a collection of k-vectors. First suppose $k\leqslant k_1$. Then $\mathscr{A}$ will be a union of blocks of consecutive members of the list $(0, k)$, $(1, k-1)$, $(2, k-2), \ldots, (k, 0)$. Show that if $\mathscr{B}$ is any such block other than the collection of all k-vectors, and $\mathscr{B}_0$ is any block containing $(0, k)$, then $|\Delta\mathscr{B}|-|\mathscr{B}|\geqslant|\Delta\mathscr{B}_0|-|\mathscr{B}_0|$. Show that this conclusion is also true if $k>k_1$. Then show that $|\Delta\mathscr{A}|-|\mathscr{A}|\geqslant|\Delta C\mathscr{A}|-|C\mathscr{A}|$, and deduce the theorem.

9.3 Denote the rank numbers of $S(k_1, \ldots, k_n)$ by N_i. Show that if $N_l<m<N_{l+1}$ then there exists an antichain $\mathscr{A}$, $|\mathscr{A}|=m$, of the form $F_{l+1}(j)\cup\{(l)S-\Delta F_{l+1}(j)\}$. (Clements 1974)

9.4 Is it true that if a set of k-vectors is compressed then it is i-compressed for all i?

9.5 Prove that the set T' in the proof of Theorem 9.3.2 is an ideal.

9.6 Show that if $\mathscr{A}$ is a collection of k-subsets of an n-set S with

$|\mathscr{A}| \leqslant \binom{n-1}{k}$ then

$$|\Delta\mathscr{A}| \geqslant \frac{\binom{n-1}{k-1}}{\binom{n-1}{k}} |\mathscr{A}|$$

by using an idea in the proof of Lemma 9.4.2.

9.7 Given a multiset $S(k_1, \ldots, k_n)$ and a positive integer l, let M denote the largest integer such that $t \leqslant M$ implies $|\Delta F_l(t)| \geqslant t$.

(i) Use the marriage theorem to show that if $X_1, \ldots, X_t$ are $t \leqslant M$ l-subsets then there exist t distinct $(l-1)$-subsets $Y_1, \ldots, Y_t$ such that $Y_i \subset X_i$ for each i.

(ii) Show that if $t > M$ then no such conclusion is possible.

(iii) Generalize to the problem of finding t distinct $(l-k)$-subsets $Z_1, \ldots, Z_t$ such that $Z_i \subset X_i$ for each i.

(Clements 1970)

10 Theorems for multisets

10.1 Intersecting families

In this section we shall attempt to extend to multisets some of the intersection results already obtained for sets. We begin with the problem with which we started Chapter 1. Now this problem was concerned with an intersecting family $\mathcal{A}$ of subsets of a set, and we are therefore faced immediately with the following question: how do we extend the concept of intersection to multisets? It has often been helpful to think of subsets of a multiset $S(k_1, \ldots, k_n)$ as vectors $\boldsymbol{x} = (x_1, \ldots, x_n)$ with $0 \leqslant x_i \leqslant k_i$ for each i. In the case $k_1 = \ldots = k_n = 1$ the non-disjointness of two sets A, B corresponds to their representing vectors $\boldsymbol{x}$ and $\boldsymbol{y}$ having the property that $x_i = y_i = 1$ for some i. If we think of this as saying that $\min(x_i, y_i) > 0$ for some i, we obtain a natural extension of the intersecting property which we shall call property I_1.

Definition 10.1.1 A collection $\mathcal{A}$ of vectors of $S(k_1, \ldots, k_n)$ is said to possess *property* I_1 if, for all $\boldsymbol{x}, \boldsymbol{y} \in \mathcal{A}$, $\min(x_i, y_i) > 0$ for some i.

In terms of divisors of a number, property I_1 requires that if $\mathcal{A}$ is any collection of divisors of m with property I_1, then g.c.d.$(a, b) > 1$ for all $a, b \in \mathcal{A}$. Here g.c.d. denotes the greatest common divisor.

There is, however, another natural way of extending the intersection property to multisets. To have $x_i = y_i = 1$ for some i in the case when $k_1 = \ldots = k_n = 1$ is to have $x_i + y_i > k_i$ for some i. We therefore make the following definition.

Definition 10.1.2 A collection $\mathcal{A}$ of vectors of $S(k_1, \ldots, k_n)$ is said to possess *property* I_2 if, for all $\boldsymbol{x}, \boldsymbol{y} \in \mathcal{A}$ ($\boldsymbol{x} = \boldsymbol{y}$ allowed!), $x_i + y_i > k_i$ for some i.

In terms of divisors of m, $\mathcal{A}$ has property I_2 provided that, for all $a, b \in \mathcal{A}$, ab does not divide m.

With these two definitions we have obtained two *different* properties I_1 and I_2, both of which reduce to the same intersection property in the special case of $k_1 = \ldots = k_n = 1$. We shall look at both of I_1 and I_2, beginning with I_1.

Recall that Theorem 1.1.1 asserts that an intersecting family of subsets of an n-set has at most 2^{n-1} members, and that any maximal intersecting family (i.e. an intersecting family contained in no larger intersecting family) has exactly 2^{n-1} members. Is there a similar result for multisets and property I_1? We find that we can obtain a bound $f(k_1, \ldots, k_n)$ corresponding to 2^{n-1}, but it is not true that every maximal I_1-family has as many as $f(k_1, \ldots, k_n)$ members.

Whether or not two vectors $\boldsymbol{x}$, $\boldsymbol{y}$ satisfy the condition that $\min(x_i, y_i) > 0$ for some i really depends only on which components x_i, y_i are zero and which are non-zero, and not upon their actual numerical values. For this reason, the idea of the *support* of a vector is useful.

Definition 10.1.3 The *support* $s(\boldsymbol{x})$ of the vector $\boldsymbol{x} = (x_1, \ldots, x_n)$ is given by $s(\boldsymbol{x}) = \{i : x_i \neq 0\}$.

Lemma 10.1.1 A collection $\mathscr{A}$ of vectors has property I_1 if and only if $s(\mathscr{A}) = \{s(\boldsymbol{x}) : \boldsymbol{x} \in \mathscr{A}\}$ is an intersecting family.

Proof $\min(x_i, y_i) > 0$ for some $i \Leftrightarrow x_i > 0, y_i > 0$ for some $i \Leftrightarrow i \in s(\boldsymbol{x}) \cap s(\boldsymbol{y})$ for some $i \Leftrightarrow s(\boldsymbol{x}) \cap s(\boldsymbol{y}) \neq \varnothing$. □

It is clear that if $\mathscr{A}$ is a collection which is maximal with respect to I_1 and if $\mathscr{A}$ contains a vector with support J, then $\mathscr{A}$ will contain all vectors with support J. If J is any subset of $\{1, \ldots, n\}$, the number of vectors in $S(k_1, \ldots, k_n\}$ with support J will be $\prod_{i \in J} k_i$. We denote this number by $w(J)$ and call it the *weight* of J. Thus

$$w(J) = \prod_{i \in J} k_i.$$

Theorem 10.1.2 (Erdös and Schönheim 1970) Let $\mathscr{A}$ be a collection of vectors of $S(k_1, \ldots, k_n)$ with property I_1. Then $|\mathscr{A}| \leqslant f(k_1, \ldots, k_n)$ where

$$f(k_1, \ldots, k_n) = \frac{1}{2} \sum_{J \subseteq \{1, \ldots, n\}} \max\{w(J), w(J')\},$$

J' being the complement of J in $\{1, \ldots, n\}$.

Proof

$$|\mathscr{A}| = \tfrac{1}{2}\sum_J \{|x \in \mathscr{A} : s(x) = J| + |x \in \mathscr{A} : s(x) = J'|\}$$

$$\leqslant \tfrac{1}{2}\sum_J \max\{w(J), w(J')\}$$

since J and J' cannot both occur as supports of vectors in $\mathscr{A}$. □

It is straightforward to construct a collection of $\mathscr{A}$ with property I_1 and such that $|\mathscr{A}| = f(k_1, \ldots, k_n)$. For any J with $w(J) > w(J')$ put into $\mathscr{A}$ all vectors x with $s(x) = J$. For all J with $w(J) = w(J')$ put into $\mathscr{A}$ all vectors x with support J (or J') according as $1 \in J$ (or J'). Observe that $w(J)w(J') = k_1 \ldots k_n$ and that, if $J_1 \cap J_2 = \varnothing$, then $w(J_1)w(J_2) \leqslant k_1 \ldots k_n$. It follows that if $w(J_1) > w(J_1')$ and $w(J_2) \geqslant w(J_2')$ then J_1 and J_2 are not disjoint. Thus $s(\mathscr{A})$ is an intersecting family, and $\mathscr{A}$ has property I_1.

Example 10.1.1 Consider $S(2, 2, 5)$:

$$w(\{1, 2, 3\}) = 2.2.5 = 20 > w(\varnothing)$$
$$w(\{1, 3\}) = 10 > w(\{2\})$$
$$w(\{2, 3\}) = 10 > w(\{1\})$$
$$w(\{3\}) = 5 > w(\{1, 2\})$$

so that, if $\mathscr{A}$ is any collection of vectors with property I_1,

$$|\mathscr{A}| \leqslant \tfrac{1}{2}\{20 + 20 + 10 + 10 + 10 + 10 + 5 + 5\} = 45.$$

An example of a collection of size 45 is, by the above construction, the set of all vectors in $S(2, 2, 5)$ with supports $\{1, 2, 3\}$, $\{1, 3\}$, $\{2, 3\}$, $\{3\}$, i.e. all vectors with third component $\geqslant 1$.

Unlike the situation for sets, where any intersecting family can be extended to one of size $f(1, \ldots, 1) = 2^{n-1}$, there exist maximal I_1-collections of size less than $f(k_1, \ldots, k_n)$ which cannot be extended. For example, in $S(2, 2, 5)$ considered in the example above, the collection of all vectors x with $x_1 \geqslant 1$ is maximal, but has size $36 < 45$. It is clear that in $S(k_1, \ldots, k_n)$ we can similarly obtain a maximal I_1-collection of size $k_1(k_2 + 1) \ldots (k_n + 1)$. We now show that, if $k_1 \leqslant \ldots \leqslant k_n$, then every collection of size less than $k_1(k_2 + 1) \ldots (k_n + 1)$ *can* be extended.

Theorem 10.1.3 (Erdös, Herzog, and Schönheim 1970). Let $\mathscr{A}$ be a collection of vectors of $S(k_1, \ldots, k_n)$, $k_1 \leqslant k_2 \leqslant \ldots \leqslant k_n$, with property I_1, which cannot be extended to a larger collection with property I_1. Then

$$|\mathscr{A}| \geqslant k_1 \prod_{i \geqslant 2} (k_i + 1).$$

Proof Let $s(\boldsymbol{x})$ denote the support of $\boldsymbol{x}$ and let $\mathscr{B} = s(\mathscr{A}) = \{s(\boldsymbol{x}) : \boldsymbol{x} \in \mathscr{A}\}$. Then $\mathscr{B}$ is, by Lemma 10.1.1, an intersecting family. Since $\mathscr{A}$ cannot be extended, neither can $\mathscr{B}$, so that, by Theorem 1.1.1, $|\mathscr{B}| = 2^{n-1}$. Let $\mathscr{B}'$ denote the set of complements of members of $\mathscr{B}$. Then, since $|\mathscr{B}| = 2^{n-1}$ and $\mathscr{B}$ cannot contain a set and its complement, $\mathscr{B} \cup \mathscr{B}'$ is a partition of the set of all subsets of $\{1, \ldots, n\}$, and we have

$$|\mathscr{A}| = \sum_{J \in \mathscr{B}} w(J) = \sum_{J' \in \mathscr{B}'} \frac{k_1 \ldots k_n}{w(J')}. \tag{10.1}$$

Therefore we must prove that

$$\sum_{J \in \mathscr{B}'} \frac{k_1 \ldots k_n}{w(J)} \geqslant k_1 \prod_{i \geqslant 2} (1 + k_i). \tag{10.2}$$

If $(1, 0, \ldots, 0) \in \mathscr{A}$ then every vector with $x_1 \geqslant 1$ will be in $\mathscr{A}$, and $|\mathscr{A}| \geqslant k_1(k_2 + 1) \ldots (k_n + 1)$ is trivially true; therefore we now suppose that $(1, 0, \ldots, 0) \notin \mathscr{A}$, i.e. $\{1\} \notin \mathscr{B}$. We then have $\{1\} \in \mathscr{B}'$. Consider the partition of $\mathscr{B}' : \mathscr{B}' = \mathscr{B}'_{(1)} \cup \mathscr{B}'_1$ where $\mathscr{B}'_{(1)}$ consists of those sets in $\mathscr{B}'$ which do not contain 1, and $\mathscr{B}'_1$ consists of those sets in $\mathscr{B}'$ which do contain 1. Then, from (10.1),

$$|\mathscr{A}| = \sum_{J \in \mathscr{B}'_{(1)}} \frac{k_1 \ldots k_n}{w(J)} + \sum_{J \in \mathscr{B}'_1} \frac{k_1 \ldots k_n}{w(J)}. \tag{10.3}$$

Now

$$\begin{aligned} \sum_{J \in \mathscr{B}'_{(1)}} \frac{k_1 \ldots k_n}{w(J)} + \sum_{J \in \mathscr{B}'_1} w(J) &= \sum_{\substack{I \in \mathscr{B} \\ 1 \in I}} w(I) + \sum_{\substack{J \in \mathscr{B}' \\ 1 \in J}} w(J) \\ &= k_1 \sum_{J \subseteq \{2, \ldots, n\}} w(J) = k_1 \prod_{i \geqslant 2} (1 + k_i), \end{aligned}$$

so that, by (10.3), proving (10.2) is equivalent to proving that

$$\sum_{J \in \mathscr{B}'_1} \frac{k_1 \ldots k_n}{w(J)} \geqslant \sum_{J \in \mathscr{B}'_1} w(J)$$

i.e.

$$\sum_{I\in\mathscr{D}} \frac{k_1 \ldots k_n}{k_1 w(I)} \geq k_1 \sum_{I\in\mathscr{D}} w(I)$$

where $\mathscr{D}$ is the set of all members of $\mathscr{B}_1'$ with the element 1 removed. But we can rewrite this as

$$\sum_{I\in\mathscr{D}} \frac{k_2 \ldots k_n}{w(I)} \geq k_1 \sum_{I\in\mathscr{D}} w(I)$$

i.e.

$$k_1 \sum_{I\in\mathscr{D}} w(I) \leq \sum_{J\in\mathscr{D}'} w(J) \tag{10.4}$$

where $\mathscr{D}$ is a collection of subsets of $\{2, \ldots, n\}$.

We now observe that $\mathscr{B}'$ is an ideal, so that $\mathscr{D}$ is an ideal of subsets of $\{2, \ldots, n\}$. Also, $\mathscr{D}$ has the property that $X \in \mathscr{D} \Rightarrow X' \notin \mathscr{D}$ where X' denotes the complement of X in $\{2, \ldots, n\}$. We can therefore make use of Theorem 6.4.1 which now tells us that if $\mathscr{D} = \{D_1, \ldots, D_r\}$ then there is a permutation ϕ of $\{1, \ldots, r\}$ such that $D_i \subset D'_{\phi(i)}$ for each i. Corresponding to each term $w(I)$ in the sum on the left of (10.4) there is therefore a term $w(J)$ in the sum on the right with $I \subset J$, so that $w(J) \geq k_2 w(I)$, k_2 being the smallest of the k_i, $2 \leq i \leq n$. We therefore have

$$k_1 \sum_{I\in\mathscr{D}} w(I) \leq \sum_{I\in\mathscr{D}} k_2 w(I) \leq \sum_{J\in\mathscr{D}'} w(J),$$

thus proving (10.4). □

We now turn our attention to I_2, the second extension of the intersection property. In this case we obtain a result very similar to Theorem 1.1.1. The argument involves the following collections $\mathscr{B}_i$ of vectors $\boldsymbol{x} = (x_1, \ldots, x_n)$ in $S(k_1, \ldots, k_n)$:

$$\mathscr{B}_1 = \{\boldsymbol{x} : x_1 > \tfrac{1}{2}k_1\}$$

$$\mathscr{B}_2 = \{\boldsymbol{x} : x_1 = \tfrac{1}{2}k_1, x_2 > \tfrac{1}{2}k_2\}$$

$$\vdots$$

$$\mathscr{B}_i = \{\boldsymbol{x} : x_1 = \tfrac{1}{2}k_1, \ldots, x_{i-1} = \tfrac{1}{2}k_{i-1}, x_i > \tfrac{1}{2}k_i\}$$

$$\vdots$$

$$\mathscr{B}_n = \{\boldsymbol{x} : x_i = \tfrac{1}{2}k_i \text{ for all } i < n, x_n > \tfrac{1}{2}k_n\}.$$

It is clear that $\mathcal{B} = \mathcal{B}_1 \cup \ldots \cup \mathcal{B}_n$ possesses property I_2.

Theorem 10.1.4 (Greene and Kleitman 1978) Let $\mathcal{A}$ be a collection of vectors of $S(k_1, \ldots, k_n)$, $k_1 \leqslant \ldots \leqslant k_n$, possessing property I_2. then

$$|\mathcal{A}| \leqslant |\mathcal{B}| = \left[\tfrac{1}{2} \prod_i (1 + k_i)\right].$$

Proof Define $S^*(k_1, \ldots, k_n)$ to be $S(k_1, \ldots, k_n)$ if k_i is odd for at least one i, and to be $S(k_1, \ldots, k_n) - \{(\tfrac{1}{2}k_1, \ldots, \tfrac{1}{2}k_n)\}$ if all the k_i are even. Note that if $\boldsymbol{x} \in S^*$, then $\boldsymbol{x} \in \mathcal{B} \Leftrightarrow \boldsymbol{x}^c = (k_1 - x_1, \ldots, k_n - x_n) \in S^* - \mathcal{B}$. If $\mathcal{A}$ satisfies the conditions of the theorem then, for any $\boldsymbol{x}$, $\boldsymbol{x}$ and $\boldsymbol{x}^c$ cannot both be in $\mathcal{A}$. Thus, if some k_i is odd,

$$|\mathcal{A}| \leqslant |\mathcal{B}| = \tfrac{1}{2}\,|S^*(k_1, \ldots, k_n)| = \tfrac{1}{2} \prod_i (1 + k_i),$$

whereas if all the k_i are even,

$$|\mathcal{A}| \leqslant |\mathcal{B}| = \tfrac{1}{2}\left(\prod_i (1 + k_i) - 1\right).$$

□

Having extended Theorem 1.1.1 to multisets, it is natural to turn next to Theorem 6.1.4 and to seek extensions corresponding to I_1 and I_2. We make the following definitions.

Definition 10.1.4 A collection of vectors in $S = S(k_1, \ldots, k_n)$ has *property* U_1 if, for all $\boldsymbol{x}$, $\boldsymbol{y}$ in the collection, $\max(x_i, y_i) < k_i$ for some i. Similarly it has *property* U_2 if, for all $\boldsymbol{x}$, $\boldsymbol{y}$ in the collection, $x_i + y_i < k_i$ for some i.

In the language of divisibility, a collection $\mathcal{A}$ of divisors of m has property U_1 if the least common multiple l.c.m.(a, b) of any two members a, b of $\mathcal{A}$ is less than m, and has property U_2 if, for all $a, b \in \mathcal{A}$, m does not divide ab.

Theorem 10.1.5 (Anderson 1976) If $\mathcal{A}$ is a collection of vectors of $S(k_1, \ldots, k_n)$ with properties I_1 and U_1, then

$$|\mathcal{A}| \leqslant \frac{\{f(k_1, \ldots, k_n)\}^2}{\prod_i (1 + k_i)},$$

where f is the function of Theorem 10.1.2.

Proof Imitate the proof of Theorem 6.1.4. □

Theorem 10.1.6 (Daykin *et al.* 1979) If $\mathscr{A}$ is a collection of vectors of $S(k_1, \ldots, k_n)$ with properties I_2 and U_2, then

$$|\mathscr{A}| \leqslant \frac{[\frac{1}{2}\prod_i (1+k_i)]^2}{\prod_i (1+k_i)}.$$

Proof Again imitate the proof of Theorem 6.1.4, using Theorem 10.1.4. □

Neither of these bounds is best possible in all cases. An exact bound for $S(k, k, \ldots, k)$ and I_1 and U_1 has been found by Engel and Gronau (1985). For I_2 and U_2, see Exercise 10.3.

10.2 Antichains in multisets

In Chapter 8 we obtained a number of results concerning the parameters of antichains of subsets of an n-set. In this present section we shall see that many of the results for sets extend in a natural way to multisets. The underlying reasons for this include the facts that the LYM inequality and normalized matching property hold for multisets, and that the generalized Macaulay theorem can be used in much the same way as the Kruskal–Katona theorem was used in Chapter 8. Throughout this section we consider $S(k_1, \ldots, k_n)$ where $k_1 \leqslant \ldots \leqslant k_n$.

First of all we consider squashed antichains. If we go through the proof of Theorem 8.1.1 we find that, on replacing $\binom{n}{i}$ by N_i, the ith rank number of the multiset, and squashed ordering by lexicographic ordering, the proof goes through in exactly the same way. Similarly the proofs of Theorem 8.1.2 and Corollary 8.1.3 can be followed and suitably adapted. So, corresponding to any antichain in the multiset we can construct a squashed antichain with the same parameters and the property that, for each i, the vectors of rank i in it, together with the vectors of rank i in the shadow of members of greater rank, constitute an initial segment of the lexicographic ordering of the vectors of rank i. The remaining results of Section 8.1 then hold also. Summarizing, we have the following theorem.

Theorem 10.2.1 (Clements 1973) There exists an antichain in $S(k_1, \ldots, k_n)$ with parameters p_i if and only if $p_i + n_i \leqslant N_i$ holds for all $i \leqslant g$, where g is the largest value of i for which $p_i \neq 0$, and where $\{n_i\}$ is defined by $n_g = 0$ and $n_{i-1} = |\Delta F_i(p_i + n_i)|$, where, as usual,

$F_k(m)$ denotes the first m k-vectors in the lexicographic ordering. Further, if these conditions are satisfied, there is an antichain called the *squashed antichain* with parameters p_i such that, at each rank level i, the p_i vectors, together with those vectors in the shadow of members of the antichain of greater rank, constitute an initial segment in the lexicographic ordering. This antichain minimizes $|\Delta^{(r)}\mathcal{A}|$, the number of r-vectors contained in at least one member of $\mathcal{A}$, over all antichains $\mathcal{A}$ with parameters p_i. □

Our next observation is that the proof of Theorem 8.2.3 made use of squashed antichains, the Kruskal–Katona theorem and the subadditivity result of Theorem 7.4.2. Since subadditivity still holds in multisets (Corollary 9.2.4) and the Clements–Lindström theorem takes over from Kruskal–Katona, the proof of Theorem 8.2.3 immediately yields the following theorem.

Theorem 10.2.2 (Clements 1977a) Suppose that there exists an antichain in $S(k_1, \ldots, k_n)$ with parameters p_i, and let $K = \sum_i k_i$. Then there exists another antichain with parameters q_i given by

$$q_i = \begin{cases} 0 & \text{if } i < \frac{1}{2}K \\ p_i & \text{if } i = \frac{1}{2}K \\ p_i + p_{K-i} & \text{if } i > \frac{1}{2}K. \end{cases}$$ □

Using this result we can now extend Theorem 8.2.4.

Theorem 10.2.3 (Clements and Gronau 1981) Let $\mathcal{A}$ be an antichain with parameters p_i in $S(k_1, \ldots, k_n)$ where $K = \sum_i k_i$ is even, such that $\mathcal{A}$ contains no more than w vectors of rank $\frac{1}{2}K$, where w is the number of vectors of rank $\frac{1}{2}K$ in $S(k_1, \ldots, k_n)$ with first component less than k_1. Then

$$\sum_{i \neq K/2} \frac{p_i}{|(i)S|} + \frac{p_{K/2}}{|(\frac{1}{2}K - 1)S|} \leq 1$$

where $|(i)S|$ denotes the rank number N_i.

Proof As in the proof of Theorem 8.2.4, there is a squashed antichain $\mathcal{A}_3$ with parameters r_i given by $r_i = 0$ if $i > \frac{1}{2}K$, $r_{K/2} = p_{K/2}$, and $r_i = p_i + p_{K-i}$ if $i < \frac{1}{2}K$. Now since $p_{K/2} \leq w$, all the vectors of rank $\frac{1}{2}K$ in $\mathcal{A}_3$ must have first component $\leq k_1 - 1$; they can therefore be thought of as $\frac{1}{2}K$-vectors in $S' = S(k_1 - 1, k_2, \ldots, k_n)$. Applying the normalized matching property in S' to them, we find that

$$|\Delta F_{K/2}(p_{K/2})| \geqslant \frac{|(\frac{1}{2}K-1)S'|}{|(\frac{1}{2}K)S'|} p_{K/2} = p_{K/2}.$$

We can therefore construct another antichain $\mathscr{A}_4$ by replacing the $p_{K/2}$ sets of size $\frac{1}{2}K$ in $\mathscr{A}_3$ by $p_{K/2}$ of the sets in their shadow. The antichain $\mathscr{A}_4$ so constructed has parameters t_i where $t_i = 0$ for $i \geqslant K/2$, $t_{K/2-1} = p_{K/2-1} + p_{K/2} + p_{K/2+1}$ and $t_i = p_i + p_{K-i}$ for each $i < \frac{1}{2}K - 1$. We now apply the LYM inequality to $\mathscr{A}_4$ to obtain

$$\sum_{i<K/2-1} \frac{p_i + p_{K-i}}{|(i)S|} + \frac{p_{K/2-1} + p_{K/2} + p_{K/2+1}}{|(\frac{1}{2}K-1)S|} \leqslant 1.$$

Since $|(i)S| = |(K-i)S|$ for each i, we thus have

$$\sum_{i \neq K/2} \frac{p_i}{|(i)S|} + \frac{p_{K/2}}{|(\frac{1}{2}K-1)S|} \leqslant 1. \qquad \square$$

Corollary 10.2.4 If $\mathscr{A}$ is a complement-free antichain in $S(k_1, \ldots, k_n)$ where K is even, then $|\mathscr{A}| \leqslant |(\frac{1}{2}K-1)S|$. $\square$

We now turn our attention to antichains which, in strong contrast, consist entirely of sets and their complements. In the case of ordinary sets, we have seen the result of Bollobás in Corollary 5.2.3, and we have widened the class of antichains by considering the condition $p_i = p_{n-i}$ in Section 8.2. Gronau has extended some of these results to multisets.

Definition 10.2.1 An antichain $\mathscr{A}$ in $S = S(k_1, \ldots, k_n)$ is called a *c-antichain* if $\boldsymbol{x} = (x_1, \ldots, x_n) \in \mathscr{A} \Rightarrow \boldsymbol{x}^c = (k_1 - x_1, \ldots, k_n - x_n) \in \mathscr{A}$.

Following the construction of $\mathscr{B}$ in Section 10.1, we define subsets of $S(k_1, \ldots, k_n)$, $k_1 \leqslant \ldots \leqslant k_n$, as follows. Let t be the smallest value of i for which k_{i+1} is odd; if all k_i are even, define $t = n$. Then define

$$\mathscr{D}_0 = \{\boldsymbol{x} \in S : x_1 < \tfrac{1}{2}k_1\}$$

$$\mathscr{D}_i = \{\boldsymbol{x} \in S : x_j = \tfrac{1}{2}k_j \text{ for each } j \leqslant i,\ x_{i+1} < \tfrac{1}{2}k_{i+1}\} \qquad (1 \leqslant i \leqslant \min(t, n-1))$$

$$\mathscr{D}_n = \{(\tfrac{1}{2}k_1, \ldots, \tfrac{1}{2}k_n)\} \text{ if } t = n.$$

Then $\mathscr{D} = \mathscr{D}_0 \cup \ldots \cup \mathscr{D}_{n-1}$ consists of the first half of the lexicographic ordering of $S - \mathscr{D}_n$. In the particular case when k_1 is odd, i.e. $t = 0$, we have the simpler situation where $S = \mathscr{D}_0 \cup \mathscr{D}_0'$. Note further that all vectors in $\mathscr{D}_i$ have rank $\geqslant \frac{1}{2}(k_1 + \ldots + k_i)$.

If $\mathscr{A}$ is a c-antichain then, apart from $x = (\frac{1}{2}k_1, \ldots, \frac{1}{2}k_n)$, which may or may not be in $\mathscr{A}$, the vectors in $\mathscr{A}$ can be put into disjoint pairs of complementary vectors x, x^c. Form the antichain $\mathscr{A}^+$ by choosing from each such pair x, x^c a vector of rank $\geq \frac{1}{2}K$, and then put $\mathscr{A}^- = \{x^c : x \in \mathscr{A}^+\}$.

Lemma 10.2.5 Let $\mathscr{A}$ be a c-antichain in $S(k_1, \ldots, k_n)$, and let $(\mathscr{A}^+)^*$ denote the squashed antichain corresponding to $\mathscr{A}^+$. Then

$$(\mathscr{A}^+)^* \subseteq \bigcup_{i \leq \min(t, n-1)} \mathscr{D}_i.$$

Proof From the way that squashed antichains are constructed it is clear that $(\mathscr{A}^+)^* \subseteq \mathscr{A}^*$. Let y be the last vector of $(\mathscr{A}^+)^*$ in the lexicographic order, and let z be the first vector of $\mathscr{B} = \mathscr{A}^* - (\mathscr{A}^+ \cup \mathscr{D}_n)^*$. Then $y < z$ (Exercise 10.11). Therefore the whole of $(\mathscr{A}^+)^*$ lies in the first half of the lexicographic ordering of $S - \mathscr{D}_n$, as was claimed by the lemma. □

Theorem 10.2.6 (Gronau 1980) Let $\mathscr{A}$ be a c-antichain and let q_i be the parameters of $\mathscr{A}^+$. Let $w = \min(t, n-1)$. Then there exist non-negative integers $q_{i,j}$, $\frac{1}{2}K \leq i \leq K$, $0 \leq j \leq w$, such that

(i) $q_i = \sum_j q_{i,j}$ for each i

(ii) for each $j = 0, \ldots, w$,

$$q_{i,j} = 0 \quad \text{if } i \geq K - \tfrac{1}{2} \sum_{l=1}^{j+1} k_l$$

and

$$\sum_{\frac{1}{2}K \leq i < K - \frac{1}{2}\sum_{l=1}^{j+1} k_l} \frac{q_{i,j}}{|(i)\mathscr{D}_j|} \leq 1.$$

The numbers $q_{i,j}$ are in fact given by $q_{i,j}$ = number of vectors of rank i in $(\mathscr{A}^+)^* \cap \mathscr{D}_j$.

Proof Consider the following partition of $(\mathscr{A}^+)^*$:

$$(\mathscr{A}^+)^* = (\mathscr{A}^+)_0^* \cup \ldots \cup (\mathscr{A}^+)_w^*$$

where $(\mathscr{A}^+)_j^* = (\mathscr{A}^+)^* \cap \mathscr{D}_j$. Each $(\mathscr{A}^+)_j^*$ is an antichain in $\mathscr{D}_j$ and (Exercise 10.8) each $\mathscr{D}_j$ is a LYM poset.

Let $q_{i,j}$ denote the number of vectors of rank i in $(\mathscr{A}^+)_j^*$. Since any vector in $\mathscr{D}_j$ has rank less than $K - \frac{1}{2}(k_1 + \ldots + k_{j+1})$, we have $q_{i,j} = 0$ whenever $i \geq K - \frac{1}{2}\sum_{l=1}^{j+1} k_l$. Also, since $(\mathscr{A}^+)^*$ consists only of vectors of ranks $\geq \frac{1}{2}K$, $q_{i,j} = 0$ whenever $i < \frac{1}{2}K$. The LYM inequality for the antichain $(\mathscr{A}^+)_j^*$ in $\mathscr{D}_j$ therefore gives

$$\sum_{\frac{1}{2}K \leqslant i < K - \frac{1}{2}\sum_{l=1}^{j+1} k_l} \frac{q_{i,j}}{|(i)\mathscr{D}_j|} \leqslant 1. \qquad \square$$

In the case when k_1 is odd, i.e. $t = 0$, we obtain

$$\sum_{\frac{1}{2}K \leqslant i < K - \frac{1}{2}k_1} \frac{q_i}{|(i)\mathscr{D}_0|} \leqslant 1 \tag{10.5}$$

where the q_i are the parameters of the antichain $\mathscr{A}^+$, i.e. where $q_{\frac{1}{2}K} = \frac{1}{2}p_{\frac{1}{2}K}$ and where $q_i = p_i$ for each $i > \frac{1}{2}K$. Similarly

$$\sum_{\frac{1}{2}K \leqslant i < K - \frac{1}{2}k_1} \frac{q_i'}{|(i)\mathscr{D}_0|} \leqslant 1 \tag{10.6}$$

where the q_i' are the parameters of the antichain $(\mathscr{A}^-)^c$. But $q_i' = q_{K-i}$, and $\frac{1}{2}K \leqslant K - i < K - \frac{1}{2}k_1 \Leftrightarrow \frac{1}{2}k_1 < i \leqslant \frac{1}{2}K$; therefore on adding (10.5) and (10.6) we obtain the following corollary.

Corollary 10.2.7 Let $\mathscr{A}$ be a c-antichain with parameters p_i in $S(k_1, \ldots, k_n)$ where k_1 is odd. Then

$$\sum_{\frac{1}{2}k_1 < i < \frac{1}{2}K} \frac{p_i}{|(K-i)\mathscr{D}_0|} + \sum_{\frac{1}{2}K \leqslant i < K - \frac{1}{2}k_1} \frac{p_i}{|(i)\mathscr{D}_0|} \leqslant 2. \qquad \square$$

This reduces to Theorem 5.2.3 when $k_n = 1$.

10.3 Intersecting antichains

Recall that the Erdös–Ko–Rado theorem is concerned with antichains of subsets of an n-set with two further properties: the sets in the antichain are all of size $\leqslant k$, where $k \leqslant \frac{1}{2}n$, and they are pairwise non-disjoint. Having extended the concept of intersecting families to multisets in two different ways by means of the properties I_1 and I_2, it is natural to look at possible extensions of EKR to these situations. In the case of property I_2 we are able to obtain extensions.

Lemma 10.3.1 Let $\mathscr{A}$ be an antichain with property I_2 in the multiset $S(k_1, \ldots, k_n)$ with every vector in $\mathscr{A}$ having rank $\leqslant \frac{1}{2}K$. Then $\mathscr{A} \cup \mathscr{A}^c$ is a c-antichain, where $\mathscr{A}^c = \{\boldsymbol{x}^c : \boldsymbol{x} \in \mathscr{A}\}$.

Proof Suppose $\boldsymbol{x} \in \mathscr{A}$ and $\boldsymbol{y}^c \in \mathscr{A}^c$, with $x_i \leqslant k_i - y_i$ for each i. We would then have $x_i + y_i \leqslant k_i$ for each i, contradicting the fact that $\mathscr{A}$

has property I_2. Therefore $\mathscr{A} \cup \mathscr{A}^c$ is an antichain. Clearly it is a c-antichain. □

We now obtain an extension of Theorem 5.2.2

Theorem 10.3.2 (Gronau 1980) Let $\mathscr{A}$ be an antichain with parameters p_i in $S(k_1, \ldots, k_n)$ where k_1 is odd. Suppose that $\mathscr{A}$ has property I_2 and that all vectors in $\mathscr{A}$ have rank $\leqslant \frac{1}{2}K$. Then

$$\sum_{\frac{1}{2}k_1 \leqslant i \leqslant \frac{1}{2}K} \frac{p_i}{|(K-i)\mathscr{D}_0|} \leqslant 1.$$

Proof Since $\mathscr{A} \cup \mathscr{A}^c$ is a c-antichain by the lemma, we can apply (10.5), noting that $(\mathscr{A} \cup \mathscr{A}^c)^+ = \mathscr{A}^c$. This gives

$$\sum_{\frac{1}{2}K \leqslant i \leqslant K - \frac{1}{2}k_1} \frac{q_i}{|(i)\mathscr{D}_0|} \leqslant 1$$

where the q_i are the parameters of $\mathscr{A}^c$; thus

$$\sum_{\frac{1}{2}K \leqslant i \leqslant K - \frac{1}{2}k_1} \frac{p_{K-i}}{|(i)\mathscr{D}_0|} \leqslant 1.$$

Finally put $K - i$ in place of i to obtain the required form of the result. □

If we restrict this result to the case $k_1 = \ldots = k_n = 1$, so that $\mathscr{A}$ is an intersecting antichain with $|A| \leqslant \frac{1}{2}n$ for each $A \in \mathscr{A}$, the theorem gives

$$\sum_{i \leqslant n/2} \frac{p_i}{\binom{n-1}{n-i}} \leqslant 1, \qquad \text{i.e. } \sum_{i \leqslant n/2} \frac{p_i}{\binom{n-1}{i-1}} \leqslant 1.$$

We thus obtain theorem 5.2.2.

We now return to the general case, no longer assuming that k_1 is odd, and obtain a generalization of the EKR theorem first stated by Greene and Kleitman (1978). Recall the definition of the sets $\mathscr{B}_i$ in Section 10.1.

Theorem 10.3.3 Let $\mathscr{A}$ be an antichain of k-vectors in $S(k_1, \ldots, k_n)$ with property I_2, where $k \leqslant \frac{1}{2}K$. Then $|\mathscr{A}|$ is at most the number of k-vectors in $\mathscr{B} = \mathscr{B}_1 \cup \ldots \cup \mathscr{B}_n$.

Proof By Lemma 10.3.1, $\mathscr{C} = \mathscr{A} \cup \mathscr{A}^c$ is a c-antichain, so we can apply Theorem 10.2.6 to it to obtain, for each $j \leq \min(t, n-1)$,

$$\sum \frac{q_{i,j}}{|(i)\mathscr{D}_j|} \leq 1,$$

where the sum is over all i such that $\frac{1}{2}K \leq i < K - \frac{1}{2}\sum_{l=1}^{j+1} k_l$ and where $q_{i,j}$ is the number of vectors of rank i in $(\mathscr{C}^+)^* \cap \mathscr{D}_j = (\mathscr{A}^c)^* \cap \mathscr{D}_j$. Since only $i = K - k$ contributes, we obtain $q_{K-k,j} \leq |(K-k)\mathscr{D}_j|$ for all j for which $\mathscr{D}_j$ can contain a $(K-k)$-vector. Therefore $|\mathscr{A}^c| \leq \sum_j |(K-k)\mathscr{D}_j|$. But $\boldsymbol{x} \in \mathscr{D}_j$ and $|\boldsymbol{x}| = i \Leftrightarrow \boldsymbol{x}^c \in \mathscr{B}_{j+1}$ and $|\boldsymbol{x}^c| = K - i$, so that $|(K-k)\mathscr{D}_j| = |(k)\mathscr{B}_{j+1}|$. Thus $|\mathscr{A}| = |\mathscr{A}^c| \leq \sum_j |(k)\mathscr{B}_{j+1}| = |(k)\mathscr{B}|$. □

Exercises 10

10.1 Let $f(k_1, \ldots, k_n)$ be as in Theorem 10.1.2, and suppose that $k_n > k_1 k_2 \ldots k_{n-1}$.

(i) Show that $f(k_1, \ldots, k_n) = k_n(1+k_1)\ldots(1+k_{n-1})$.

(ii) Verify that in this case the bound of Theorem 10.1.5 is $\left(\frac{k_n}{1+k_n}\right)^2 \prod_{i=1}^{n} (1+k_i)$, and exhibit an example of a collection $\mathscr{A}$ for which

$$|\mathscr{A}| = \frac{k_n^2 - 1}{(1+k_n)^2} \prod_{i=1}^{n} (1+k_i).$$

10.2 Show that

$$f(k, k, \ldots, k) = \tfrac{1}{2} \sum_{i=0}^{n} \binom{n}{i} k^{\max(i, n-i)}$$

$$= \sum_{i<n/2} \binom{n}{i} k^{n-i} + \frac{1}{2}\binom{n}{n/2} k^{n/2}.$$

10.3 In the case where at least two of the k_i are odd, exhibit a collection $\mathscr{A}$ which attains the bound of Theorem 10.1.6.

10.4 Let $\mathscr{A}$ be a collection of vectors of $S(k_1, \ldots, k_n)$ such that, for all $\boldsymbol{x}, \boldsymbol{y} \in \mathscr{A}$, if $\min(x_i, y_i) = 0$ for each i then $\max(x_i, y_i) = k_i$ for each i. Use the ideas of the proof of Theorem 10.1.2 to prove that, if not all the k_i are 1, then $|\mathscr{A}| \leq f(k_1, \ldots, k_n)$.

(Li 1978)

10.5 Deduce Theorem 5.2.3 from Corollary 10.2.7.

10.6 Show that $|(i)\mathscr{D}_j| \leq |([\frac{1}{2}(K+1)])\mathscr{D}_j|$ for each $i \geq [\frac{1}{2}(K+1)]$.

10.7 (i) Show that if $\mathscr{A}$ is a c-antichain in $S(k_1, \ldots, k_n)$ then $|\mathscr{A}| \leqslant |(\frac{1}{2}K)S| = N_{K/2}$ if $K = \sum_i k_i$ is even, whereas $|\mathscr{A}| \leqslant 2\sum_{j=0}^{t} |(\frac{1}{2}(K+1))\mathscr{D}_j|$ if K is odd.
(ii) Show that these bounds can be attained.
(iii) Deduce that $|\mathscr{A}| \leqslant 2\binom{n-1}{[\frac{1}{2}n]-1}$ when $k_n = 1$.
(Gronau 1980)

10.8 In the proof of Theorem 10.2.6 it was asserted that $\mathscr{D}_j$ was a LYM poset. Prove this assertion.

10.9 Extend Theorem 10.3.3 to the case where each vector in $\mathscr{A}$ has rank $\leqslant k$.

10.10 Let $\mathscr{A}$ be a collection of divisors of m such that if $a, b \in \mathscr{A}$ then a and b have at most l prime factors in common, counted according to multiplicity. Show that $|\mathscr{A}| \leqslant \sum_{i=0}^{l+1} N_i$. Re-express this result in terms of vectors in $S(k_1, \ldots, k_n)$.
(Woodall, private communication)

10.11 Prove that $y < z$ in the proof of Lemma 10.2.5.

11 The Littlewood–Offord problem

11.1 Early results

In 1943 Littlewood and Offord published the following result. Let $z_1, \ldots, z_n$ be complex numbers such that $|z_i| \geqslant 1$ for each i; then the number of sums $\sum_{i=1}^n \varepsilon_i z_i$ ($\varepsilon_i = \pm 1$) lying inside any circle of radius r cannot exceed

$$\frac{Cr^{2^n} \log n}{n^{1/2}} \tag{11.1}$$

for some constant C. Two years later Erdös showed, by a very nice application of Sperner's theorem, that the log term can be omitted, thus giving the best order of magnitude possible for the bound. In this chapter we shall explore variations on this problem, showing that some generalizations of Sperner's theorem are central to the development. After giving the basic ideas and the early results in this section, we shall discuss M-part Sperner theorems in Section 11.2 and then show their relevance to the Littlewood–Offord problem in Section 11.3.

There are, in fact, two equivalent formulations of the result of Littlewood and Offord described above. If $\sum_{i=1}^n z_i$ is added to each sum, the differences between sums remain unchanged, and the coefficients of the z_i are now $\delta_i = 0$ or 2. If we scale everything by a factor of $\frac{1}{2}$, we obtain the following: the number of sums $\sum_{i=1}^n \delta_i z_i$ ($\delta_i = 0$ or 1) lying inside any circle of *diameter* r cannot exceed the bound (11.1). This alternative formulation with coefficients 0 or 1 instead of ± 1 is sometimes more convenient.

We start our survey by considering combinations of real numbers, noting that the one-dimensional analogue of a disc of radius 1 is an interval of length 2.

Theorem 11.1.1 (Erdös 1945) Let $x_1, \ldots, x_n$ be real numbers such that $|x_i| \geqslant 1$ for each i, and let J be any interval of length 2 open at at least one end. Then the number of sums $\sum_{i=1}^n \varepsilon_i x_i$ ($\varepsilon_i = \pm 1$) lying in J is at most $\binom{n}{[n/2]}$.

Proof Without loss of generality we can assume that all the x_i are positive; for any negative x_i can be replaced by $-x_i$. We now associate to each sum $\sum_{i=1}^{n} \varepsilon_i x_i$ the set $A = \{i : \varepsilon_i = 1\}$. If A_1 and A_2 are two such sets, and $A_1 \subset A_2$, then the corresponding sums would differ by at least 2 and so they could not both be in J. It follows that the sets A corresponding to sums in J must form an antichain; the result is therefore immediate from Sperner's theorem. □

Given any interval of length $2r$, open at at least one end, we can split it up into r intervals of length 2. Applying the above theorem to each of these intervals we obtain the following corollary.

Corollary 11.1.2 If the x_i are as in Theorem 11.1.1 and J is any interval of length $2r$ open at at least one end, then the number of sums $\sum_{i=1}^{n} \varepsilon_i x_i (\varepsilon_i = \pm 1)$ lying in J is at most $r\binom{n}{[n/2]}$. □

This bound is not, however, the best possible, for we can replace it by the sum of the r largest binomial coefficients $\binom{n}{i}$.

Theorem 11.1.3 (Erdös 1945) Let $x_1, \ldots, x_n$ be real numbers, $|x_i| \geqslant 1$ for each i, and let J be any interval of length $2r$, open at at least one end. Then the number of sums $\sum_{i=1}^{n} \varepsilon_i x_i$ $(\varepsilon_i = \pm 1)$ lying in J is at most

$$\sum_{i=1}^{r} \binom{n}{[(n+i)/2]}.$$

Proof As in the proof of Theorem 11.1.1, assume that each x_i is positive, and associate to each sum $\sum_{i=1}^{n} \varepsilon_i x_i$ the set $A = \{i : \varepsilon_i = 1\}$. If A_1 and A_2 are two such sets and $A_1 \subset A_2$ and $|A_2 - A_1| \geqslant r$, then the corresponding sums will differ by at least $2r$, so that at most one of them can be in J. The result now follows from Exercise 2.9. □

We now return to the case of complex numbers. We shall use j rather than i as a suffix so as to avoid confusion with i where $i^2 = -1$!

Theorem 11.1.4 (Erdös 1945) Let $z_1, \ldots, z_n$ be complex numbers with $|z_j| \geqslant 1$ for each j. Then the number of sums $\sum_{j=1}^{n} \varepsilon_j z_j$ $(\varepsilon_j = \pm 1)$

lying inside a circle of radius r is at most

$$\frac{Cr2^n}{n^{1/2}}$$

for some constant C.

Proof If $|z| \geqslant 1$ where $z = x + iy$, x and y real, then either $|x| \geqslant \frac{1}{2}$ or $|y| \geqslant \frac{1}{2}$. Thus either at least half of the $z_j = x_j + iy_j$ have $|x_j| \geqslant \frac{1}{2}$ or at least half of them have $|y_j| \geqslant \frac{1}{2}$. Thus, on replacing each z_j by iz_j if necessary, we can assume that $|x_j| \geqslant \frac{1}{2}$ for at least half of the z_j. Further, since z_j can be replaced by $-z_j$ if x_j is negative, we can suppose that at least half of the z_j have real part $x_j \geqslant \frac{1}{2}$.

Therefore suppose now that $z_1, \ldots, z_t$ have real part $\geqslant \frac{1}{2}$, $t \geqslant \frac{1}{2}n$. Consider the 2^t sums $\sum_{j=1}^{n} \varepsilon_j z_j$ where $\varepsilon_{t+1}, \ldots, \varepsilon_n$ have been given fixed values. We require that the sums $\sum_{j=1}^{t} \varepsilon_j z_j$ lie inside a circle of radius r, and hence that the sums $\sum_{j=1}^{t} \varepsilon_j x_j$ lie inside an interval of length $2r$. By Corollary 11.1.2, the number of such sums is therefore at most $2r\binom{t}{[\frac{1}{2}t]} = \frac{Cr2^t}{t^{1/2}}$ for some constant C; this is because the sums $\sum_{j=1}^{t} \varepsilon_j(2x_j)$ lie inside a region of length $4r$, and $2x_j \geqslant 1$ for each j.

Thus for each choice of $\varepsilon_{t+1}, \ldots, \varepsilon_n$ we obtain at most $Cr2^t/t^{1/2}$ sums lying inside the circle. But there are 2^{n-t} ways of choosing $\varepsilon_{t+1}, \ldots, \varepsilon_n$. Therefore the total number of sums lying inside the circle is at most

$$\frac{Cr2^t2^{n-t}}{t^{1/2}} = \frac{Cr2^n}{t^{1/2}} = \frac{C'r2^n}{n^{1/2}}$$

where $C \leqslant C' \leqslant 2^{1/2}C$ since $\frac{1}{2}n \leqslant t \leqslant n$. □

In the case $r = 1$ the bound given by Theorem 11.1.4 is a constant multiple of $\binom{n}{[n/2]}$. In fact, we can obtain $\binom{n}{[n/2]}$ itself as the best possible upper bound. This improvement, conjectured by Erdös, was proved by Katona and Kleitman independently. Since the proof depends upon a generalization of Sperner's theorem, we now turn to such generalizations.

11.2 *M*-part Sperner theorems

In this section we present some generalizations of the fundamental theorem of Sperner. We consider the set S partitioned into M disjoint non-empty sets: $S = X_1 \cup \ldots \cup X_M$. A colourful interpretation is to think of the elements of S being coloured by M colours, X_i being the set of those elements coloured by the ith colour. A generalization of the antichain condition is then as follows: a collection $\mathscr{A}$ of subsets of S has the property that if $A, B \in \mathscr{A}$ and $A \subset B$ then $B - A$ is not contained in any of the parts X_i. In terms of colours this property requires that there do not exist sets $A, B \in \mathscr{A}$ with $A \subset B$ and $B - A$ monochromatic. The antichain condition is of course given by $M = 1$.

Rather surprisingly, although the condition with $M = 2$ is less stringent than the antichain condition with $M = 1$, the upper bound $\binom{n}{[n/2]}$ still holds in the case $M = 2$.

Theorem 11.2.1 (Two-part Sperner theorem) (Katona 1966*b*, Kleitman 1965). If $|S| = n$, $S = X_1 \cup X_2$, $X_1 \neq \varnothing$, $X_2 \neq \varnothing$, $X_1 \cap X_2 = \varnothing$, and if $\mathscr{A} = \{A_1, \ldots, A_m\}$ is a collection of subsets of S such that there are no i, j with $A_i \subset A_j$ and $A_j - A_i \subseteq X_1$ or X_2, then $|\mathscr{A}| \leqslant \binom{n}{[n/2]}$.

Proof The set of subsets of X_1 can be partitioned into symmetric chains, as can the set of subsets of X_2. Each subset of S is uniquely expressible in the form $E \cup F$ with $E \subseteq X_1$ and $F \subseteq X_2$. If

$$C : E_1 \subset E_2 \subset \ldots \subset E_p$$

and

$$C' : F_1 \subset F_2 \subset \ldots \subset F_q$$

are, respectively, symmetric chains of subsets of X_1, X_2, then the subsets of S of the form $E_i \cup F_j$ can be displayed in a 'symmetric rectangle' as follows:

$$\begin{array}{cccc} E_1 \cup F_1 & E_1 \cup F_2 & \ldots\ldots\quad\ldots & E_1 \cup F_q \\ E_2 \cup F_1 & E_2 \cup F_2 & \ldots\ldots\quad\ldots & E_2 \cup F_q \\ \vdots & & & \\ E_p \cup F_1 & E_p \cup F_2 & \ldots\ldots\quad\ldots & E_p \cup F_q. \end{array}$$

In this way the whole of $\mathscr{P}(S)$ is partitioned into symmetric rectangles. Note that no two members of $\mathscr{A}$ can appear in the same row or column of any such rectangle; therefore the number of members of $\mathscr{A}$ occurring in any such $p \times q$ rectangle is at most

$\min(p, q)$. However, by Exercise 11.2, $\min(p, q)$ is just the number of subsets of S of size $[n/2]$ in the rectangle. Thus, considering all of the symmetric rectangles, the number of sets in $\mathscr{A}$ is at most the total number of subsets of S of size $[n/2]$, i.e. $\binom{n}{[n/2]}$. □

It would have been nice if there were a corresponding three-part Sperner theorem, but unfortunately the bound $\binom{n}{[n/2]}$ does not hold in this case.

Example 11.2.1 Take $S = \{1\} \cup \{2\} \cup \{3\}$, i.e. colour the three elements of $\{1, 2, 3\}$ differently. Let $\mathscr{A}$ be the set of all 2-subsets of S, together with $\varnothing$. Then $\mathscr{A}$ does not contain A, B with $A \subset B$ and $B - A$ monochromatic, but $|\mathscr{A}| = 4 > \binom{3}{1}$.

An example for general n is given in Exercise 11.1. If the bound $\binom{n}{[n/2]}$ is to be valid in a three-part theorem, some extra conditions have to be put on the set $\mathscr{A}$. Several such conditions have been found. For example, Griggs and Kleitman (1977) have shown that if $S = X_1 \cup X_2 \cup X_3$ then the following additional condition on $\mathscr{A}$ will, along with the absence of A, B such that $A \subset B$ and $B - A$ monochromatic, ensure that $|\mathscr{A}| \leq \binom{n}{[n/2]}$: $\mathscr{A}$ does not contain four sets $A_1, \ldots, A_4$ such that $A_1 \cap X_1 = A_4 \cap X_1 \neq A_2 \cap X_1 = A_3 \cap X_1$, $A_1 \cap X_2 = A_3 \cap X_2 \supset A_2 \cap X_2 = A_4 \cap X_2$, and $A_1 \cap X_3 = A_2 \cap X_3 \supset A_3 \cap X_3 = A_4 \cap X_3$.

If the n-set S is k-coloured, how large can a collection $\mathscr{A}$ of subsets of S be which has the property that there are no A, B with $A \subset B$ and $B - A$ monochromatic? Let $f(n, k)$ denote the largest possible such $|\mathscr{A}|$, taken over all k-colourings of S; then $f(n, 1) = f(n, 2) = \binom{n}{[n/2]} < f(n, 3)$. The question is: how large can $f(n, k)$ be?

Theorem 11.2.2 (Graham and Chung: described in Griggs, Odlyzko, and Shearer (in press)). As $n \to \infty$,

$$f(n, k) \leq \{1 + o(1)\} k^{1/2} \binom{n}{[n/2]}.$$

Proof Let $\mathcal{A}$ be a collection achieving the bound $f(n, k)$, and let $n_k = |X_k|$ where X_k is the largest of the k colour classes X_i, $S = X_1 \cup \ldots \cup X_k$. For each subset T of the $n - n_k$ other elements of S let $\mathcal{A}(T) = \{A \cap X_k : A \in \mathcal{A}, A - X_k = T\}$. Then $\mathcal{A}(T)$ is an antichain of subsets of X_k and so, by Sperner's theorem, $|\mathcal{A}(T)| \leq \binom{n_k}{[n_k/2]}$. Now $n_k \geq n/k$, so, on using the approximation

$$\binom{n}{[n/2]} \approx \left(\frac{2}{\pi n}\right)^{1/2} 2^n$$

which we obtain from Stirling's approximation, we have

$$\begin{aligned} f(n, k) = |\mathcal{A}| &\leq 2^{n-n_k}\binom{n_k}{[\frac{1}{2}n_k]} = 2^{n-n_k}\left(\frac{2}{\pi n_k}\right)^{1/2} 2^{n_k}\{1 + o(1)\} \\ &= \left(\frac{2}{\pi n_k}\right)^{1/2} 2^n\{1 + o(1)\} \\ &\leq k^{1/2}\left(\frac{2}{\pi n}\right)^{1/2} 2^n\{1 + o(1)\} \leq k^{1/2}\binom{n}{[n/2]}\{1 + o(1)\}. \qquad \square \end{aligned}$$

It has recently been proved that

$$d_k = \lim_{n \to \infty} \frac{f(n, k)}{\binom{n}{[n/2]}}$$

exists. In fact, we have the following theorem which we state without proof.

Theorem 11.2.3 (Griggs, Odlyzko, and Shearer, to be published) As $n \to \infty$,

$$f(n, k)n^{1/2}2^{-n} \to e_k \qquad \text{where } e_k \sim \left(\frac{k}{2 \log k}\right)^{1/2}. \qquad \square$$

For our next results, we replace the antichain condition by a condition involving chains.

Theorem 11.2.4 (Sali 1983) If $|S| = n$ and $S = X_1 \cup \ldots \cup X_M$ where the X_i are non-empty and disjoint, and if $\mathcal{A}$ is a collection of subsets of S such that $\mathcal{A}$ does not contain $l + 1$ different members $A_1, \ldots, A_{l+1}$ satisfying $A_1 \subset \ldots \subset A_{l+1}$ and $A_{l+1} - A_1 \subseteq X_i$ for some i, then $|\mathcal{A}| \leq Ml\binom{n}{[n/2]}$.

Theorem 11.2.5 (Griggs 1980*b*) Under the conditions of Theorem 11.2.4, $|\mathscr{A}| \leq 2^{M-2} l \binom{n}{[n/2]}$.

Note that Griggs' result reduces to Theorem 11.2.1 when $M = 2$ and $l = 1$, but that Sali's result is stronger than that of Griggs if $M > 4$. Both have applications to the Littlewood–Offord problem. We shall now give a proof of Sali's result, making use once again of symmetric chains.

Proof of Theorem 11.2.4 The basic idea is to note that for each i the set of subsets of X_i has a symmetric chain decomposition; just as, in the proof of Theorem 11.2.1, the set of subsets of S was partitioned into symmetric rectangles (products of two chains), so here the set of subsets of S can be partitioned into 'symmetric bricks' (products of m chains). Each brick will be of the form $\mathscr{B} = C_1 \times C_2 \times \ldots \times C_M$ where C_i is one of the symmetric chains of subsets of X_i. If we can prove that the number of members of $\mathscr{A}$ in $\mathscr{B}$ is at most $Mlw(\mathscr{B})$ where $w(\mathscr{B})$ is the number of sets in $\mathscr{B}$ of size $[\frac{1}{2}n]$, then the theorem will follow on considering all possible bricks $\mathscr{B}$.

Now we can think of $\mathscr{B} = C_1 \times \ldots \times C_M$ as the set of vectors $(x_1, \ldots, x_M)$ with integer components satisfying $1 \leq x_i \leq k_i$ where k_i is the size of C_i. Therefore $\mathscr{B}$ is isomorphic to the set of subsets of a multiset, or to the set of divisors of $\Pi_i p_i^{k_i-1}$; thus $\mathscr{B}$ itself has a symmetric chain decomposition. Recall the structure of the chains in the symmetric chain decomposition presented in Chapter 3. Each divisor is obtained from the previous divisor in the chain by including one more prime; the primes are added not in any order but in such a way that in each chain the prime p_i is added (as many times as necessary) before p_{i+1} is added. Each symmetric chain of subsets in $\mathscr{B}$ therefore consists of M parts (some of them possibly empty) such that in the ith part the sets (vectors) differ only in the ith component. There can therefore be at most l members of $\mathscr{B}$ in each part and hence at most Ml members of $\mathscr{B}$ in each chain; thus there are at most $Mlw(\mathscr{B})$ members of $\mathscr{A}$ in $\mathscr{B}$, where $w(\mathscr{B})$ is the number of chains in $\mathscr{B}$. But, by the symmetry of the chains, each chain in $\mathscr{B}$ contains exactly one member of size $[\frac{1}{2}n]$. Thus each symmetric brick $\mathscr{B}$ contains at most $Mlw(\mathscr{B})$ members of $\mathscr{A}$ where $w(\mathscr{B})$ is the number of members of $\mathscr{B}$ of size $[\frac{1}{2}n]$. Considering each $\mathscr{B}$ in turn, we find that the total number of members of $\mathscr{A}$ is $\leq Ml\binom{n}{[n/2]}$. □

11.3 Littlewood–Offord results

Having obtained some generalizations of Sperner's theorem in the previous section, we now apply these to the Littlewood–Offord problem. First of all we obtain the best possible improvement of Theorem 11.1.4 in the case $r = 1$.

Theorem 11.3.1 (Katona 1966*b*, Kleitman 1965) Let $z_1, \ldots, z_n$ be complex numbers with $|z_j| \geqslant 1$ for each j. Then the number of sums $\sum_{j=1}^{n} \varepsilon_j z_j$ $(\varepsilon_j = \pm 1)$ lying inside a circle of unit radius is at most $\binom{n}{[n/2]}$.

Proof By replacing z_j by $-z_j$ if necessary, we can assume that each z_j has real part $\geqslant 0$. To each of the sums $\sum_j \varepsilon_j z_j$ we can associate two sets, namely $A = \{j : \varepsilon_j = 1,\ z_j$ in the first quadrant$\}$ and $B = \{j : \varepsilon_j = 1,\ z_j$ in the fourth quadrant$\}$. If two sums give rise to the same A but to two different sets B_1, B_2 such that $B_1 \subset B_2$, then the sums differ by a sum of complex numbers all in the fourth quadrant and all of size $\geqslant 2$; the sums therefore cannot both lie inside a unit circle. So the sets $C = A \cup B$ corresponding to sums in a unit circle must satisfy the conditions of the two-part Sperner theorem 11.2.1; thus the number of sums is at most $\binom{n}{[n/2]}$. □

Theorems 11.1.1 and 11.3.1, both with bound $\binom{n}{[n/2]}$, deal with sums of vectors in one and two dimensions respectively whose sums lie in a 'unit ball' in that particular dimension. What about three dimensions? Since the two-part Sperner theorem does not extend to three parts, a new idea is needed. The ingenuity of Kleitman provides the idea we need; his argument in fact deals with *any* dimension!

Theorem 11.3.2 (Kleitman 1970) Let $\boldsymbol{x}_1, \ldots, \boldsymbol{x}_n$ be n vectors in d-dimensional space, with $|\boldsymbol{x}_i| \geqslant 1$ for each i. Then the number of sums $\sum_{i=1}^{n} \varepsilon_i \boldsymbol{x}_i$ $(\varepsilon_i = \pm 1)$ lying in a unit ball is at most $\binom{n}{[n/2]}$.

Proof The idea behind the proof is to imitate the construction of a symmetric chain decomposition of the sets of subsets of an n-set into $\binom{n}{[n/2]}$ chains; here we shall partition the set of linear combinations

$\sum_{i=1}^{n} \varepsilon_i \boldsymbol{x}_i$ into $\binom{n}{[n/2]}$ blocks such that no two linear combinations in the same block can both have sums lying in a unit ball. The theorem will follow immediately from this partition into blocks.

The blocks are constructed step by step, as were the chains. Suppose that all the combinations of $\boldsymbol{x}_1, \ldots, \boldsymbol{x}_{n-1}$ have been put into blocks as required. In the chain construction, each chain for the case $n-1$ gave rise to two chains (one possibly empty) for the case n, the chains being of sizes one more and one less than the size of the chain from which they arose. We now mimic this construction procedure. Consider any one of the blocks, say $U = \{\boldsymbol{w}_1, \ldots, \boldsymbol{w}_k\}$ where the $\boldsymbol{w}_i$ are combinations of the form $\sum_{i=1}^{n-1} \varepsilon_i \boldsymbol{x}_i$, no two of which lie in a unit ball. From U we construct first of all $U' = \{\boldsymbol{w}_i + \boldsymbol{x}_n : i = 1, \ldots, k-1\}$; U' is just a translate of U. Now choose from U the vector $\boldsymbol{w}_j$ with the largest component in the direction of $\boldsymbol{x}_n$, and consider the blocks

$$V_1 = U' - \{\boldsymbol{w}_j + \boldsymbol{x}_n\}, \qquad V_2 = U'' \cup \{\boldsymbol{w}_j + \boldsymbol{x}_n\}$$

where $U'' = \{\boldsymbol{w}_i - \boldsymbol{x}_n : \boldsymbol{w}_i \in U\}$. These blocks are of size $|U| - 1$ and $|U| + 1$ respectively, and we can check (Exercise 11.3) that no two members of either block can both lie in a unit ball. Now the construction of the blocks is exactly parallel to that of the chains in the symmetric chain construction: each chain of size k gave rise to a chain of size $k-1$ and to another of size $k+1$, and the blocks grow in exactly the same way. It follows that the final number of blocks must be the same as the final number of chains, i.e. $\binom{n}{[n/2]}$. The proof of the theorem is therefore complete. □

Before we apply Sali's result 11.2.4 to obtain our final theorem of this chapter, let us introduce some notation.

Let $g_d(n, r)$ denote the maximum number of sums $\sum_{i=1}^{n} \varepsilon_i \boldsymbol{x}_i$ ($\varepsilon_i = 0$ or 1) lying inside a ball of diameter r, where the $\boldsymbol{x}_i$ are in d-dimensional space and where $|\boldsymbol{x}_i| \geqslant 1$ for each i. We have therefore seen that

$$g_d(n, 1) = \binom{n}{[n/2]}$$

for all $d \geqslant 1$, and that

$$g_1(n, r) = \sum_{i=1}^{r} \binom{n}{[(n+i)/2]}.$$

Further results have been obtained by Kleitman (1976a) for $r = \sqrt{2}$,

$\sqrt{3}$ and $\sqrt{5}$ in two dimensions (see also Exercise 11.7 for another related type of problem). We now prove the following theorem.

Theorem 11.3.3 (Sali 1983)

$$g_d(n, r) \leq 2^{d-1}[rd^{1/2}]\binom{n}{[n/2]}.$$

Proof (essentially Griggs 1980b) Let $\boldsymbol{x}_1, \ldots, \boldsymbol{x}_n$ be n vectors each of length ≥ 1 in d-dimensional space. Without loss of generality we can assume that each has a non-negative first component, so they can be placed into 2^{d-1} sets according to the orthants in which they lie. Let B be an open ball of diameter r, and let $\mathcal{A} = \{E \subseteq \{1, \ldots, n\} : \sum_{i \in E} \boldsymbol{x}_i \in B\}$.

Note that the sum of any k vectors $\boldsymbol{x}_i$ in the same orthant must be $\geq k/d^{1/2}$ (since, for example, all vectors of length ≥ 1 in the first orthant have component $\geq 1/d^{1/2}$ in the direction of the unit vector $(1/d^{1/2})(1, 1, \ldots, 1)$); thus, if $E_1 \subset E_2$ and all the vectors $\boldsymbol{x}_i$, $i \in E_2 - E_1$, are in the same orthant, then the difference between the sums $\sum_{i \in E_1} \boldsymbol{x}_i$ and $\sum_{i \in E_2} \boldsymbol{x}_i$ is at least $|E_2 - E_1|/d^{1/2}$. Thus if $E_1, E_2 \in \mathcal{A}$ and $E_1 \subseteq E_2$, we must have $|E_2 - E_1| \leq [rd^{1/2}]$. We now apply Theorem 11.2.4 with S as the set of sums $\sum_{i=1}^{n} \varepsilon_i \boldsymbol{x}_i$, $M = 2^{d-1}$, the set X_i consisting of the sums in the ith orthant, and $l = [rd^{1/2}]$ to obtain $|\mathcal{A}| \leq 2^{d-1}[rd^{1/2}]\binom{n}{[n/2]}$. □

Exercises 11

11.1 Show that there is no three-part extension of Theorem 11.2.1 by considering the following example. Let $X_1 = \{1, \ldots, n-2\}$, $X_2 = \{n-1\}$, $X_3 = \{n\}$, $S = X_1 \cup X_2 \cup X_3$, n odd, and define

$$\mathcal{A} = \{A \subseteq S : |A| = |A \cap X_1| = \tfrac{1}{2}(n-3)\}$$
$$\mathcal{B} = \{B \subseteq S : |B \cap X_1| = \tfrac{1}{2}(n-3),\ |B \cap X_2| = |B \cap X_3| = 1\}$$
$$\mathcal{C} = \{C \subseteq S : |C \cap X_1| = \tfrac{1}{2}(n-1),\ |C \cap X_2| = 0,\ |C \cap X_3| = 1\}$$
$$\mathcal{D} = \{D \subseteq S : |D \cap X_1| = \tfrac{1}{2}(n-1),\ |D \cap X_2| = 1,\ |D \cap X_3| = 0\}.$$

Show that if $\mathcal{F} = \mathcal{A} \cup \mathcal{B} \cup \mathcal{C} \cup \mathcal{D}$ then $|\mathcal{F}| = 2\binom{n-1}{\frac{1}{2}(n-1)} >$

$\binom{n}{\frac{1}{2}(n-1)}$, and that there do not exist $F_1, F_2 \in \mathscr{F}$ such that $F_1 \subset F_2$ and $F_2 - F_1 \subseteq X_i$ for some i. (Katona 1966b)

11.2 Verify that in the proof of Theorem 11.2.1 $\min(p, q)$ is the number of sets of size $[\frac{1}{2}n]$ occurring in the $p \times q$ symmetric rectangle.

11.3 Verify that in the proof of Theorem 11.3.2 any two vectors in V_2 are distance $\geqslant 2$ apart.

11.4 Given a colouring of a set S, we say that a chain $A_1 \subset A_2 \subset \ldots \subset A_l$ of subsets of S is *monochromatic* if $l = 1$ or $A_l - A_1$ is monochromatic. Show that if S is two-coloured then $\mathscr{P}(S)$ can be partitioned into $\binom{n}{[n/2]}$ monochromatic chains.

(Griggs, Odlyzko, Shearer)

11.5 With $f(n, k)$ as in Theorem 11.2.2, show that $f(n, n) = 2^{n-1}$.

11.6 A two-colouring of the elements of S is *balanced* if the number of elements of one colour differs by at most 1 from the number of elements of the other colour. Suppose that S has been given two different two-colourings, one arbitrary and the other inducing a balanced colouring on each block of the first. Let $\mathscr{A}$ be a collection of subsets of S which contains no A, B with $A \subset B$ and $B - A$ monochromatic with respect to the first colouring and balanced with respect to the second. Show that $|\mathscr{A}| \leqslant \binom{n}{[n/2]}$. (Greene and Kleitman 1976*b*)

11.7 (i) Suppose that the two-dimensional vectors $\boldsymbol{x}_1, \boldsymbol{x}_2$ are such that $|\boldsymbol{x}_i| \geqslant 1$ for each i, and that the angle between $\boldsymbol{x}_1$ and $\boldsymbol{x}_2$ is greater than $\pi/3$. Show that no two of $\mathbf{0}, \boldsymbol{x}_1, \boldsymbol{x}_2$ and $\boldsymbol{x}_1 + \boldsymbol{x}_2$ can be less than distance 1 apart.

(ii) Let $|\boldsymbol{x}_i| \geqslant 1$, $i = 1, \ldots, n$, and let $\boldsymbol{x}_1$ and $\boldsymbol{x}_2$ be at an angle θ, $\pi/3 < \theta < 2\pi/3$. Show that the number of sums $\sum_i \varepsilon_i x_i$ ($\varepsilon_i = 0$ or 1) in any disc of diameter 1 is at most

$$\binom{n+1}{[\frac{1}{2}(n+1)]} - 2\binom{n-1}{[\frac{1}{2}(n-1)]}.$$ (Kleitman 1976*a*)

12 Miscellaneous methods

12.1 The duality theorem of linear programming

Some problems involving subsets of a set can be solved using the duality theorem of linear programming. Suppose that we are considering the following problem:

$$\text{maximize } P = \sum_{i=1}^{n} c_i x_i = \mathbf{c}'\mathbf{x}$$

where $\mathbf{c}$ and $\mathbf{x}$ are $n \times 1$ column vectors and $\mathbf{c}'$ is the transpose of $\mathbf{c}$, subject to the conditions

$$\mathbf{x} \geqslant \mathbf{0} \quad \text{and} \quad A\mathbf{x} \leqslant \mathbf{b}$$

where A is an $m \times n$ matrix and $\mathbf{b}$ is an $m \times 1$ vector. Here we write $\mathbf{x} \geqslant \mathbf{y}$ to mean $x_i \geqslant y_i$ for each i. We define the *dual problem* to be

$$\text{minimize } Q = \sum_{i=1}^{m} b_i y_i = \mathbf{b}'\mathbf{y}$$

subject to the conditions

$$\mathbf{y} \geqslant 0 \quad \text{and} \quad A'\mathbf{y} \geqslant \mathbf{c}.$$

We observe that if $\mathbf{x}$ is any solution of the original problem and $\mathbf{y}$ is any solution of the dual problem then $A\mathbf{x} \leqslant \mathbf{b}$ and $A'\mathbf{y} \geqslant \mathbf{c}$ so that $\mathbf{x}'\mathbf{A}' \leqslant \mathbf{b}'$, whence

$$\mathbf{b}'\mathbf{y} \geqslant \mathbf{x}'A'\mathbf{y} \geqslant \mathbf{x}'\mathbf{c} = \mathbf{c}'\mathbf{x}.$$

Thus every solution of the original problem is at most as large as every solution of the dual problem, so that $\max P \leqslant \min Q$. The duality theorem asserts that we have equality.

Theorem 12.1.1 (The duality theorem) If either of the above problems has a solution then both of the problems are solvable, and $\max P = \min Q$. □

A proof of this theorem can be found in almost any book on linear programming (e.g. Ziants 1974). We shall give two illustrations of the use of this theorem in the text, and another two in the exercises. The first illustration is the original proof due to Kleitman and Milner (1973) of a result which we have already proved in a different way (Theorem 4.2.4).

Theorem 12.1.2 (Kleitman and Milner 1973) Let $\mathcal{A}$ be an antichain on an n-set and suppose that $|\mathcal{A}| \geqslant \binom{n}{k}$ where $k \leqslant \frac{1}{2}n$. Then $\sum_{A \in \mathcal{A}} |A| \geqslant k\,|\mathcal{A}|$, i.e. the average size of the sets in $\mathcal{A}$ is at least k.

Proof We first of all note that we can suppose that $|\mathcal{A}| = \binom{n}{k}$, for we can ignore the $|\mathcal{A}| - \binom{n}{k}$ largest sets in $\mathcal{A}$. Let x_i (instead of the usual p_i) denote the number of i-sets in $\mathcal{A}$; then (generously)

$$\sum_{i=0}^{n} x_i \geqslant \binom{n}{k} \tag{12.1}$$

and, by the LYM inequality,

$$\sum_{i=0}^{n} \frac{x_i}{\binom{n}{i}} \leqslant 1. \tag{12.2}$$

Our problem is to minimize $\sum_{i=0}^{n} ix_i$, i.e. to maximize

$$P = \sum_{i=0}^{n} (-i)x_i,$$

subject to the conditions $\boldsymbol{x} \geqslant 0$, (12.1) and (12.2). Now (12.1) and (12.2) can be rewritten as

$$\begin{bmatrix} \frac{1}{\binom{n}{0}} & \frac{1}{\binom{n}{1}} & \dots & \dots & \frac{1}{\binom{n}{n}} \\ -1 & -1 & \dots & \dots & -1 \end{bmatrix} \boldsymbol{x} \leqslant \begin{bmatrix} 1 \\ -\binom{n}{k} \end{bmatrix}.$$

The dual problem is therefore to minimize

$$Q = y_1 - \binom{n}{k} y_2$$

subject to the conditions $\boldsymbol{y} \geqslant 0$ and

$$\frac{y_1}{\binom{n}{j}} - y_2 \geqslant -j \quad (0 \leqslant j \leqslant n). \tag{12.3}$$

The idea is now to exhibit a specific example of a vector $\boldsymbol{y} \geqslant \mathbf{0}$ satisfying (12.3) for which $Q = -k\binom{n}{k}$. We shall then have $\min Q \leqslant -k\binom{n}{k}$ so that $\max P \leqslant -k\binom{n}{k}$; the minimum value of $\sum_{i=0}^{n} ix_i$ will then be at least $k\binom{n}{k} = k\,|\mathscr{A}|$ as required. Making an inspired guess, we therefore consider

$$y_1 = \binom{n}{k}(y_2 - k), \qquad y_2 = k + \frac{k}{n-2k+1}.$$

For this $\boldsymbol{y}$ we certainly have $y_1 - \binom{n}{k}y_2 = -k\binom{n}{k}$, so we have only to check that this $\boldsymbol{y}$ satisfies the conditions $\boldsymbol{y} \geqslant 0$ and (12.3). Certainly $\boldsymbol{y} \geqslant \mathbf{0}$ since $k \leqslant \frac{1}{2}n$ forces $y_2 > k$ which in turn makes $y_1 > 0$. It therefore remains only to prove (12.3). Now

$$\frac{y_1}{\binom{n}{j}} - y_2 + j = \frac{\binom{n}{k}(y_2 - k)}{\binom{n}{j}} - y_2 + j$$

$$= \left\{\frac{\binom{n}{k}}{\binom{n}{j}} - 1\right\}(y_2 - k) - (k - j).$$

If we define

$$\psi(n, k, i) = \frac{k - i}{\binom{n}{k}\Big/\binom{n}{i} - 1}$$

we then have

$$\psi(n, k, k-1) = \frac{k}{n - 2k + 1} = y_2 - k$$

so that

$$\frac{y_1}{\binom{n}{j}} - y_2 + j = \left\{\frac{\binom{n}{k}}{\binom{n}{j}} - 1\right\}\{\psi(n, k, k-1) - \psi(n, k, j)\}, \quad (12.4)$$

and we have to prove that the expression on the right is non-negative.

Now, for fixed n and k, $\psi(n, k, i)$ is an increasing function of i, $i \leqslant \frac{1}{2}n$ (see Exercise 12.2); therefore if $j < k$ both brackets on the right of (12.4) are non-negative, whereas if $k \leqslant j \leqslant \frac{1}{2}n$ both brackets are $\leqslant 0$. Therefore (12.3) holds for all $j \leqslant \frac{1}{2}n$. But it then follows that (12.3) hold also for $j > \frac{1}{2}n$, for if $j > \frac{1}{2}n$ then $n - j < \frac{1}{2}n$ so that

$$\frac{y_1}{\binom{n}{j}} - y_2 + j > \frac{y_1}{\binom{n}{n-j}} - y_2 + (n - j) \geqslant 0.$$

Thus (12.3) is satisfied and the theorem is proved. □

The normalized matching property for products

As our second example of the use of the duality theorem we show that the normalized matching property of two ranked posets is, under certain conditions, preserved by their product.

Let F and G be two ranked posets. The *product poset* $F \times G$ is defined to have elements (x, y) with $x \in F$, $y \in G$, and to have the order relation defined by $(x, y) \leqslant (x', y') \Leftrightarrow x \leqslant x'$ in F and $y \leqslant y'$ in G. We can define a rank function on $F \times G$ simply by defining the rank of (x, y) to be the sum of the ranks of x and y in F and G respectively. (Recall Exercise 4.7.)

Theorem 12.1.3 (Harper 1974, Hsieh and Kleitman 1973) Let F and G be ranked posets whose rank numbers are log concave, and where both F and G possess the normalized matching property. Then $F \times G$ also has the normalized matching property.

Proof (Hsieh and Kleitman 1973) Let f_p and g_p respectively denote the numbers of elements of rank p in F and G. Then the number of elements of rank k in $F \times G$ is $\sum_p f_p g_{k-p}$. Let X be a given set of elements of rank k in $F \times G$, and, for fixed p, let $x_{p,k-p}$ denote the proportion of the $f_p g_{k-p}$ elements of the form (x, y), $r(x) = p$, $r(y) = k - p$, which are in X. (Here, for convenience, we use r for the rank in either F or G.) Let $Y = \nabla X$ denote the shade of X in $F \times G$, and let $y_{p,k+1-p}$ denote the proportion of elements of $F \times G$ of rank $k + 1$ of the form (x, y), $r(x) = p$, $r(y) = k + 1 - p$, which are in Y. We then have

$$x_{p,k-p} \geqslant 0, \qquad y_{p,k+1-p} \geqslant 0. \tag{12.5}$$

Let g be any element of G. By the normalized matching property of F, the proportion of $(k + 1)$-ranked elements of Y of the form $(\ ,g)$ is at least as great as the proportion of k-ranked elements of X of the form $(\ ,g)$. Thus, on averaging over all $g \in G$ of fixed rank $k-p$, we obtain $y_{p+1,k-p} \geqslant x_{p,k-p}$. Thus

$$y_{p+1,k-p} - x_{p,k-p} \geqslant 0 \quad \text{for all } p. \tag{12.6}$$

Similarly we obtain

$$y_{p,k-p+1} - x_{p,k-p} \geqslant 0 \quad \text{for all } p. \tag{12.7}$$

To prove that $F \times G$ has the normalized matching property we have to show that

$$H = \frac{\sum_p x_{p,k-p} f_p g_{k-p}}{\sum_p f_p g_{k-p}} - \frac{\sum_p y_{p,k-p+1} f_p g_{k-p+1}}{\sum_p f_p g_{k-p+1}} \leqslant 0.$$

If H is ever positive under conditions (12.5)–(12.7) then it is in fact unbounded, for these conditions permit the replacement of the xs and ys by any scalar multiple. Therefore if H is ever positive, the linear programming problem of maximizing H under these conditions is not solvable. Thus, by Theorem 12.1.1, we can prove that $H \leqslant 0$ by showing that the dual problem has a solution. Now the dual problem is to minimize a zero linear combination of $s_1, s_2, \ldots, t_1, t_2, \ldots$ subject to the conditions $s_p \geqslant 0$, $t_p \geqslant 0$, and

$$s_p + t_p \geqslant \frac{f_p g_{k-p}}{\sum_p f_p g_{k-p}} \tag{12.8}$$

and

$$s_{p-1} + t_p \leqslant \frac{f_p g_{k-p+1}}{\sum_p f_p g_{k-p+1}}. \tag{12.9}$$

The theorem will therefore be proved if we can show that conditions (12.8) and (12.9), together with $s_p \geqslant 0$ and $t_p \geqslant 0$, are feasible. We do this by exhibiting a solution. Take

$$s_p = \frac{\sum_{r \leqslant p} f_r g_{k-r}}{\sum_r f_r g_{k-r}} - \frac{\sum_{r \leqslant p} f_r g_{k-r+1}}{\sum_r f_r g_{k-r+1}}$$

and

$$t_p = \frac{\sum_{r \leqslant p} f_r g_{k-r+1}}{\sum_r f_r g_{k-r+1}} - \frac{\sum_{r < p} f_r g_{k-r}}{\sum_r f_r g_{k-r}}.$$

These clearly satisfy (12.8) and (12.9), so we need only check that they are $\geqslant 0$. To show that $s_p \geqslant 0$ we have to show that

$$\frac{\sum_r f_r g_{k-r+1}}{\sum_r f_r g_{k-r}} \geqslant \frac{\sum_{r \leqslant p} f_r g_{k-r+1}}{\sum_{r \leqslant p} f_r g_{k-r}}$$

i.e.

$$\frac{\sum_r f_r g_{k-r}\theta_{k-r}}{\sum_r f_r g_{k-r}} \geqslant \frac{\sum_{r\leqslant p} f_r g_{k-r}\theta_{k-r}}{\sum_{r\leqslant p} f_r g_{k-r}} \tag{12.10}$$

where $\theta_i = g_{i+1}/g_i$. Now we use the given information that the g_i form a log concave sequence, i.e. $g_{i-1}g_{i+1} \leqslant g_i^2$ for each i, i.e. $\theta_i \leqslant \theta_{i-1}$, i.e. θ_i is a decreasing function of i, i.e. θ_{k-i} is an increasing function of i. The truth of (12.10) therefore follows from the following lemma, and $t_p \geqslant 0$ follows similarly from the log concavity of the f_i. □

Lemma 12.1.4 Let θ_i be an increasing non-negative function of i and let each p_i be non-negative. Then, if f is defined by

$$f(k) = \frac{\sum_{i\leqslant k} p_i\theta_i}{\sum_{i\leqslant k} p_i},$$

f is an increasing function of k.

Proof $f(k) \leqslant f(k+1) \Leftrightarrow p_0\theta_0 + \ldots + p_k\theta_k \leqslant \theta_{k+1}(p_0 + \ldots + p_k) \Leftrightarrow \sum_{i=0}^{k} (\theta_{k+1} - \theta_i)p_i \geqslant 0$. But $\theta_i \leqslant \theta_{k+1}$ for each $i \leqslant k$. □

12.2 Graph-theoretic methods

Sometimes the language of graph theory is helpful in the study of subset problems. Our first example is a nice application of the idea of a tree (a connected graph with no cycles) and the fact that in a tree the number of vertices is always one more than the number of edges.

Theorem 12.2.1 (Bondy 1972) If $\mathscr{A} = \{A_1, \ldots, A_n\}$ is a collection of n distinct subsets of an n-set S, then there is an $(n-1)$-subset S' of S such that the sets $A_i \cap S'$ are all distinct.

Proof We define a graph G with vertices labelled $A_1, \ldots, A_n$, and with A_i and A_j joined by an edge labelled x if and only if $A_i = A_j \cup \{x\}$ or $A_j = A_i \cup \{x\}$. Suppose we can show that there are at most $n-1$ distinct edge labels in G, i.e. there is some $y \in S$ which does not appear as an edge label. We then assert that $S' = S - \{y\}$ has the required property, for two sets $A_i \cap S'$ and $A_j \cap S'$ can be equal only if one of A_i, A_j is obtained from the other by removing y and this possibility has been ruled out by the non-appearance of y as an edge label.

We have therefore only to show that there are at most $n-1$ edge labels in G. If G is a tree or a disjoint union of trees, then G has at most $n-1$ edges and so the required result is immediate. Therefore suppose that G has a cycle C. It is clear that every label appearing on an edge of C must occur at least twice on C, since every element removed from a set at some stage of the cycle must be replaced. It follows that if we remove an edge from C the resulting graph G' has the same set of edge labels as G. We can therefore remove edges from cycles until no cycles are left, leaving the set of edge labels unchanged in the process. But the graph we finish up with, having no cycles, must be a tree or a disjoint union of trees, and we have already observed that such a graph can have at most $n-1$ labels. It follows therefore that the original graph also has at most $n-1$ labels. □

For generalizations of Bondy's result, see Frankl (1983).

As our second example of the use of graphs, we consider an extension of Exercise 5.1. In that exercise we showed that if $A_1, \ldots, A_m$ are subsets of an n-set S such that $A_i \cap A_j = \varnothing \Rightarrow A_i \cup A_j = S$, then $m \leqslant 2^{n-1} + \binom{n-1}{[\frac{1}{2}(n-2)]}$. Now let $\mathscr{A}$ be a collection of subsets of an n-set S such that, if $\mathscr{B}$ is any subcollection of $\mathscr{A}$, then $\bigcap_{A\in\mathscr{B}} A = \varnothing \Rightarrow \bigcup_{A\in\mathscr{B}} A = S$. Can we find the best possible upper bound for $|\mathscr{A}|$?

Theorem 12.2.2 (Li 1977) Let $\mathscr{A}$ be a collection of subsets of an n-set S such that, for any subcollection $\mathscr{B}$ of $\mathscr{A}$, $\bigcap_{A\in\mathscr{B}} A = \varnothing \Rightarrow \bigcup_{A\in\mathscr{B}} A = S$. Then $|\mathscr{A}| \leqslant 2^{n-1} + 1$.

This upper bound can be attained (see Exercise 12.1). The proof of the theorem will make use of the graph-theoretic concept of a covering: a *covering* $\mathscr{C}$ of a graph G is a set of vertices of G such that every edge of G is incident with at least one vertex of $\mathscr{C}$. The *degree* of a vertex is the number of edges incident with that vertex. A *star* is a graph in which there is a vertex which is incident with every edge, i.e. a graph which has a covering consisting of only one vertex.

Lemma 12.2.3 Let G be a graph with n vertices, each of degree $\geqslant 1$. Then G has at most $2^{n-1} + 1$ coverings.

Proof We use induction on n. The removal of an edge from G cannot decrease the number of coverings, so we assume that every edge is incident with a vertex of degree 1, and hence that every connected component of G is a star. If G is connected then G is a star, and any covering contains the common vertex or consists of all the other vertices; the number of coverings is therefore $2^{n-1}+1$. If G is not connected then we can write G as a disjoint union, $G=G_1 \cup G_2$, where G_1 has $k \geq 2$ vertices and G_2 has $n-k \geq 2$ vertices for some k, $2 \leq k \leq n-2$. We can apply the induction hypothesis to G_1 and G_2 to find that G_1 has at most $2^{k-1}+1$ coverings and G_2 at most $2^{n-k-1}+1$ coverings. Since a covering of G is obtained by combining a covering of G_1 with a covering of G_2, it follows that there are at most $(2^{k-1}+1)(2^{n-k-1}+1)$ coverings of G. But this number is $2^{n-2}+2^{k-1}+2^{n-k-1}+1 \leq 2^{n-1}+1$. □

Proof of Theorem 12.2.2 Let $\mathcal{A}$ satisfy the conditions of the theorem. Construct a graph G with vertices labelled $1, \ldots, n$ and with vertex i joined to vertex j by an edge if and only if the set $\{i, j\}$ has non-empty intersection with every set in $\mathcal{A}$. Then every member of $\mathcal{A}$ is a covering of G, so the theorem will follow from the lemma provided that we can show that G has no vertex of degree 0. Consider therefore any $i \in \{1, \ldots, n\}$. Since $i \notin \bigcup\{A \in \mathcal{A} : i \notin A\}$, we have $\bigcup\{A \in \mathcal{A} : i \notin A\} \neq S$, so that, by the conditions of the theorem, $\bigcap\{A \in \mathcal{A} : i \notin A\} \neq \varnothing$, i.e. there exists some j such that $j \in A$ whenever $i \notin A$. It follows that $\{i, j\}$ has non-empty intersection with each $A \in \mathcal{A}$; thus i is joined to j in G, and i has degree ≥ 1. □

A related problem is to obtain the largest possible value of $|\mathcal{A}|$ under the condition $\bigcap_{A \in \mathcal{B}} A = \varnothing \Rightarrow \bigcup_{A \in \mathcal{B}} A = S$ whenever $\mathcal{B}$ is a subcollection of $\mathcal{A}$ consisting of k sets, k being a given fixed positive integer. (In contrast, $\mathcal{B}$ could be of any size in Theorem 12.2.2.) We have seen that the bound obtained in Exercise 5.1 for the case $k=2$ is very different from the bound obtained in Theorem 12.2.2. For $k \geq 3$ it turns out that the bound is again $2^{n-1}+1$ (see Ein, Richmond, Shearer, and Sturtevant 1981). These authors also investigate the problem when all the sets in $\mathcal{A}$ are required to be distinct. Finally, we note that Li (1978) has extended Exercise 5.1 to multisets: if $\mathcal{A}$ is a collection of divisors of m such that if g.c.d.$(a, b)=1$ then l.c.m.$(a, b)=m$, then $|\mathcal{A}|$ is bounded by precisely the bound of Theorem 10.1.2 due to Erdös and Schönheim (1970) (see Exercise 10.4).

12.3 Using network flow

One of the basic theorems in the study of network flows is the theorem of Ford and Fulkerson (1956) which relates flow through a network to capacities of cuts. If D is a directed graph, suppose that to each arc (x, y) we assign a non-negative real number $c(x, y)$ called its *capacity*. The capacity can be thought of as the maximum amount of flow which can be sent along the arc. Suppose also that D contains two particular vertices s, t which are the *source* and the *sink* of the flow. A *flow* f of *value* v from s to t is a non-negative real function f defined on the set of arcs of D such that $0 \leqslant f(x, y) \leqslant c(x, y)$ for each arc (x, y), and, for each vertex x,

$$\sum_{y}{}_1 f(x, y) - \sum_{y}{}_2 f(y, x) = \begin{cases} v & \text{if } x = s \\ 0 & \text{if } x = s \text{ or } t \\ -v & \text{if } x = t \end{cases} \qquad (12.11)$$

where Σ_1 is the sum over all arcs (x, y) out of x, and Σ_2 is the sum over all arcs (y, x) into x. Condition (12.11) just states that the amount of flow leaving s is equal to the amount of flow arriving at t, and that, at each vertex other than the source and sink, the amount of flow in is balanced by the amount of flow out.

Let X and Y be sets of vertices of D such that $s \in X, t \in Y$ and every other vertex in D is in precisely one of X, Y. Let $\langle X, Y\rangle$ denote the set of arcs joining a vertex in X to a vertex in Y. Then $\langle X, Y\rangle$ is called a *cut* separating s and t, and we define the capacity of the cut $\langle X, Y\rangle$ to be $c(X, Y) = \Sigma_{(x,y)\in\langle X,Y\rangle}\, c(x, y)$. Intuitively, a cut is a set of arcs whose removal from the network prevents any flow from s to t. It is clear that since any flow from s to t must pass along the arcs of the cut $\langle X, Y\rangle$ the value of any flow cannot exceed the capacity of the cut. Thus the maximum flow value is less than or equal to the minimum capacity of a cut. The Ford–Fulkerson theorem asserts that in fact we have equality.

Theorem 12.3.1 (Ford and Fulkerson 1956) The maximum value of a flow from s to t is equal to the minimum capacity of a cut separating s and t. Further, if all the capacities are integers, there is a flow of maximum value in which every $f(x, y)$ is an integer. □

As our first application of the Ford–Fulkerson theorem, we give yet another proof of the marriage theorem.

Proof of Theorem 2.2.1 Suppose that we have r sets $A_1, \ldots, A_r$ such that, for each $m \leqslant r$ the union of any m of them contains at least m elements. Represent the sets $A_1, \ldots, A_r$ by vertices $x_1, \ldots, x_r$ and the elements $a_1, \ldots, a_n$ of their union by vertices $y_1, \ldots, y_n$, and join x_i to y_j precisely when $a_j \in A_i$. Introduce a source s and a sink t, and join s to each x_i and each y_i to t. Assign capacity 1 to each of the arcs (s, x_i) and (y_i, t), and capacity M to each arc (x_i, y_j), where M is any integer greater than $\max(n, r) = n$.

We want to show that each x_i can be paired with a different y_j. Suppose that we cannot find r disjoint arcs from $\{x_i\}$ to $\{y_i\}$. Then there is no flow of value r from s to t, and so there must be a cut of capacity less than r. Considering the values of the capacities, we see that this cut must consist of, say, u arcs of the form (s, x_i) and v arcs of the form (y_i, t) where $u + v < r$. There can be no arc from any of the remaining $r - u$ vertices x_i to any of the $n - v$ remaining vertices y_i, and so the only arcs from the remaining $r - u$ vertices x_i go to at most v vertices y_i. In terms of the original sets, we therefore have $r - u$ sets which contain in their union at most v, and hence fewer than $r - u$, elements. But this contradicts the hypotheses, so we must, after all, be able to find r disjoint arcs from $\{x_i\}$ to $\{y_i\}$. These r arcs yield distinct representatives for the r sets A_i. □

As our next example, we show how network ideas can be applied to the theory of ranked posets. We have seen in Theorem 2.3.4 that if $\mathscr{A}$ is a collection of subsets of an n-set S such that no $k + 1$ members of $\mathscr{A}$ form a chain, then $|\mathscr{A}|$ is at most the sum of the k largest binomial coefficients $\binom{n}{i}$. Generalizing the definition of the Sperner property, we shall say that a ranked poset P has property S_k if the size of the largest set of elements of P, with no $k + 1$ members forming a chain, is the sum of the k largest rank numbers. We have thus seen in Section 2.3 that any poset with the LYM property possesses property S_k.

While investigating conditions under which a poset P has property S_k, Griggs (1980*a*) introduced the further property T_k. We say that P has property T_k if there exist disjoint chains in P which cover the $(k + 1)$th-largest rank and each of which meets each of the k largest ranks. For example, any symmetric chain order has property T_k. Griggs showed that if P possesses property S_k for all k then it must possess property T_k for all k also. His proof used a deep result of Greene and Kleitman (1976*a*) concerning k-saturated chain partitions; this will be described in Chapter 13. However, Griggs, Sturtevant, and Saks (1980) showed in a later paper that the result can be obtained quite neatly by using a network-flow argument. We now present their proof.

Theorem 12.3.2 (Griggs 1980*a*) If the ranked poset P has property S_k for all k then it has property T_k for all k.

Proof (Griggs *et al.* 1980) Let N_i denote the ith rank number of P, and let n be the largest rank in P. Let $M_0, M_1, \ldots, M_n$ be the rank numbers N_i listed in decreasing order. We suppose that P possesses property S_k for each $k \leqslant n$. To prove that P has property T_k for each k we need only show that it has property T_k for all those value k for which $M_k > M_{k+1}$ and for $k = n$. Therefore let k have such a value.

Consider the ranked poset P' obtained from P by taking the elements of the $k+1$ ranks of P with largest rank numbers and with the order relation of P. Then P' has property S_k since P has. To show that P' has property T_k we need to show that there are M_k disjoint chains in P' which meet every rank.

We now construct a network as follows. Take a source s and a sink t, and vertices u_x and v_x corresponding to each $x \in P'$. Then include the following arcs and capacities (where K is any integer greater than $|P'|$):

(s, u_x), with capacity K, for all $x \in P'$ of rank 0 in P';
(v_x, t), with capacity K, for all $x \in P'$ of rank k in P';
(u_x, v_x), with capacity 1, for all $x \in P'$;
(v_x, u_y), with capacity K, for all $x,y \in P'$ such that y covers x.

Note that any flow from s to t will have to travel along arcs as follows: $(s, u_{x(0)})$, $(u_{x(0)}, v_{x(0)})$, $(v_{x(0)}, u_{x(1)})$, $(u_{x(1)}, v_{x(1)})$, $(v_{x(1)}, u_{x(2)}), \ldots,$ $(v_{x(k)}, t)$, where $x(i)$ is of rank i in P' and where $x(i)$ covers $x(i-1)$ for each $i = 1, \ldots, k$.

Now the set of arcs (u_x, v_x) corresponding to elements $x \in P'$ of a rank with rank number M_k forms a cut with capacity M_k. We shall show that this is the minimum possible capacity of a cut. It will then follow from the Ford–Fulkerson theorem that there is an integer-valued flow from s to t of value M_k. Because of the nature of the network, which permits a flow of at most 1 through each u_x and v_x, this flow must consist of M_k separate unit flows from s to t. Each of these follows a succession of arcs as noted above and hence corresponds to a chain in P' which meets every rank. The theorem will then be proved.

Therefore it remains to prove that there is no cut in P' with capacity less than M_k. Let C be any cut with minimum cut capacity. Since its capacity is at most $M_k < K$, it must consist of arcs $(u_{x_1}, v_{x_1}), \ldots, (u_{x_r}, v_{x_r})$, say, where r is the capacity of C. Now form a poset P'' by removing $x_1, \ldots, x_r$ from P'. No flow from s to t is possible in P'', so no chain in P'' meets all $k+1$ ranks. Thus P'' is a subset of P' containing no $k+1$ members forming a chain. Since P' has property S_k, it follows that $|P''|$ is at most the sum of the k largest rank numbers in P', i.e. $|P''| \leqslant M_0 + \ldots + M_{k-1}$. But $|P''| + r = |P'| =$

$M_0 + \ldots + M_k$, so we must have $r \geq M_k$. Thus the capacity of C is at least M_k as required. □

The converse result is not in general true, but can be proved if the rank numbers N_i are unimodal (see Griggs 1980*a*).

Exercises 12

12.1 Exhibit a collection $\mathscr{A}$ of $2^{n-1}+1$ subsets of an n-set S satisfying the conditions of Theorem 12.2.2.

12.2 (i) Show that if $0 < u_1 \leq u_2 \leq \ldots \leq u_{p+1} < 1$ and $1 < v_1 \leq v_2 < \ldots \leq v_{q+1}$ then

$$f(i) = \frac{i}{1 - u_1 u_2 \ldots u_i}$$

increases with i, and

$$g(i) = \frac{i}{v_1 v_2 \ldots v_i - 1}$$

decreases as i increases.

(ii) Let $\psi(n, k, i)$ be as in the proof of Theorem 12.1.2. Put $a_i = \binom{n}{i} \Big/ \binom{n}{i-1}$. Verify that $a_1 > a_2 > \ldots > a_n$ and that $a_i > 1$ if $i \leq \frac{1}{2}n$. Verify also that

$$\psi(n, k, i) = \frac{k-i}{a_k a_{k-1} \ldots a_{i+1} - 1}$$

if $i < k$ and

$$\psi(n, k, i) = \frac{i-k}{1 - 1/a_i \ldots a_{k+1}}$$

if $i > k$.

(iii) Use (i) to deduce that ψ is an increasing function of i.

12.3 (Preliminary to 12.4). Let $\mathscr{A}$ be an antichain on an n-set and let $\mathscr{A}^* = \{B : B \subseteq A \text{ for some } A \in \mathscr{A}\}$. Let u_i denote the number of i-sets in $\mathscr{A}$ and let $x_i = u_i \Big/ \binom{n}{i}$. If x_i^* denotes the corresponding number for $\mathscr{A}^*$, show that $\Sigma_i x_i \leq 1$ and $x_j^* \geq \Sigma_{i \geq j} x_i$ and hence $|\mathscr{A}^*| \geq \Sigma_i x_i \left\{ \Sigma_{j \leq i} \binom{n}{j} \right\}$.

12.4 Prove Kleitman's result that if $\mathscr{A}$ is an antichain on an n-set such that $|\mathscr{A}| \geqslant \binom{n}{k}$, $k \leqslant \frac{1}{2}n$, then $|\mathscr{A}^*| \geqslant \sum_{i \leqslant k} \binom{n}{i}$ (hints follow). By 12.3 it suffices to prove that the minimum value of $\sum_i x_i \left\{ \sum_{j \leqslant i} \binom{n}{j} \right\}$ subject to $\boldsymbol{x} \geqslant \mathbf{0}$, $\sum_i (-x_i) \geqslant -1$ and $\sum_i x_i \binom{n}{i} \geqslant \binom{n}{k}$ is at least $\sum_{i \leqslant k} \binom{n}{i}$.

(i) Show that the dual problem is to maximize $-u + \binom{n}{k} v$ subject to $u \geqslant 0$, $v \geqslant 0$, $-u + \binom{n}{i} v \leqslant \sum_{j \leqslant i} \binom{n}{j}$.

(ii) Note that if there is a solution for which $-u + \binom{n}{k} v = \sum_{i \leqslant k} \binom{n}{i}$, then $\sum_{i \leqslant k} \binom{n}{i}$, being a solution of the dual problem, will be a lower bound for solutions of the original problem.

(iii) Letting $m_k = \sum_{i \leqslant k} \binom{n}{i}$, show that $-u + \binom{n}{k} v = m_k$ requires

$$\frac{m_k - m_j}{\binom{n}{k} - \binom{n}{j}} \leqslant v \leqslant \frac{m_i - m_k}{\binom{n}{i} - \binom{n}{k}} \quad (1 \leqslant j < k < i \leqslant \tfrac{1}{2}n)$$

and use the log concavity of binomial coefficients to show that v, u exist.

(Odlyzko, quoted in Greene and Kleitman 1978)

12.5 Extend 12.4 to multisets.

12.6 Let $\mathscr{A}$ be a collection of subsets of an n-set such that there are no three distinct sets A, B, C in $\mathscr{A}$ for which $A \cup B = C$. Show that $|\mathscr{A}| \leqslant \binom{n}{[n/2]} \left\{ 1 + 0\left(\frac{1}{n^{1/2}} \right) \right\}$ as follows.

(i) Recall from Exercise 5.8 that if $\mathscr{A}$ contains x_i sets of size i then, for each $j \leqslant \frac{1}{2}n$,

$$\sum_{k=j}^{2j} \frac{j}{k} \frac{x_k}{\binom{n}{k}} \leqslant 1. \tag{12.12}$$

Therefore the problem is to maximize $\sum_i x_i$ subject to (12.12) and $\boldsymbol{x} \geqslant 0$.

(ii) The dual problem is to minimize $y_1 + \ldots + y_{[n/2]}$ subject to $y \geqslant 0$ and

$$\sum_{k=\frac{1}{2}j}^{\min(j,[n/2])} \frac{k}{j\binom{n}{j}} y_k \geqslant 1 \qquad (j = 1, \ldots, n). \tag{12.13}$$

(iii) The required result will follow if we can exhibit values of the y_i satisfying $y \geqslant 0$ and (12.13) for which $\Sigma_i\, y_i = \binom{n}{[n/2]}\left\{1 + O\left(\frac{1}{n^{1/2}}\right)\right\}$.

(iv) Take

$$y_{2m} = \binom{n}{2m} - \binom{n}{2m-1} + \frac{1}{2m}\binom{n}{2m-1}$$

$$y_{2m+1} = \binom{n}{2m+1} - \binom{n}{2m} + \frac{1}{2m+1}\binom{n}{2m} + \frac{m}{2m+1} y_m,$$

and verify that this choice of y solves (iii).

(Kleitman 1976*b*)

12.7 Use Theorem 12.1.3 to show that $C(n, k)$, the set of all subsets of an n-set S which intersect a given k-subset T of S, has the normalized matching property. (This gives another proof that $C(n, k)$ has the Sperner property (see Theorem 3.5.1).)

(Griggs 1982)

12.8 Let $A_1 \subset A_2 \subset \ldots \subset A_r$ be a chain of subsets of S, and let a_i and b_i, $i = 1, \ldots, r$, be given non-negative integers such that $a_i \leqslant b_i$ for each i. Let $P = \{X \in S : a_i \leqslant |X \cap A_i| \leqslant b_i\}$ for each i.

(i) Show that P is the product of $\mathcal{P}(A_r')$ and the poset P' which consists of those subsets of A_r which are in P.

(ii) Prove that P' is a LYM poset with log concave rank numbers.

(iii) Deduce that P is a LYM poset with log concave rank numbers.

(West, Harper, and Daykin 1983)

13 Lattices of antichains and saturated chain partitions

13.1 Antichains

In this final chapter we look at antichains in a general poset. We begin by showing that a partial ordering can be defined on the set $\mathfrak{A}$ of antichains of a poset P, under which $\mathfrak{A}$ is a lattice containing the set of maximum-sized antichains as a sublattice. This is essentially work of Dilworth (1960), although we shall not necessarily follow his development. We shall then study some recent work, initially due to Greene and Kleitman (1976*a*), which extends the fundamental Theorem 3.2.1 of Dilworth on chain partitions of a poset.

Let $\mathfrak{A}(P)$ denote the set of antichains on the finite poset P, and, for $\mathscr{A}, \mathscr{B} \in \mathfrak{A}(P)$, write

$$\mathscr{A} \leqslant \mathscr{B} \Leftrightarrow \text{for all } A \in \mathscr{A} \text{ there exists } B \in \mathscr{B} \text{ such that } A \leqslant B.$$

Then clearly

$$\mathscr{A} \leqslant \mathscr{A} \quad \text{for all} \quad \mathscr{A} \in \mathfrak{A}(\mathrm{P}). \tag{13.1}$$

Next suppose that $\mathscr{A} \leqslant \mathscr{B}$ and $\mathscr{B} \leqslant \mathscr{A}$. If $A \in \mathscr{A}$, there exists $B \in \mathscr{B}$ such that $A \leqslant B$, and, for this B, there exists $A_1 \in \mathscr{A}$ such that $B \leqslant A_1$. We therefore have $A \leqslant B \leqslant A_1$, whence $A = A_1 = B$ since $\mathscr{A}$ is an antichain. Thus $A \in \mathscr{B}$. Thus every member of $\mathscr{A}$ is in $\mathscr{B}$, and similarly every member of $\mathscr{B}$ is in $\mathscr{A}$, so that $\mathscr{A} = \mathscr{B}$. It follows that

$$(\mathscr{A} \leqslant \mathscr{B} \quad \text{and} \quad \mathscr{B} \leqslant \mathscr{A}) \Rightarrow \mathscr{A} = \mathscr{B}. \tag{13.2}$$

Next suppose that $A \leqslant \mathscr{B}$ and $\mathscr{B} \leqslant \mathscr{C}$ where $\mathscr{A}, \mathscr{B}, \mathscr{C} \in \mathfrak{A}(P)$. If $A \in \mathscr{A}$ then there exists $B \in \mathscr{B}$ such that $A \leqslant B$, and for this B there exists $C \in \mathscr{C}$ such that $B \leqslant C$. Therefore, given $A \in \mathscr{A}$, there exists $C \in \mathscr{C}$ such that $A \leqslant C$; thus $\mathscr{A} \leqslant \mathscr{C}$. We have therefore proved that

$$(\mathscr{A} \leqslant \mathscr{B} \quad \text{and} \quad \mathscr{B} \leqslant \mathscr{C}) \Rightarrow \mathscr{A} \leqslant \mathscr{C}. \tag{13.3}$$

The following is now immediate from (13.1), (13.2), and (13.3).

Theorem 13.1.1 $(\mathfrak{A}(P), \leqslant)$ is a poset. □

We now exhibit a relation between antichains and ideals. This is really just a reworking of the relation between antichains and increasing functions into $\{0, 1\}$ discussed in Section 3.4.1. For $\mathscr{A} \in \mathfrak{A}(P)$ define

$$\bar{\mathscr{A}} = \{X \in P : X \leqslant A \text{ for some } A \in \mathscr{A}\}.$$

Then $\bar{\mathscr{A}}$ is clearly an ideal. We now prove the following lemma.

Lemma 13.1.2 $\bar{\mathscr{A}} \subseteq \bar{\mathscr{B}} \Leftrightarrow \mathscr{A} \leqslant \mathscr{B}$.

Proof Suppose first that $\mathscr{A} \leqslant \mathscr{B}$. If $X \in \bar{\mathscr{A}}$ then there exists A such that $X \leqslant A$. Since $\mathscr{A} \leqslant \mathscr{B}$ it then follows that $A \leqslant B$ for some $B \in \mathscr{B}$. We then have $X \leqslant B$, so that $X \in \bar{\mathscr{B}}$. Therefore

$$\mathscr{A} \leqslant \mathscr{B} \Rightarrow \bar{\mathscr{A}} \subseteq \bar{\mathscr{B}}. \tag{13.4}$$

Conversely, suppose that $\bar{\mathscr{A}} \subseteq \bar{\mathscr{B}}$. Then, if $A \in \mathscr{A}$, it follows that $A \in \bar{\mathscr{A}}$ and so $A \in \bar{\mathscr{B}}$, so that there exists $B \in \bar{\mathscr{B}}$ such that $A \leqslant B$. Thus $\mathscr{A} \leqslant \mathscr{B}$. We have therefore proved that

$$\bar{\mathscr{A}} \subseteq \bar{\mathscr{B}} \Rightarrow \mathscr{A} \leqslant \mathscr{B} \tag{13.5}$$

and the lemma follows from (13.4) and (13.5). □

Corollary 13.1.3 $\bar{\mathscr{A}} = \bar{\mathscr{B}} \Leftrightarrow \mathscr{A} = \mathscr{B}$.

Proof $\bar{\mathscr{A}} = \bar{\mathscr{B}} \Leftrightarrow \bar{\mathscr{A}} \subseteq \bar{\mathscr{B}}$ and $\bar{\mathscr{B}} \subseteq \bar{\mathscr{A}} \Leftrightarrow \mathscr{A} \leqslant \mathscr{B}$ and $\mathscr{B} \leqslant \mathscr{A} \Leftrightarrow \mathscr{A} = \mathscr{B}$. □

This connection between antichains and ideals sheds light on Theorem 13.1.1, for the partial order $\leqslant$ on $\mathfrak{A}(P)$ reflects the partial order of inclusion on the set of ideals $\mathscr{A}$. We have

(i) $\mathscr{A} \leqslant \mathscr{A}$ since $\bar{\mathscr{A}} \subseteq \bar{\mathscr{A}}$

(ii) $\mathscr{A} \leqslant \mathscr{B}$ and $\mathscr{B} \leqslant \mathscr{A} \Rightarrow \bar{\mathscr{A}} \subseteq \bar{\mathscr{B}}$ and $\bar{\mathscr{B}} \subseteq \bar{\mathscr{A}} \Rightarrow \bar{\mathscr{A}} = \bar{\mathscr{B}} \Rightarrow \mathscr{A} = \mathscr{B}$

(iii) $\mathscr{A} \leqslant \mathscr{B}$ and $\mathscr{B} \leqslant \mathscr{C} \Rightarrow \bar{\mathscr{A}} \subseteq \bar{\mathscr{B}}$ and $\bar{\mathscr{B}} \subseteq \bar{\mathscr{C}} \Rightarrow \bar{\mathscr{A}} \subseteq \bar{\mathscr{C}} \Rightarrow \mathscr{A} \leqslant \mathscr{C}$.

We also note that the correspondence between antichains and ideals is in fact bijective, since every ideal in P is an $\bar{\mathscr{A}}$ for some antichain $\mathscr{A}$, namely the set of maximal elements of the ideal. We now use this correspondence to show that $(\mathfrak{A}(P), \leqslant)$ is more than a poset—it is a distributive lattice. In what follows we let $\max(\mathscr{F})$ denote the set of maximal elements of the ideal $\mathscr{F}$. The fact that $\max(\mathscr{F}) = \mathscr{A} \Leftrightarrow \mathscr{F} = \bar{\mathscr{A}}$ is proved as Exercise 13.1.

Lemma 13.1.4 The set of ideals of a finite poset P, under inclusion, is a distributive lattice.

Proof Let $\mathcal{M}$ and $\mathcal{N}$ be ideals of P. We first show that $\mathcal{M} \cap \mathcal{N}$ is an ideal. Let $Y \in \mathcal{M} \cap \mathcal{N}$ and $X \leqslant Y$. Since $Y \in \mathcal{M}$, $X \in \mathcal{M}$ also; similarly $X \in \mathcal{N}$. Thus $X \in \mathcal{M} \cap \mathcal{N}$, and we have shown that $\mathcal{M} \cap \mathcal{N}$ is an ideal. Note that $\mathcal{M} \cap \mathcal{N} \subseteq \mathcal{M}$ and $\mathcal{M} \cap \mathcal{N} \subseteq \mathcal{N}$, so certainly $\mathcal{M} \cap \mathcal{N}$ is a lower bound for $\mathcal{M}$ and $\mathcal{N}$. If $\mathcal{R}$ is any ideal such that $\mathcal{R} \subseteq \mathcal{M}$ and $\mathcal{R} \subseteq \mathcal{N}$ then $\mathcal{R} \subseteq \mathcal{M} \cap \mathcal{N}$; so $\mathcal{M} \cap \mathcal{N}$ is in fact the greatest lower bound $\mathcal{M} \wedge \mathcal{N}$. Similarly we can show that $\mathcal{M} \cup \mathcal{N}$ is an ideal which is the least upper bound $\mathcal{M} \vee \mathcal{N}$. Thus the set of ideals is a lattice.

Finally, to show that the distributive law holds we have to show that $\mathcal{M} \vee (\mathcal{N} \wedge \mathcal{K}) = (\mathcal{M} \vee \mathcal{N}) \wedge (\mathcal{M} \vee \mathcal{K})$, i.e. $\mathcal{M} \cup (\mathcal{N} \cap \mathcal{K}) = (\mathcal{M} \cup \mathcal{N}) \cap (\mathcal{M} \cup \mathcal{K})$. But this is true by the distributive law of set theory. □

It now follows from the one–one correspondence between antichains and ideals, with the corresponding connection between their partial ordering relations given by Lemma 13.1.2, that $(\mathfrak{A}(P), \leqslant)$ is also a distributive lattice. Since the greatest lower bound of $\bar{\mathcal{A}}$ and $\bar{\mathcal{B}}$ is $\bar{\mathcal{A}} \cap \bar{\mathcal{B}}$, the greatest lowest bound of $\mathcal{A}$ and $\mathcal{B}$ in $\mathfrak{A}(P)$ must be the antichain corresponding to $\bar{\mathcal{A}} \wedge \bar{\mathcal{B}}$, i.e. $\mathcal{A} \cap \mathcal{B} = \max(\bar{\mathcal{A}} \cap \bar{\mathcal{B}})$. A similar situation holds for least upper bounds, and we have the following theorem.

Theorem 13.1.5 (Dilworth 1960) $(\mathfrak{A}(P), \leqslant)$ is a distributive lattice. If $\mathcal{A}, \mathcal{B} \in \mathfrak{A}(P)$, $\mathcal{A} \wedge \mathcal{B} = \max(\bar{\mathcal{A}} \cap \bar{\mathcal{B}})$ and $\mathcal{A} \vee \mathcal{B} = \max(\bar{\mathcal{A}} \cup \bar{\mathcal{B}})$. □

We note that $\max(\bar{\mathcal{A}} \cup \bar{\mathcal{B}}) = \max(\mathcal{A} \cup \mathcal{B})$ (see Exercise 13.2). Thus $\mathcal{A} \vee \mathcal{B} = \max(\mathcal{A} \cup \mathcal{B})$, a nice result. In contrast, however, $\mathcal{A} \wedge \mathcal{B}$ is *not* in general the set of minimal elements of $\mathcal{A} \cup \mathcal{B}$.

13.2 Maximum-sized antichains

Having shown that the antichains of a poset P form a lattice, we now turn our attention to the maximum-sized antichains of P and show that they form a sublattice. We shall denote by $d(P)$ the size of the largest antichain(s) in P: by Dilworth's theorem $d(P)$ is also the minimum number of chains into which P can be decomposed.

We begin by constructing further antichains from any two given antichains, not necessarily of maximum size. If $\mathcal{A}$ and $\mathcal{B}$ are antichains, then $\mathscr{A} \cup \mathscr{B}$ will not in general be an antichain. The members of $\mathscr{A} \cup \mathscr{B}$ fall into four categories:

(i) members of $\mathscr{A} \cap \mathscr{B}$;
(ii) $X \in \mathscr{A} \cup \mathscr{B}$: $X \subset Y$ for some $Y \in \mathscr{A} \cup \mathscr{B}$;
(iii) $X \in \mathscr{A} \cup \mathscr{B}$: $X \supset Y$ for some $Y \in \mathscr{A} \cup \mathscr{B}$;
(iv) $X \in \mathscr{A} \cup \mathscr{B}$: $X \not\subset Y$ and $Y \not\subset X$ for all $Y(\neq X) \in \mathscr{A} \cup \mathscr{B}$; $X \notin \mathscr{A} \cap \mathscr{B}$.

We note that $\mathscr{A} \vee \mathscr{B}$, the set of maximal elements of $\mathscr{A} \cup \mathscr{B}$, consists of those elements in categories (i), (iii), and (iv). Now define $\mathscr{A} \nabla \mathscr{B}$ by

$$\mathscr{A} \nabla \mathscr{B} = \text{set of minimal elements of } \mathscr{A} \cup \mathscr{B}, \tag{13.6}$$

and note that $\mathscr{A} \nabla \mathscr{B}$ consists of all the elements of $\mathscr{A} \cup \mathscr{B}$ except those in category (iii). Clearly $\mathscr{A} \nabla \mathscr{B}$ is an antichain. Another antichain is $\mathscr{A} \triangle \mathscr{B}$ defined by

$$\mathscr{A} \triangle \mathscr{B} = ((\mathscr{A} \cup \mathscr{B}) - (\mathscr{A} \vee \mathscr{B})) \cup (\mathscr{A} \cap \mathscr{B}). \tag{13.7}$$

This is the set of all non-maximal elements of $\mathscr{A} \cup \mathscr{B}$ together with the elements of $\mathscr{A} \cap \mathscr{B}$; it consists of those elements in categories (i) and (ii). It is clear that

$$\mathscr{A} \triangle \mathscr{B} \subseteq \mathscr{A} \nabla \mathscr{B}. \tag{13.8}$$

We can also show easily (see Exercise 13.3) that

$$\mathscr{A} \triangle \mathscr{B} \subseteq \mathscr{A} \wedge \mathscr{B}, \tag{13.9}$$

and it is clear by considering categories (i)–(iv) that

$$|\mathscr{A} \vee \mathscr{B}| + |\mathscr{A} \triangle \mathscr{B}| = |\mathscr{A} \cup \mathscr{B}| + |\mathscr{A} \cap \mathscr{B}|. \tag{13.10}$$

Lemma 13.2.1 If $\mathscr{A}$ and $\mathscr{B}$ are maximum-sized antichains, then so are all of $\mathscr{A} \vee \mathscr{B}$, $\mathscr{A} \wedge \mathscr{B}$, $\mathscr{A} \triangle \mathscr{B}$, and $\mathscr{A} \nabla \mathscr{B}$. Further,

$$\mathscr{A} \wedge \mathscr{B} = \mathscr{A} \triangle \mathscr{B} = \mathscr{A} \nabla \mathscr{B}.$$

Proof (Green and Kleitman 1976*a*). By (13.10),

$$|\mathscr{A} \vee \mathscr{B}| + |\mathscr{A} \triangle \mathscr{B}| = |\mathscr{A} \cup \mathscr{B}| + |\mathscr{A} \cap \mathscr{B}| = |\mathscr{A}| + |\mathscr{B}| = 2d(P).$$

It therefore follows from (13.8) and (13.9) that

$$|\mathscr{A} \vee \mathscr{B}| + |\mathscr{A} \nabla \mathscr{B}| \geq 2d(P)$$

and

$$|\mathscr{A} \vee \mathscr{B}| + |\mathscr{A} \wedge \mathscr{B}| \geq 2d(P).$$

But none of $|\mathscr{A} \vee \mathscr{B}|$, $|\mathscr{A} \nabla \mathscr{B}|$, and $|\mathscr{A} \wedge \mathscr{B}|$ can be larger than $d(P)$, and so they must all be equal to $d(P)$. It follows that $\mathscr{A} \vee \mathscr{B}$, $\mathscr{A} \nabla \mathscr{B}$, and $\mathcal{A} \wedge \mathcal{B}$ are all maximum-sized antichains, and (13.10) then shows that $\mathcal{A} \triangle \mathcal{B}$ is also of maximum size. Finally, (13.8) and (13.9) show that $\mathcal{A} \wedge \mathcal{B}$ and $\mathcal{A} \nabla \mathcal{B}$ are both equal to $\mathcal{A} \triangle \mathcal{B}$. □

Corollary 13.2.2 If $\mathcal{A}$ and $\mathcal{B}$ are maximum-sized antichains then $\mathcal{A} \wedge \mathcal{B}$ is the set of minimal elements of $\mathcal{A} \cup \mathcal{B}$.

Proof By Lemma 13.2.1, $\mathcal{A} \wedge \mathcal{B} = \mathcal{A} \nabla \mathcal{B}$. □

Collecting information together we can now state the following theorem.

Theorem 13.2.3 (Dilworth 1960) The antichains of maximum size $d(P)$ in $(\mathfrak{A}(P), \preccurlyeq)$ form a sublattice $(\mathcal{S}(P), \preccurlyeq)$. For $\mathcal{A}$, $\mathcal{B} \in \mathcal{S}(P)$, $\mathcal{A} \vee \mathcal{B}$ is the set of maximal elements in $\mathcal{A} \cup \mathcal{B}$ and $\mathcal{A} \wedge \mathcal{B}$ is the set of minimal elements of $\mathcal{A} \cup \mathcal{B}$. □

We note that the fact that $\mathcal{A} \triangle \mathcal{B} = \mathcal{A} \nabla \mathcal{B}$ for maximum-sized antichains $\mathcal{A}$, $\mathcal{B}$ reflects the fact that category (iv) is then empty; anything in (iv) could be added to one of the antichains and increase its size.

Since $(\mathcal{S}(P), \preccurlyeq)$ is a finite lattice it must have a 'greatest' member $\mathcal{A}^*$ such that $\mathcal{A} \preccurlyeq \mathcal{A}^*$ for all antichains of maximum size. We can use the existence of $\mathcal{A}^*$ to obtain yet another proof of Sperner's theorem (Freese 1974). $\mathcal{A}^*$ must be invariant under every automorphism of $(\mathcal{P}(S), \subseteq)$. If $A \in \mathcal{A}^*$, $|A| = k$, then A will be mapped to all the k-subsets of S by permutations of the elements of S, and so all k-subsets of S must belong to $\mathcal{A}^*$. Clearly $\mathcal{A}^*$ cannot then have any other members, so $|\mathcal{A}^*| = \binom{n}{k} \leqslant \binom{n}{[n/2]}$.

13.3 Saturated chain partitions

Recall that Dilworth's theorem asserts that if $d_1(P)$ is the size of the largest antichain in P then there exists a partition of P into $d_1(P)$ chains. It is clear that at least $d_1(P)$ chains are needed to cover the whole of P; Dilworth showed that no more than $d_1(P)$ are needed.

We generalize the concept of an antichain by defining a *k-union* of P to be a subset A of P such that no $k+1$ members of A form a chain. Thus a 1-union is just an antichain. If A has the maximum size of all k-unions in P we say that A is a *Sperner k-union*. Thus a Sperner 1-union is just a maximum-sized antichain. We denote the size of a Sperner k-union by $d_k(P)$.

Lemma 13.3.1 A set $\mathscr{A}$ of elements of a poset P is a k-union if and only if it is the union of k antichains.

Proofs Any union of k antichains is trivially a k-union. Conversely, suppose that $\mathscr{A}$ is a k-union. For $A \in \mathscr{A}$, let $\delta(A)$ denote the size of the largest chain in $\mathscr{A}$ whose bottom member is A, and let

$$\mathscr{A}_i = \{A \in \mathscr{A} : \delta(A) = i\}.$$

Then the sets $\mathscr{A}_1, \ldots, \mathscr{A}_k$ form a partition of $\mathscr{A}$ into k antichains (some of which may be empty). □

There are other ways of obtaining k antichains from a k-union, but the method used here is particularly natural in that it leads to an ordering on the set of k-unions, as we shall briefly explain later.

Note that if P is partitioned into chains then each chain can meet a given k-union in at most k elements (and fewer than k if the size of the chain is less than k). Accordingly, if $\mathscr{C} = \{C_1, \ldots, C_q\}$ is any partition of P into chains, $d_k(P) \leqslant \beta_k(\mathscr{C})$, where

$$\beta_k(\mathscr{C}) = \sum_{i=1}^{q} \min(k, |C_i|).$$

Any chain partition $\mathscr{C}$ for which $d_k(P) = \beta_k(\mathscr{C})$ is called a *k-saturated* chain partition. Dilworth's theorem therefore asserts that there exists a 1-saturated chain partition of P. Greene and Kleitman (1976*a*) proved that k-saturated partitions exist for all $k \geqslant 1$. In fact they proved the stronger result that, for each $k \geqslant 1$, there exists a chain partition which is simultaneously k-saturated and $(k+1)$-saturated. In general, there is no chain partition which is simultaneously k-saturated for all $k \geqslant 1$. The example given by Greene and Kleitman is the poset P with six elements $a, \ldots, f$ and the partial ordering $a \leqslant b \leqslant c \leqslant d$, $e \leqslant c$, $b \leqslant f$. For this poset, $d_1(P) = 2$, $d_2(P) = 4$, $d_3(P) = 5$, $d_4(P) = 6$. The only partition of P into 2 chains is $\{a, b, f\} \cup \{c, d, e\}$; this is 1-saturated and 2-saturated but not 3-saturated. The partition $\{a, b, c, d\} \cup \{e\} \cup \{f\}$ is 2-saturated and 3-saturated but not 1-saturated. However, there do exist posets which have a chain partition which is k-saturated for all k; the set of subsets of a set is one example.

We are going to give a proof of this result of Greene and Kleitman, using the elegant arguments due to Saks (1979) and Perfect (1984). First we make a formal statement of the theorem.

Theorem 13.3.2 (Greene and Kleitman 1976*a*) If P is a poset and $k \geq 1$ then there exists a chain partition of P which is simultaneously k-saturated and $(k+1)$-saturated.

Proof We shall consider product posets such as $P \times [\boldsymbol{k}]$ where $[\boldsymbol{k}] = \{1, 2, \ldots, k\}$ has the usual ordering $<$ and where $P \times [\boldsymbol{k}]$ consists of all pairs (x, i), $x \in P$, $i \in [\boldsymbol{k}]$, witn $(x, i) \leq (y, j) \Leftrightarrow x \leq y$ in P and $i \leq j$. We shall call any element of the form (x, i) an element *at level i*.

Now there is an immediate connection between antichains (1-unions) in $P \times [\boldsymbol{k}]$ and k-unions in P. If $\mathscr{A}$ is any antichain in $P \times [\boldsymbol{k}]$ we can consider $\mathscr{A}_1 \cup \mathscr{A}_2 \cup \ldots \cup \mathscr{A}_k$ where $\mathscr{A}_i = \{x \in P : (x, i) \in \mathscr{A}\}$. Each $\mathscr{A}_i$ is an antichain in P and, further, the $\mathscr{A}_i$ are all disjoint. Thus corresponding to each antichain of $P \times [\boldsymbol{k}]$ we obtain a k-union *of the same size* in P. It follows that

$$d_1(P \times [\boldsymbol{k}]) \leq d_k(P). \qquad (13.11)$$

The key to the proof is to consider certain chain partitions of $P \times [\boldsymbol{k+1}]$ which we shall call *special.* For any chain partition $\mathscr{C}$, call an element x *maximal in* $\mathscr{C}$ if x is the 'top' member of the chain in $\mathscr{C}$ in which it lies, and let $M(\mathscr{C})$ denote the set of all elements which are maximal in $\mathscr{C}$. Note that $|M(\mathscr{C})|$ is therefore the number of chains in $\mathscr{C}$. If x and y are in the same chain in $\mathscr{C}$, if $y < x$, and if no element in the chain lies between x and y, we say that x *covers* y in $\mathscr{C}$. Also, if T is a subset of $P \times [\boldsymbol{k+1}]$, the *projection* of T on P is the set of all $x \in P$ such that $(x, i) \in T$ for some $i \leq k+1$. If $\mathscr{C}$ is a chain partition of P, denote by $M_i(\mathscr{C})$ the projection on P of the maximal elements of $\mathscr{C}$ at level i. Having made all these definitions we now say that a chain partition $\mathscr{C}$ of $P \times [\boldsymbol{k+1}]$ is special if

(i) $M_1(\mathscr{C}) \supseteq M_2(\mathscr{C}) \supseteq \ldots \supseteq M_k(\mathscr{C})$

(ii) if $a \in M_{k+1}(\mathscr{C})$ and $a \notin M_k(\mathscr{C})$, then $(a, k+1)$ covers (a, k) in $\mathscr{C}$.

Thus, in a special partition $\mathscr{C}$, if (p, i) is the top element of a chain $(i \leq k)$, so are $(p, i-1), \ldots, (p, 1)$, and if $(p, k+1)$ is at the top of a chain but (p, k) is not, then $(p, k+1)$ covers (p, k) in its chain. Note also that (ii) states that

$$M_{k+1}(\mathscr{C}) = M^*_{k+1}(\mathscr{C}) \cup M'_{k+1}(\mathscr{C})$$

where $M^*_{k+1}(\mathscr{C}) = M_{k+1}(\mathscr{C}) \cap M_k(\mathscr{C})$ and where $M'_{k+1}(\mathscr{C})$ consists of elements p such that $(p, k+1)$ covers (p, k) in $\mathscr{C}$.

The proof will proceed as follows. We shall show (Theorem 13.3.3) that there exists a minimum chain partition (i.e. a chain partition containing $d_1(P \times [\mathbf{k+1}])$ chains) of $P \times [\mathbf{k+1}]$ which is special and which has $m_k = d_1(P \times [\mathbf{k+1}]) - d_1(P \times [\mathbf{k}])$ chains consisting entirely of elements at level 1. Suppose that $\mathscr{C} = \{C_1, \ldots, C_{m_k}, \ldots, C_q\}$ is such a partition, where $q = d_1(P \times [\mathbf{k+1}])$ and where $C_1, \ldots, C_{m_k}$ are the chains consisting of elements at level 1. Let us now define $\mathscr{C}' = \{C'_1, \ldots, C'_q\}$ where $C'_i = \{(x, j) : (x, j+1) \in C_i\}$. Then $C'_1, \ldots, C'_{m_k}$ are all empty. Let $\mathscr{C}''$ denote the set of non-empty members of $\mathscr{C}'$; then $\mathscr{C}''$ is a chain partition of $P \times [\mathbf{k}]$ which, by the definition of m_k, is minimum and which is also special because $\mathscr{C}$ was special.

Returning to $\mathscr{C}$, we can associate with $\mathscr{C}$ a chain decomposition $\mathscr{C}^*$ of P which we obtain from $\mathscr{C}$ by restricting the chains of $\mathscr{C}$ to level $k+1$ and taking projections onto P. The set of maximal elements of $\mathscr{C}^*$ is just $M_{k+1}(\mathscr{C})$, so that we obtain $|M_{k+1}(\mathscr{C})|$ chains in $\mathscr{C}^*$. Now

$$\begin{aligned}\beta_1(\mathscr{C}) = |M(\mathscr{C})| = \sum_{i=1}^{k+1} |M_i(\mathscr{C})| &\geq k\,|M_k(\mathscr{C})| + |M_{k+1}(\mathscr{C})| \\ &\geq k\,|M^*_{k+1}(\mathscr{C})| + |M^*_{k+1}(\mathscr{C})| + |M'_{k+1}(\mathscr{C})| \\ &= (k+1)\,|M^*_{k+1}(\mathscr{C})| + |M'_{k+1}(\mathscr{C})|\end{aligned}$$

whereas

$$\beta_{k+1}(\mathscr{C}^*) = \sum_{C \in \mathscr{C}^*} \min(|C|, k+1) \leq |M'_{k+1}(\mathscr{C})| + (k+1)\,|M^*_{k+1}(\mathscr{C})|$$

since all the chains in $\mathscr{C}^*$ with maximal element in $M'_{k+1}(\mathscr{C})$ are, by (ii), of size 1. Combining these inequalities with (13.11), we obtain

$$\beta_{k+1}(\mathscr{C}^*) \leq \beta_1(\mathscr{C}) = d_1(P \times [\mathbf{k+1}]) \leq d_{k+1}(P).$$

However, since $\mathscr{C}^*$ is a chain decomposition of P, we trivially must have $d_{k+1}(P) \leq \beta_{k+1}(\mathscr{C}^*)$, so we must have $d_{k+1}(P) = \beta_{k+1}(\mathscr{C}^*)$. $\mathscr{C}^*$ is therefore $(k+1)$-saturated. If we now repeat the argument with $\mathscr{C}''$ in place of $\mathscr{C}$ we obtain the same $\mathscr{C}^*$ as before and conclude that $\mathscr{C}^*$ is also k-saturated. We therefore obtain a chain partition of P which is simultaneously k-saturated and $(k+1)$-saturated. The proof of Theorem 13.3.2 will therefore be complete as soon as we have established the following theorem.

Theorem 13.3.3 There exists a minimum chain partition of $P \times [\mathbf{k}+\mathbf{1}]$ which is special and which has $m_k = d_1(P \times [\mathbf{k}+\mathbf{1}]) - d_1(P \times [\mathbf{k}])$ chains consisting only of elements at level 1.

The first step in the proof is to omit the 'special' requirement.

Lemma 13.3.4 If $r > d_1(P \times [k])$ then all antichains of size r in the poset $P \times [\mathbf{k}+\mathbf{1}]$ contain at least $r - d_1(P \times [\mathbf{k}])$ elements at level 1.

Proof Let $\mathcal{A}$ be an antichain of size r in $P \times [\mathbf{k}+\mathbf{1}]$, and suppose that it contains m elements at level 1. If $\mathcal{A}'$ denotes the set of these elements at level 1, then $\mathcal{A} - \mathcal{A}'$ is an antichain in $R = (P \times [\mathbf{k}+\mathbf{1}]) - (P \times [1])$. But R is isomorphic to $P \times [\mathbf{k}]$, so that $|\mathcal{A} - \mathcal{A}'| \leqslant d_1(P \times [\mathbf{k}])$. Thus $r - m \leqslant d_1(P \times [\mathbf{k}])$, i.e. $m \geqslant r - d_1(P \times [\mathbf{k}])$ as required. □

Lemma 13.3.5 There exists a minimum chain partition of $P \times [\mathbf{k}+\mathbf{1}]$ which contains at least (and hence exactly) m_k chains consisting entirely of elements at level 1.

Proof Let $\mathscr{C}$ be a minimum chain partition of $P \times [\mathbf{k}+\mathbf{1}]$, and suppose that it contains s chains consisting entirely of elements at level 1. We shall show that if $s < m_k$ we can increase s by 1. Suppose that $\mathscr{C} = \{C_1, \ldots, C_s, C_{s+1}, \ldots, C_q\}$ where $q = d_1(P \times [\mathbf{k}+\mathbf{1}])$ and where $C_1, \ldots, C_s$ consist entirely of elements at level 1. Let $P' = (P \times [\mathbf{k}+\mathbf{1}]) - (C_1 \cup \ldots \cup C_s)$. P' is covered by $q - s$ chains. It cannot be covered by fewer, for otherwise its $t < q - s$ covering chains, together with $C_1, \ldots, C_s$, would form a chain covering of $P \times [\mathbf{k}+\mathbf{1}]$ of $s + t < q$ chains. We can therefore apply Dilworth's theorem to P' to deduce that any maximum-sized antichain in P' has size $q - s$. Now by Theorem 13.2.3 the maximum-sized antichains form a lattice, so there is a 'greatest' antichain $\mathcal{A}^+$ such that $\mathcal{A} \leqslant \mathcal{A}^+$ for all maximum-sized antichains $\mathcal{A}$. By Lemma 13.3.4, $\mathcal{A}^+$ contains at least $q - s - d_1(P \times [\mathbf{k}]) = m_k - s$ elements at level 1. If $s < m_k$, choose any such element a. Without loss of generality we can assume that $a \in C_{s+1}$. Put $C_{s+1} = C_{s+1}^- \cup C_{s+1}^+$ where C_{s+1}^- consists of all elements of C_{s+1} up to a, and put $P'' = P' - C_{s+1}^-$, noting that C_{s+1}^- consists entirely of elements at level 1.

We next observe that P'' cannot contain an antichain of size $q - s$. For if $\mathcal{A}$ were such an antichain then $\mathcal{A}$ would contain an element b of C_{s+1}^+ and, since $\mathcal{A} \leqslant \mathcal{A}^+$, there would have to be an element

$c \in \mathscr{A}^+$ such that $b \leqslant c$, but then $a < b \leqslant c$ where a, $c \in \mathscr{A}^+$, contradicting the fact that $\mathscr{A}^+$ is an antichain. Since P'' does not accordingly contain an antichain of size $q - s$, P'' can, by Dilworth's theorem, be covered by $q - s - 1$ chains, say $C''_{s+2}, \ldots, C''_q$.

Now replace $\mathscr{C}$ by $\mathscr{C}' = \{C_1, \ldots, C_s, C^-_{s+1}, C''_{s+2}, \ldots, C''_q\}$. This also is a minimum chain partition of $P \times [\mathbf{k+1}]$, and it contains $s + 1$ chains, $C_1, \ldots, C_s, C^-_{s+1}$, which consist entirely of elements at level 1. If $s + 1 < m_k$, repeat the above procedure; after $m_k - s$ steps a minimum chain partition with m_k chains of elements at level 1 will be obtained. □

Proof of Theorem 13.3.3 The chain partition whose existence is guaranteed by Lemma 13.3.5 possesses some of the properties required by the theorem, but it may not be special. We are therefore going to alter it by applying two types of transformation which will eventually change the chain partition into a special one. The two operations involved are those of *insertion* and *transfer*. If z covers x in a chain partition and w is in a different chain but satisfies $x < w < z$, then we can remove w from its chain and insert it between x and z. Again, if x is the maximal element of a chain C_1 of the chain partition and $x < y$ where y is in another chain C_2, then we can make y cover x by transferring the part of the chain C_2 from y upwards to the top of C_1. Note that neither of these operations increases the number of chains in the chain partition.

Let $\mathscr{C}$ be the minimum chain partition given by Lemma 13.3.5. First of all we perform a sequence of insertions so as to obtain a minimum chain decomposition $\mathscr{C}_1$ in which every element (a, j) which is not a top element of a chain in $\mathscr{C}_1$ is covered either by another element at level j or by $(a, j+1)$. To do this, we take the levels j in turn, in increasing order, looking at each (a, j) which is not at the top of a chain. If (a, j) is covered by (b, i) where $b > a$ and $i > j$ or by $(a, i+1)$ where $i > j$, then insert $(a, j+1), \ldots, (a, i)$ into the chain. This operation does not affect coverings already dealt with since it operates only at levels greater than j. Note also that it does not affect the m_k chains consisting only of elements at level 1.

Having obtained $\mathscr{C}_1$ we now apply to it a sequence of transfers which lead to a special chain partition. If (a, j) is maximal in $\mathscr{C}_1$ with $1 < j \leqslant k + 1$, but $(a, j-1)$ is not maximal, being covered by (b, i) for some $i \leqslant k$, then $(b, i+1) > (a, j)$ and we can transfer the part of the chain containing $(b, i+1)$ from $(b, i+1)$ upwards to the chain containing (a, j). Such a transfer does not affect the m_k chains containing only elements at level 1. We continue doing such transfers until no further transfers are possible. Assuming for the moment that this process will indeed terminate after a finite number of transfers have taken place, we shall finally obtain a chain partition $\mathscr{C}_2$ such that if (a, j) is maximal in $\mathscr{C}_2$ then either $(a, j-1)$ is maximal also or $(a, j-1)$ is covered by an element at level $k+1$. In the latter case we must, by the property of $\mathscr{C}_1$, have $j-1=k$, i.e. $j=k+1$, and (a, k) must be covered by $(a, k+1)$. Thus $\mathscr{C}_2$ is special.

Why does this transfer process eventually terminate? To see why, define, for a given chain partition $\mathscr{C}$ of $P\times[\mathbf{k+1}]$, numbers $v_1, \ldots, v_k$ as follows: let v_i denote the number of elements a of P such that (a, i) is covered by some (b, j) and $(a, i+1)$ is covered by $(b, j+1)$. Each time a transfer of the type described is used to cover an element at level i, v_{i-1} increases by 1 and v_j is unchanged for each $j\leqslant i-1$ (although v_i may increase). Thus every transfer results in a chain partition for which the vector $(v_1, \ldots, v_k)$ is lexicographically greater than before. Since $v_i\leqslant|P|$ for each i, such increases in the lexicographic order can occur only a finite number of times. The proof of Theorem 13.3.3, and hence that of Theorem 13.3.2, is now complete. □

As mentioned earlier, the proof of Theorem 13.3.2 presented here is the combined work of Saks and Perfect. Saks used these ideas in 1979 to prove that k-saturated chain partitions exist for every k, and then Perfect exploited his method to prove the stronger result of Theorem 13.3.2 in her 1984 paper.

We now give an application of the existence of a k-saturated chain partition by presenting Griggs' original proof of Theorem 12.3.2. Recall that if the rank numbers N_i of a ranked poset P are, in decreasing order of magnitude, $M_0, M_1, \ldots, M_n$, then P is said to have property T_k if there exists a set of M_k disjoint chains in P, each of which intersects each of the $k+1$ largest ranks. Theorem 12.3.2 asserted that if, for each k, the size of the largest k-union in P is the sum of the k largest rank numbers, then P possesses property T_k for each k.

Proof of Theorem 12.3.2 (Griggs 1980*a*) As in the first proof, let P' be the ranked poset consisting of the $k+1$ largest ranks in P, ordered as in P. Then P' also has the property that the size of its largest k-union is $M_0+\ldots+M_{k-1}$. Now by Theorem 13.3.2 there exists a k-saturated chain partition $\mathscr{C}=\{C_1, C_2, \ldots\}$ of P', for which

$$\sum_i \min(|C_i|, k)=d_k(P')=M_0+\ldots+M_{k-1}=|P'|-M_k.$$

We therefore have

$$|P'| - M_k = \sum_{|C_i| \leq k} |C_i| + ku$$

where u is the number of chains in $\mathscr{C}$ of size $k+1$. Since $|P'| = \Sigma_i |C_i|$ we therefore have $u(k+1) - M_k = ku$, whence $u = M_k$. Thus $\mathscr{C}$ has M_k chains of size $k+1$, and these chains must intersect each of the $k+1$ largest ranks of P. □

13.4 The lattice of k-unions

Greene and Kleitman (1976*a*) have extended the ideas of Section 13.1 to the set of k-unions. It turns out that an ordering can be defined on the set of k-unions in such a way that the resulting poset is in fact a lattice. The definition of the ordering depends upon the ideas in the proof of Lemma 13.3.1. There each k-union $\mathscr{A}$ is partitioned into k 1-unions (antichains) $\mathscr{A}_1, \ldots, \mathscr{A}_k$ where $\mathscr{A}_i$ is the set of all $A \in \mathscr{A}$ from which the largest chain starting at A has size i. These antichains are in fact totally ordered by the ordering of Theorem 13.1.1: we have

$$\mathscr{A}_k \leqslant \mathscr{A}_{k-1} \leqslant \ldots \leqslant \mathscr{A}_1.$$

Thus to each k-union $\mathscr{A}$ we can associate a k-tuple $(\mathscr{A}_1, \ldots, \mathscr{A}_k)$ of antichains satisfying

(i) $\mathscr{A}_i \cap \mathscr{A}_j = \varnothing \quad (i \neq j)$

(ii) $\mathscr{A}_k \leqslant \ldots \leqslant \mathscr{A}_1$.

If now $\mathscr{A}$ and $\mathscr{B}$ are k-unions, we write $\mathscr{A} \leqslant \mathscr{B}$ if and only if $\mathscr{A}_i \leqslant \mathscr{B}_i$ for each $i = 1, \ldots, k$. It can be shown that $\mathfrak{A}_k$, the set of k-unions, is a lattice in which

$$\mathscr{A} \vee \mathscr{B} = \bigcup_i (\mathscr{A}_i \vee \mathscr{B}_i).$$

The formula for $\mathscr{A} \wedge \mathscr{B}$ turns out to be much more complicated. The lattice is not in general distributive but, surprisingly, the set of all Sperner k-unions forms a sublattice which *is* distributive.

Exercises 13

13.1 Show that if $\mathscr{A}$ is an antichain and $\mathscr{F}$ is an ideal in P then $\max(\mathscr{F}) = \mathscr{A} \Leftrightarrow \mathscr{F} = \bar{\mathscr{A}}$.

13.2 Prove that, if $\mathscr{A}$ and $\mathscr{B}$ are antichains then $\max(\bar{\mathscr{A}} \cup \bar{\mathscr{B}}) = \max(\mathscr{A} \cup \mathscr{B})$.

13.3 Show that, in the notation of Section 13.2, $\mathscr{A} \triangle \mathscr{B} \subseteq \mathscr{A} \wedge \mathscr{B}$.

13.4 Give an example to show that there need not be any inclusion relations between $\mathscr{A} \wedge \mathscr{B}$ and $\mathscr{A} \nabla \mathscr{B}$.

13.5 For $\mathscr{A}, \mathscr{B} \in \mathscr{S}(P)$ let

$$\mathscr{A}_{\mathscr{B}} = \{A \in \mathscr{A} : A \geqslant B \quad \text{for some } B \in \mathscr{B}\},$$
$$\mathscr{A}^{\mathscr{B}} = \{A \in \mathscr{A} : A \leqslant B \quad \text{for some } B \in \mathscr{B}\}.$$

Show that

(i) $\mathscr{A} \leqslant \mathscr{B} \Leftrightarrow \mathscr{A}^{\mathscr{B}} = \mathscr{A} \Leftrightarrow \mathscr{B}_{\mathscr{A}} = \mathscr{B}$;

(ii) $\mathscr{A}_{\mathscr{B}} \cup \mathscr{A}^{\mathscr{B}} = \mathscr{A}$;

(iii) $\mathscr{A}_{\mathscr{B}} \cap \mathscr{B}_{\mathscr{A}} = \mathscr{A}^{\mathscr{B}} \cap \mathscr{B}^{\mathscr{A}} = \mathscr{A} \cap \mathscr{B}$;

(iv) $\mathscr{A}_{\mathscr{B}} \cap \mathscr{A}^{\mathscr{B}} = \mathscr{B}_{\mathscr{A}} \cap \mathscr{B}^{\mathscr{A}} = \mathscr{A} \cap \mathscr{B}$. (Freese 1974)

13.6 Let $\mathscr{C}$ be a chain partition of P, let α_k denote the number of chains of $\mathscr{C}$ of size $\geqslant k$, and let S_k denote the set of elements of P contained in chains of $\mathscr{C}$ of size less than k. Show that $\mathscr{C}$ is k-saturated if and only if $d_k(P) = |S_k| + k\alpha_k$.

13.7 Show that $d_k(P) = \min_{S \subseteq P} \{|S| + kd_1(P - S)\}$.

13.8 Show that the antichains of maximum size in P have non-empty intersection if and only if any two of them have non-empty intersection.

13.9 Let $\Delta_k(P) = d_k(P) - d_{k-1}(P)$. Show that if $\mathscr{C}$ is a k-saturated chain partition of P with α_k chains of size $\geqslant k$ then $\Delta_{k+1}(P) \leqslant \alpha_k \leqslant \Delta_k(P)$. Deduce that $\Delta_1 \geqslant \Delta_2 \geqslant \Delta_3 \ldots$.

13.10 Prove that if $\Delta_k(P) = \Delta_{k+1}(P)$ and $\mathscr{C}$ is k-saturated then $\mathscr{C}$ is $(k+1)$-saturated. (Green and Kleitman 1976*a*)

Hints and solutions

Exercises 1

1.1 Since $A \cup B = S \Leftrightarrow A' \cap B' = \varnothing$, apply Theorem 1.1.1 to the A_i'.

1.2 Yes: e.g. if n is odd take all subsets of size $\geqslant \frac{1}{2}(n+1)$.

1.3 Follow the proof of Sperner's theorem, noting that $p_i = 0$ whenever $i > h$.

1.4 (a) 6; (b) 20. (Include the empty antichain and the antichain $\{\varnothing\}$.)

1.5 For each k-subset Y there are $2^k - 1$ choices of X. Use the binomial theorem to evaluate $\Sigma_k \binom{n}{k}(2^k - 1)$. Alternatively, every $x \in S$ is in 0, 1 or 2 of the sets in a pair $X \subset Y$, and not every x belongs to 0 or 2 of them.

1.6 For any $B \in \mathscr{B}$ there is a subset B_1 of S which is comparable with B but with no other member of $\mathscr{B}$ (otherwise B would not be needed in $\mathscr{B}$). Let B^* be the larger of B and B_1 and let $\mathscr{B}^* = \{B^* : B \in \mathscr{B}\}$. Prove that $\mathscr{B}^*$ is an antichain. To eliminate the possibility $A^* \subseteq B^*$ consider four cases: (i) $A^* = A$, $B^* = B_1$ (here we would have B_1 and A comparable); (ii) $A^* = A$, $B^* = B$; (iii) $A^* = A_1$, $B^* = B$; (iv) $A^* = A_1$, $B^* = B_1$.

1.7 The sets of indices corresponding to sums in I must form an antichain. (This is followed up in Chapter 11.)

1.8 Apply Theorem 1.3.1 to $A_1, \ldots, A_m, B_1', \ldots, B_m'$.

1.9 The same as 1.8 with the A_i and B_i interchanged.

1.10 For each i there is a set T_i, $|T_i| = t$, $T_i \cap A_j = \varnothing \Leftrightarrow i = j$. Apply Theorem 1.3.1.

1.11 $X_i \not\subseteq X_j$ means that there exists k such that $k \in X_i$, $k \notin X_j$, i.e. $x_i \in A_k$, $x_j \notin A_k$.

1.12 (i) Every member of $\mathscr{A}$ intersects every member of $b(\mathscr{A})$ and so contains a member of $b(b(\mathscr{A}))$. However, if $A \in b(b(\mathscr{A}))$, then A intersects every set in $b(\mathscr{A})$, so A' contains no member of $b(\mathscr{A})$, so there must be some $A_1 \in \mathscr{A}$ such that $A' \cap A_1 = \varnothing$, i.e. $A_1 \subseteq A$. Thus every member of $b(b(\mathscr{A}))$ contains a member

of $\mathscr{A}$. Thus if $X \in \mathscr{A}$, then $X \supseteq Y$ for some $Y \in b(b(\mathscr{A}))$, and $Y \supseteq Z$ for some $Z \in \mathscr{A}$. Since $\mathscr{A}$ is an antichain, $X = Y = Z$; therefore every set in $\mathscr{A}$ is in $b(b(\mathscr{A}))$ and vice versa.
(ii) Take $S = \{1, \ldots, 4\}$ and let $\mathscr{A}$ consist of the two sets $\{1, 2\}$ and $\{2, 3\}$. Then $c(\mathscr{A}) = \{\{2, 4\}, \{1, 3, 4\}\}$ and $c(c(\mathscr{A})) = \{\{1, 2\}, \{2, 3\}, \{4\}\}$.

1.13 $b(\mathscr{A})$ consists of $\{a, b, c\}$, $\{c, d, e\}$, $\{b, d, f\}$, $\{a, e, f\}$.

1.14 The sets of vertices of minimal paths from x to y form a Menger antichain.

Exercises 2

2.1 Count the pairs (E, B) with $E \subset B$, $E \in \Delta\mathscr{B}$, $B \in \mathscr{B}$, in two different ways to obtain $k\,|\mathscr{B}| \leqslant (n - k + 1)\,|\Delta\mathscr{B}|$. Alternatively, apply the first part of the lemma to the collection $\mathscr{B}' = \{B' : B \in \mathscr{B}\}$.

2.2 Apply the normalized matching property.

2.3 If $\mathscr{A}$ contains $\alpha_i\binom{n}{i}$ sets of size i then, by repeated use of the normalized matching property, $\alpha_{n-i} \geqslant \alpha_i$ for all $i \leqslant \frac{1}{2}n$; therefore the average size of the sets of sizes i and $n - i$ is $\geqslant \frac{1}{2}n$.

2.4 By Corollary 2.1.2 any m sets of size k are contained in at least m sets of size $k + 1$; therefore apply Theorem 2.2.1.

2.5 Let $S_i = \{j : A_i \cap B_j \neq \varnothing\}$, and use Theorem 2.2.1 to show that the sets S_i possess an s.d.r.. Then remove the s.d.r. from the A_i and B_i to obtain a similar problem with m replaced by $m - 1$; hence repeat the process.

2.6 The set of all maximal chains is a regular covering.

2.7 1,3,4,3,1.

2.8 Immediate from Corollary 2.3.3.

2.9 (a) If there are sets in $\mathscr{A}$ of size $h < \frac{1}{2}(n - r)$, consider the smallest such h and replace the sets by an equal number of sets of size $h + r$; this is possible by normalized matching since $N_{h+r} \geqslant N_h$. In this way push all the sets into the middle r ranks.

2.10 For each $X \subseteq S$ take $\lambda(X) = u(X)\binom{n}{|X|}$ where $u(X)$ is the number of collections $\mathscr{A}_i$ containing X. Also take R as the union of the $\mathscr{A}_i$. The left-hand side in Theorem 2.3.2 is then $\Sigma_i\,|\mathscr{A}_i|$. On the right, note that, for any chain C, if C contains an X occurring in more than one $\mathscr{A}_i$ then $\lambda(X) \leq t\binom{n}{|X|}$ and X is the only member of $R \cap C$, whereas if each X in C occurs in at most one $\mathscr{A}_i$ then $\sum_{X \in R \cap C} \lambda(X) \leq \sum_{i=0}^{n} \binom{n}{i} = 2^n$.

2.12 (a) Induction step: $S(k+1, 2) = S(k, 1) + 2S(k, 2) = 1 + 2(2^{k-1} - 1) = 2^k - 1$.
(b) Induction step: $S(k+1, 3) = S(k, 2) + 3S(k, 3) = 2^{k-1} - 1 + \frac{3}{2}(3^{k-1} - 2^k + 1) = \frac{1}{2}(3^k - 2^{k+1} + 1)$.

2.13 If we require $1, \ldots, k$ to be in different blocks, there are then k choices for each of the other $n-k$ elements; thus there are k^{n-k} such partitions. Alternatively, use $S(n, k) \geq kS(n-1, k)$ from Theorem 2.4.1. For the second inequality use induction:

$$S(n+1, k) \leq \binom{n-1}{k-2}(k-1)^{n-k+1} + k\binom{n-1}{k-1}k^{n-k}$$
$$\leq \binom{n}{k-1}k^{n-k+1}.$$

2.15 Given a permutation of $1, \ldots, n$ take the first k_1 elements as k_1 blocks of size 1, the next k_2 pairs of elements as k_2 blocks of size 2, and so on. This procedure will give all partitions but with duplications. Firstly, there are $k_i!$ orderings of the blocks of size i all of which give the same partition; secondly in each block of size i there are $i!$ orderings of the elements, all of which give the same partition. Therefore we have to divide n by $k_i!$ and by $(i!)^{k_i}$ for each i.

2.17 Each surjection f determines a unique partition of X into k blocks, where the ith block consists of those $x \in X$ such that $f(x) = i$. Further, each partition arises from $k!$ surjections.

2.18 k^n = number of mappings from an n-set to a k-set Y

$$= \sum_{A \subseteq Y} |A|!\, S(n, |A|) \quad \text{(on considering the image } A \text{ of each } f)$$
$$= \sum_i i! \binom{k}{i} S(n, i) = \sum_{i=0}^{k} S(n, i)[k]_i.$$

For $n=3$, $1[k]_1 + 3[k]_2 + 1[k]_3 = k + 3k(k-1) + k(k-1)(k-2) = k^3$.

2.19 If, for each h, any h A_is contain more than h elements in their union, the first part of the proof of Theorem 2.2.1 shows that all the A_i have the required property. If $|A_{i_1} \cup \ldots \cup A_{i_h}| = h$, where h is minimal, then we claim that all of these A_{i_j} have the required property, for if, say, $a \in A_{i_1}$ is chosen to represent A_{i_1}, and t of the remaining A_{i_j} had less than t elements other than a in their union, then they originally had exactly t elements in their union, contradicting the minimality of h.

2.20 Use 2.19 and induction, noting that once an element has been chosen to represent one set, it has to be deleted from the other sets which now have $k-1$ effective elements.

Exercises 3

3.1 The size of a chain beginning with a set of size k is $n+1-2k$. For the second part, note that $n+i$ is always odd for a chain of subsets of an n-set of size i to exist. The given expression is 0 if $n+i$ is even. If $n+i$ is odd, put $i=n+1-2k$ to obtain

$$\binom{n}{k}-\binom{n}{k-1}=\binom{n}{n-k}-\binom{n}{n-k+1}$$

$$=\binom{n}{(n+i-1)/2}-\binom{n}{(n+1+i)/2}=\text{given expression.}$$

3.2 The contribution to the sum from each of the chains is a sum of the form $\pm(1-1+1-\ldots)$, and so is $1, -1$, or 0.

3.3 $\bigcup \mathscr{A}_i$ can contain at most r members of each chain. Using 3.1,

$$\left|\bigcup \mathscr{A}_i\right| \leqslant \sum_{j=1}^{n+1}\left\{\binom{n}{[(n+j)/2]}-\binom{n}{[(n+j+1)/2]}\right\}\min(r,j)$$

$$\leqslant \sum_{j=1}^{n+1}\binom{n}{[(n+j)/2]}\{\min(r,j)-\min(r,j-1)\}$$

$$=\sum_{j=1}^{r}\binom{n}{[(n+j)/2]}.$$

3.4 Take B_i as the h-subset occurring in the same chain of a given symmetric chain decomposition as A_i'. Then $B_i \subset A_i'$.

3.5 The number of antichains consisting of r subsets of size $[n/2]$ is $\binom{m}{r}$ where $m=\binom{n}{[n/2]}$. Summing over r gives $2^{\binom{n}{[n/2]}}$.

3.6 Either there is a chain of size $\geqslant n^{1/2}$ or every chain decomposition of $\{A_i\}$ needs more than $n^{1/2}$ chains, in which case there is an antichain of more than $n^{1/2}$ sets. But a chain and an antichain have the required property.

3.7 (a) For each $x \in P$ let $f(x) =$ size of longest chain ending at x. Then $f(x) = f(y) \Rightarrow x, y$ incomparable. Thus if the largest chain size is k, the antichains $\mathscr{A}_i = \{x \in P : f(x) = i\}$, $i = 1, \ldots, k$, form a partition of P. This proves that the maximum size of the chain $\geqslant$ minimum number of antichains partitioning P. The other way round is trivial.
(b) If there is no antichain of size $s+1$ then the minimum number of antichains into which P can be partitioned is $\geqslant (rs+1)/s > r$, so that the maximum chain size is greater than r.
(c) Given a sequence $a_1, \ldots, a_{rs+1}$, define $a_i \lessdot a_j$ if $i < j$ and $a_i < a_j$, and apply (b) to the poset consisting of $a_1, \ldots, a_{rs+1}$ under $\lessdot$.

3.8 Think how basic elements are paired with previous elements. There are $\binom{n}{[n/2]}$ sets of basic elements (one for each chain) and each yields a zigzag which does not go below the x-axis from $(0, 0)$ to (n, p), $p = n - 2$ (number of basic elements). A chain of length $n - 2k + 1$ has k basic elements, so that $p = n - 2k$, i.e. $k = \frac{1}{2}(n - p)$, and there are $\binom{n}{k} - \binom{n}{k-1}$ such chains.

3.9 To include the sets of size $[n/2]$ the sequence must be of length $\geqslant \binom{n}{[n/2]}$.

3.10 In a chain of size $n + 1 - 2k$ the largest set has size $n - k$. By 3.1, the sum of sizes of the largest sets is

$$\sum_{k=1}^{\lambda} (n-k)\left\{\binom{n}{k} - \binom{n}{k-1}\right\} + n$$

$$= \binom{n}{0} + \binom{n}{1} + \ldots + \binom{n}{\lambda - 1} + (n - \lambda)\binom{n}{\lambda}$$

where $\lambda = [\frac{1}{2}n]$. Divide by (n/λ) and use the facts that

$$\sum_{i=0}^{\lambda} \binom{n}{i} = 2^{n-1}\left\{1 + O\left(\frac{1}{n^{1/2}}\right)\right\}$$

and $(n/\lambda) \sim (2/\pi n)^{1/2} 2^n$. Thus the average size of each U_i and V_i in the proof of Theorem 3.4.2 is $\frac{1}{2}k + O(n^{1/2})$, and this gives a bound asymptotically half that of the previous one.

3.11 For $n = 2$, $\mathscr{P}(S) = \{\varnothing, \{1\}\} \cup \{\{2\}, \{1, 2\}\} = \{\varnothing, \{2\}\} \cup \{\{1\},$

$\{1, 2\}\}$. For $n = 3$, $\mathscr{P}(S) = \{\emptyset, \{1\}, \{1, 2\}\} \cup \{\{2\}, \{2, 3\}, \{1, 2, 3\}\} \cup \{\{3\}, \{1, 3\}\} = \{\{2\}, \{1, 2\}\} \cup \{\{1\}, \{1, 3\}, \{1, 2, 3\}\} \cup \{\emptyset, \{3\}, \{2, 3\}\}$.

3.12 Let $f(n) = \binom{n}{m+1} - \binom{n}{m}$. Then $d_{m+1} - d_m = f(n) - f(n-k)$.
If $1 \leq m+1 \leq \frac{1}{2}(n-k)$ then $f(n) > f(n-k)$ is easily verified.
If $[\frac{1}{2}(n-k)] < m+1 \leq [\frac{1}{2}n]$, $f(n) > 0$ and $f(n-k) < 0$.
If $[\frac{1}{2}n] \leq m \leq n$, use $d_m = \sum_{j=1}^{k} \binom{n-j}{m-1}$ to obtain $d_m - d_{m+1} = \sum_{j=1}^{k} \left(\binom{n-j}{m-1} - \binom{n-j}{m} \right)$ where each bracket is ≥ 0.

3.13 Consider the $k\,.\,(n-1)!$ permutations of $\{1, \ldots, n\}$ beginning with an element of T, $|T| = k$. If $|A \cap T| = j$, the number of such permutations starting with the set $A \in C(n, k)$ is

$$j(|A|-1)!\,(n-|A|)! = \frac{|A \cap T|\,(n-1)!}{\binom{n-1}{|A|-1}}.$$

Since $\mathscr{A}$ is an antichain, such permutations arising from different $A \in \mathscr{A}$ are all different. Therefore

$$\sum_{A \in \mathscr{A}} \frac{|A \cap T|\,(n-1)!}{\binom{n-1}{|A|-1}} \leq k(n-1)!$$

as required. For $C(n, k, h)$, a similar argument gives

$$\sum_{A \in \mathscr{A}} \frac{1}{\binom{n-h}{|A|-h}} \cdot \binom{|A \cap T|}{h} \leq \binom{k}{h}.$$

3.14 Use induction and $|L(m, n)| = \sum_{j=0}^{n} |L(m-1, j)|$.

3.15 (i) Suppose $A_1, A_2 \in g(\mathscr{A})$, $A_2 \subset A_1$. Then there exist $B_1, B_2 \in \mathscr{A}$ with $B_i \cap \{1, \ldots, n-1\} = A_i$. But then $B_2 \cap \{1, \ldots, n-1\} \subset A_1$, contradicting the definition of $g(\mathscr{A})$.
(ii) (a) If $A \cup \{n\} \in \mathscr{A}$, choose $D \in \mathscr{A}$ such that $D \subseteq A \cup \{n\}$ and $|D \cap \{1, \ldots, n-1\}|$ is minimal. Let $B = D \cap \{1, \ldots, n-1\}$. Then $B \subseteq A$ and $B \in g(\mathscr{A})$. Conversely, given $B \in g(\mathscr{A})$, choose $D \in \mathscr{A}$ such that $D \cap \{1, \ldots, n-1\} = B$.
(b) If $A \in \mathscr{A}$, use (a) to obtain $B \in g(\mathscr{A})$ with $B \subseteq A$. If $C \in g(\mathscr{A})$, $C \cap A = \emptyset$, choose $D \in \mathscr{A}$ with $D \cap \{1, \ldots, n-1\} = C$ to obtain $D \cap A = \emptyset$, which is a contradiction. Con-

versely, suppose $B \in g(\mathscr{A})$, $B \subseteq A$, and there is no $C \in g(\mathscr{A})$ with $C \cap A = \varnothing$. By (a), $A \cup \{n\} \in \mathscr{A}$. Suppose $A \notin \mathscr{A}$; then $\{1, \ldots, n\} - A \in \mathscr{A}$, and by (a) we can choose $C \in g(\mathscr{A})$ with $C \subseteq \{1, \ldots, n\} - A$. But then $C \cap A = \varnothing$.

Exercises 4

4.1 (a) $\binom{n}{k-1}\binom{n}{k+1} \leqslant \binom{n}{k}^2$

$$\Leftrightarrow k!\,k!\,(n-k)!\,(n-k)!$$
$$\leqslant (k-1)!\,(k+1)!\,(n-k-1)!\,(n-k+1)!$$
$$\Leftrightarrow k(n-k) \leqslant (k+1)(n-k+1),$$

which is true.

(b) Assume that the $S(n, k)$ are log concave, and consider $n+1$.

$$S(n+1, k+1)S(n+1, k-1)$$
$$= \{S(n-k) + (k+1)S(n, k+1)\}\{S(n, k-2)$$
$$+ (k-1)S(n, k-1)\}$$
$$= S(n, k-2)S(n, k) + (k^2-1)S(n, k-1)S(n, k+1)$$
$$+ (k+1)S(n, k-2)S(n, k+1) + (k-1)S(n, k-1)S(n, k)$$
$$\leqslant \{S(n, k-1)\}^2 + k^2\{S(n, k)\}^2 + 2kS(n, k-1)S(n, k)$$
$$= \{S(n, k-1) + kS(n, k)\}^2 = \{S(n+1, k)\}^2.$$

4.2 Rank numbers are 1, 3, 6, 9, 11, 12, 12, 12, 11, 9, 6, 3, 1. $L = 5$; $2k_1 - K + 1 = 3$.

4.3 If $\mathscr{A}$ is a collection of divisors of m such that no $k+1$ members of $\mathscr{A}$ form a chain, then $|\mathscr{A}|$ is at most the sum of the k largest rank numbers. The proof is the same; it uses normalized matching which still holds.

4.4 Identical result.

4.5 If $N_{i-1} = N_i$, Corollary 4.2.2 gives $|\Delta\mathscr{B}| \geqslant |\mathscr{B}|$, whence the result. If $N_{i-1} < N_i$ then $|\Delta\mathscr{B}|/N_{i-1} \geqslant |\mathscr{B}|/N_i$ so that

$$\frac{N_{i-1} - |\Delta\mathscr{B}|}{N_{i-1}} \leqslant \frac{N_i - |\mathscr{B}|}{N_i} \leqslant \frac{N_i - |\mathscr{B}|}{N_{i-1}}$$

whence $N_{i-1} - |\Delta\mathscr{B}| \leqslant N_i - |\mathscr{B}|$ as required.

4.6 (a) $N_{\lambda-r}\sum_{i=0}^{\lambda-r}(K-2i)N_i \leq N_{\lambda-r}\sum_{i=0}^{\lambda-r-1}(K-2i-2)N_i$

$$+(2r-1)N_{\lambda-r}^2+2N_{\lambda-r}\sum_{i=0}^{\lambda-r}N_i$$

where the first term on the right is $\leq N_{\lambda-r-1}\sum_{i=0}^{\lambda-r-1}(K-2i-2)N_{i+1}\leq N_{\lambda-r-1}\sum_{j=0}^{\lambda-r}(K-2j)N_j$. Therefore

$$N_{\lambda-r}\sum_{i=0}^{\lambda-r}(K-2i)N_i \leq N_{\lambda-r-1}\sum_{i=0}^{\lambda-r-1}(K-2i)N_i$$

$$+2N_{\lambda-r}\sum_{i=0}^{\lambda-r}N_i+(2r-1)N_{\lambda-r}^2+(2r+1)N_{\lambda-r-1}N_{\lambda-r}$$

whence the result.

(b) $N_\lambda\sum_{i=0}^{\lambda}(K-2i)N_i \leq 2\sum_{r=0}^{\lambda-1}N_{\lambda-r}\sum_{i=0}^{\lambda-r}N_i+2\sum_{r=0}^{\lambda-1}rN_{\lambda-r}^2$

$$+2\sum_{r=0}^{\lambda-1}rN_{\lambda-r-1}N_{\lambda-r}+KN_0$$

$$\leq 2\sum_{j=0}^{\lambda}N_j\left(\sum_{i=0}^{j}N_i\right)+2\sum_{j=0}^{\lambda-1}N_j\left(\sum_{i=j+1}^{\lambda}N_i\right)$$

$$+2\sum_{i=0}^{\lambda-1}N_i\left(\sum_{j=i}^{\lambda-1}N_j\right)$$

$$\leq 2\sum_{j=0}^{\lambda}N_j\left(\sum_{i=0}^{\lambda}N_i\right)+2\sum_{i=0}^{\lambda-1}N_i\left(\sum_{j=i}^{\lambda-1}N_j\right)$$

$$\leq 2\left(\sum_{i=0}^{\lambda}N_i\right)^2+2\left(\sum_{i=0}^{\lambda}N_i\right)^2<4\tau^2(m).$$

(c) Since (cf. Exercise 3.1) there are $\binom{K}{k}-\binom{K}{k-1}$ chains of size $K+1-2k$,

$$J(m)=\binom{K+1}{2}+\sum_{k=1}^{\lambda}\binom{K+1-2k}{2}(N_k-N_{k-1}).$$

Since $\binom{x+1}{2}-\binom{x-1}{2}=2x-1$, we obtain

$$J(m)=\sum_{k=0}^{\lambda-1}(2K-4k-1)N_k+\binom{K+1-2\lambda}{2}N_\lambda$$

$$\leq 2\sum_{k=0}^{\lambda}(K-2k)N_k;$$

therefore use (b).

(d) Consider the case $n = 2\lambda$.

$$J(m) = \sum_{k=0}^{\lambda-1} (2n-4k-1)\binom{n}{k} = (2n-1)\sum_{k=0}^{\lambda-1}\binom{n}{k} - 4\sum_{k=0}^{\lambda-1} k\binom{n}{k}$$

$$= (2n-1)\sum_{k=0}^{\lambda-1}\binom{n}{k} - 4n\sum_{k=1}^{\lambda-1}\binom{n-1}{k-1}$$

$$= (2n-1)\left\{2^{n-1} - \tfrac{1}{2}\binom{n}{\lambda}\right\} - 4n\left\{2^{n-2} - \binom{n-1}{\lambda-1}\right\}$$

$$= 4n\binom{n-1}{\lambda-1} - (n-\tfrac{1}{2})\binom{n}{\lambda} - 2^{n-1} \sim n\left(\frac{2}{\pi n}\right)^{1/2} 2^n$$

$$= \left(\frac{2}{\pi}\right)^{1/2} n^{1/2} 2^n.$$

Deal with n odd similarly.

4.7 (ii) Multiply out the product of the two expressions, which must be $\geqslant 0$.

(iv) The poset of divisors of $m = p_1^{k_1} \ldots p_n^{k_n}$ is the product of the n chains C_i where $C_i = \{1, p_i, p_i^2, \ldots, p_i^{k_i}\}$. Each C_i trivially has log concave rank numbers.

Exercises 5

5.1 Remove the smaller set from each complementary pair of sets, to be left with an intersecting family. By EKR the number of sets removed is at most $\binom{n-1}{[n/2]-1}$; apply Theorem 1.1.1 to the rest.

5.2 All $[n/2]$ – subsets of $\{1, \ldots, n\}$ containing 1.

5.3 If n is even, apply Sperner to the antichain $\mathscr{A} \cup \mathscr{A}'$. If n is odd, apply EKR to the sets of size less than $\frac{1}{2}n$ in $\mathscr{A} \cup \mathscr{A}'$.

5.4 If $h \leqslant \frac{1}{2}n$ then the result is immediate from 5.1.2. Now suppose $h > \frac{1}{2}n$. Suppose that $|A_i| < n-h$, $1 \leqslant i \leqslant m_0$, and $|A_i| \geqslant n-h$, $m_0 < i \leqslant m$. Then $m_0 \leqslant \sum_{i=g}^{n-h-1}\binom{n-1}{i-1}$ by 5.1.2. For sets of size between $n-h$ and h, no set and its complement can both be A_is, so $m - m_0 \leqslant \frac{1}{2}\sum_{i=n-h}^{h}\binom{n}{i}$. Use $\binom{n}{r} = \binom{n-1}{r-1} + \binom{n-1}{r}$ to deduce $m - m_0 \leqslant \sum_{i=n-h}^{h}\binom{n-1}{i-1}$.

5.5 Let k be such that $f(k)\binom{n-1}{k-1}$ takes its maximum value. By Theorem 5.2.2,

$$1 \geqslant \sum_{i=1}^{m} \frac{1}{\binom{n-1}{|A_i|-1}} = \sum_{i=1}^{m} \frac{f(|A_i|)}{f(|A_i|)\binom{n-1}{|A_i|-1}} \geqslant \frac{1}{f(k)\binom{n-1}{k-1}} \sum_{i=1}^{m} f(|A_i|).$$

5.6 Let p be the size of the smallest set A_i. Replace all sets A_i of size p by those sets of size $p+1$ containing them. The resulting collection of sets is also an intersecting antichain, and, by Lemma 5.2.5, $\Sigma_i 1\Big/\binom{n-1}{|A_i|-1}$ has not decreased. Continue until only sets of size $[\frac{1}{2}n]$ are left; the sum is then $\leqslant 1$ by EKR.

5.7 The first part is clear from the conditions for 2-independence in the text. To prove 5.2.4, replace each set containing a chosen element of S by its complement to obtain an intersecting antichain $\mathscr{B}$ of subsets of an $(n-1)$-set. By Milner's theorem,

$$|\mathscr{A}| = |\mathscr{B}| \leqslant \binom{n-1}{[(n+1)/2]} = \binom{n-1}{[(n-2)/2]}.$$

5.8 (i) $A_i \cup A_j \neq C$ is equivalent to $(C-A_i) \cap (C-A_j) \neq \varnothing$, so by EKR we have

$$m \leqslant \binom{k-1}{k-h-1} = \frac{k-h}{k}\binom{k}{h}.$$

(ii) Let $\mathscr{B}$ be the set of all h-subsets of C in $\mathscr{A}$. By (i), $\mathscr{B}$ contains at most $\dfrac{k-h}{k}\dbinom{k}{h}$ sets. Hence the proportion of saturated chains through C from rank h to rank $2h$ and containing no member of $\mathscr{B}$ is at least h/k. Hence the proportion of saturated chains through C containing no smaller member of $\mathscr{A}$ is $\geqslant h/k$; hence their number is

$$\geqslant \frac{h}{k}\frac{k!}{h!}\frac{(n-k)!}{(n-2h)!}$$

Summing over all such C we obtain

$$\frac{n!}{h!\,(n-2h)!} \geqslant \sum_{k=h}^{2h} p_k \frac{h}{k}\frac{k!}{h!}\frac{(n-k)!}{(n-2h)!}.$$

5.9 Take $\mathscr{A} = \mathscr{B} \cup \mathscr{C}$ where $\mathscr{B}$ is the set of all $(q-1)$-subsets which contain all of $1, \ldots, k$, and where $\mathscr{C}$ is the set of all q-subsets which do not contain all of $1, \ldots, k$.

Exercises 6

6.1 (a) Take $\mathscr{U} = \mathscr{P}(S) - \mathscr{A}$, $\mathscr{D} = \mathscr{B}$ to obtain $(2^n - |\mathscr{A}|)\,|\mathscr{B}| \geq 2^n(|\mathscr{B}| - |\mathscr{B} \cap \mathscr{A}|)$.
(b) Apply (a) to the downsets $\mathscr{A}'$, $\mathscr{B}'$ where $\mathscr{A}' = \{X : X' \in \mathscr{A}\}$.

6.2 As in Exercise 2.3. Alternatively, use Theorem 6.4.1; the average size in each pair must be $\leq \frac{1}{2}n$.

6.3 Take $S = \{1, \ldots, n\}$ and define $\mathscr{A}_1$ to consist of the one set $\{1\}$, $\mathscr{A}_2$ to consist of all 2-sets containing 1 together with $\{2, \ldots, n\}$, $\mathscr{A}_3$ to consist of all 3-sets containing 1 together with all $(n-2)$-subsets of $\{2, \ldots, n\}, \ldots, \mathscr{A}_k$ to consist of all sets of size $\geq k$ containing 1 and all $(n-k+1)$-subsets of $\{2, \ldots, n\}$.

6.4 $\mathscr{H} \cup \mathscr{I}$ is an upset and $\mathscr{J} \cup \mathscr{K}$ is a downset, so $(|\mathscr{H}| + |\mathscr{I}|)(|\mathscr{K}| + |\mathscr{J}|) \geq 2^n\,|\mathscr{I}|$. But $|\mathscr{H}| + |\mathscr{I}| + |\mathscr{J}| + |\mathscr{K}| = 2^n$, so that $|\mathscr{I}|\,|\mathscr{J}| \leq |\mathscr{H}|\,.\,|\mathscr{K}|$ on substituting for 2^n. Thus $|\mathscr{I}|\,|\mathscr{J}| \leq |\mathscr{H}|\,|\mathscr{K}| \leq (\frac{1}{2}(|\mathscr{H}| + |\mathscr{K}|))^2 = (\frac{1}{2}(2^n - |\mathscr{I}| - |\mathscr{J}|))^2$, whence $|\mathscr{I}|^{1/2} + |\mathscr{J}|^{1/2} \leq 2^{n/2}$. But $\mathscr{P} \subseteq \mathscr{I}$, $\mathscr{Q} \subseteq \mathscr{J}$.

6.5 Take $\mathscr{P} = \mathscr{A}$, $\mathscr{Q} = \mathscr{A}'$ to obtain $|\mathscr{A}|^{1/2} + |\mathscr{A}|^{1/2} \leq 2^{n/2}$.

6.6 (i) Let $X \in (\text{mid}\,\mathscr{A})^+$. Then $X \supseteq Y$ where $Y \in \text{mid}\,\mathscr{A}$. $Y \in \mathscr{A}^+$, so that $Y \supseteq A$ for some $A \in \mathscr{A}$. So $X \supseteq A$ and $X \in \mathscr{A}^+$. The other way round is trivial.
(ii) As in (i), $\mathscr{A}^- = (\text{mid}(\mathscr{A}))^-$. But $\mathscr{C}(\mathscr{A}) = \mathscr{A}^+ \cup \mathscr{A}^-$.
(iii) Apply Kleitman's lemma to $\mathscr{U} = \mathscr{A}^+$ and $\mathscr{D} = \mathscr{A}^-$, and use $\frac{1}{2}(|\mathscr{A}^+| + |\mathscr{A}^-|) \geq (|\mathscr{A}^+|\,|\mathscr{A}^-|)^{1/2}$.
(iv) $|\mathscr{A}^+ \cup \mathscr{A}^-| = |\mathscr{A}^+| + |\mathscr{A}^-| - |\mathscr{A}^+ \cap \mathscr{A}^-|$, so use (iii).
(v) This follows from (iv) since $|\text{mid}\,\mathscr{A}| \geq |\mathscr{A}|$, and $f(x) = 2.2^{n/2}x^{1/2} - x$ is an increasing function of x for $0 < x < 2^n$.

6.7 (i) $X \in \mathscr{A} \vee \mathscr{B} \Rightarrow X = A \cup B$ for some $A \in \mathscr{A}$, $B \in \mathscr{B} \Rightarrow X = A \cup B$ where $A \cup B \in \mathscr{A} \cap \mathscr{B} \Rightarrow X \in \mathscr{A} \cap \mathscr{B}$. Conversely, $X \in \mathscr{A} \cap \mathscr{B} \Rightarrow X \in \mathscr{A}$ and $X \in \mathscr{B} \Rightarrow X = X \cup X \in \mathscr{A} \vee \mathscr{B}$.
(ii) $X \in \mathscr{A} \wedge \mathscr{B} \Rightarrow X = A \cap B$ for some $A \in \mathscr{A}$, $B \in \mathscr{B} \Rightarrow X = A \cap B$ where $A \cap B \in \mathscr{A} \cap \mathscr{B} \Rightarrow X \in \mathscr{A} \cap \mathscr{B}$. Conversely, $X \in \mathscr{A} \cap \mathscr{B} \Rightarrow X \in \mathscr{A}$ and $X \in \mathscr{B} \Rightarrow X = X \cap X \in \mathscr{A} \wedge \mathscr{B}$.

6.8 Put $\mathscr{A} = \mathscr{U} \cap \mathscr{D}$, $\mathscr{B} = \mathscr{P}(S)$, and note that $\mathscr{A} \vee \mathscr{B} \subseteq \mathscr{U}$, $\mathscr{A} \wedge \mathscr{B} \subseteq \mathscr{D}$.

6.9 Obtain

$$\sum_i f(i)g(i)\mu(i) \sum_j \mu(j) + \sum_i \mu(i) \sum_j \mu(j)f(j)g(j)$$
$$\geqslant \sum_i f(i)\mu(i) \sum_j g(j)\mu(j) + \sum_i g(i)\mu(i) \sum_j f(j)\mu(j).$$

6.10 $\alpha(A)\beta(B) = \mu_1(A)\mu_2(B)h(B) \leqslant \mu_1(A \cup B)\mu_2(A \cap B)h(A \cup B) = \varphi(A \cup B)\delta(A \cap B)$, so by Theorem 6.2.1,

$$\sum_{A \subseteq S} \mu_1(A) \sum_{A \subseteq S} \mu_2(A)h(A) \leqslant \sum_{A \subseteq S} \mu_1(A)h(A) \sum_{A \subseteq S} \mu_2(A).$$

6.11 Given μ, f, g with f, g increasing, take μ_1, μ_2 as in hint and $h = f$. Then

$$\mu_1(A)\mu_2(B) = \frac{g(A)\mu(A)\mu(B)}{\Sigma_A\, g(A)\mu(A)} \leqslant \frac{g(A \cup B)\mu(A \cup B)\mu(A \cap B)}{\Sigma_A\, g(A)\mu(A)}$$
$$= \mu_1(A \cup B)\mu_2(A \cap B)$$

so that, by 6.10,

$$\sum_A \mu(A)f(A) \leqslant \sum_A \frac{g(A)\mu(A)f(A)}{\Sigma_A\, g(A)\mu(A)}$$

and FKG follows in the case $\Sigma\, \mu(A) = 1$. The general result follows by scaling.

(i) $X \in (\sigma\mathscr{A})^+ \Leftrightarrow X \supseteq B$ for some $B \in \sigma\mathscr{A} \Leftrightarrow \sigma X \subseteq \sigma B$ for some $\sigma B \in \mathscr{A}$. So $|(\sigma\mathscr{A})^+| = |\mathscr{A}^-|$.

(ii) $2^n\, |\mathscr{A}| \leqslant 2^n\, |\mathscr{A}^+ \cap \mathscr{A}^-| \leqslant |\mathscr{A}^+| \,.\, |\mathscr{A}^-| = |\mathscr{A}^+| \,.\, |(\sigma\mathscr{A})^+|$ by Kleitman. But $|\mathscr{A}^+| \,.\, |(\sigma\mathscr{A})^+| \leqslant 2^n\, |\mathscr{A}^+ \cap (\sigma\mathscr{A})^+|$ by Exercise 6.1(b). Similarly for the last part.

6.13 (i)

$$\mathscr{D} \cap \mathscr{D}' = (\mathscr{A} \cup \mathscr{B}_2 \cup \mathscr{C}) \cap (\mathscr{A}' \cup \mathscr{B}_2' \cup \mathscr{C}')$$
$$= (\mathscr{A} \cap \mathscr{A}') \cup (\mathscr{B}_2 \cap \mathscr{B}_2') \cup (\mathscr{C} \cap \mathscr{C}') \cup (\mathscr{A} \cap \mathscr{B}_2') \cup (\mathscr{A}' \cap \mathscr{B}_2)$$
$$\cup (\mathscr{A} \cap \mathscr{C}') \cup (\mathscr{A}' \cap \mathscr{C}) \cup (\mathscr{B}_2 \cap \mathscr{C}') \cup (\mathscr{B}_2' \cap \mathscr{C})$$

where $\mathscr{A} \cap \mathscr{A}' = \mathscr{B}_2 \cap \mathscr{B}_2' = \varnothing$ (complement-free), $\mathscr{A} \cap \mathscr{B}_2' = \mathscr{A}' \cap \mathscr{B}_2 = \varnothing$ (since $\mathscr{A} \cap \mathscr{B}_2' = (\mathscr{A}_1 \cap \mathscr{B}_2') \cup (\mathscr{A}_2 \cap \mathscr{B}_2') = \mathscr{A}_1 \cap \mathscr{B}_2' = \mathscr{B}_1' \cap \mathscr{B}_2' = \varnothing$). Also $\mathscr{C} \cap \mathscr{C}' = \varnothing$; for $A_i \cup B_i = (A_j \cup B_j)' = (A_j' \cap B_j') \Rightarrow A_i \subseteq A_j'$ where $A_i \in \mathscr{A}_1$ and $A_j' \in \mathscr{B}_1$, contradicting $A \nsubseteq B$. Similarly for other intersections.

(ii) $\mathscr{A} \cap \mathscr{C} \neq \varnothing \Rightarrow A = A_1 \cup B_1$ for some $A \in \mathscr{A}$, $A_1 \in \mathscr{A}_1$, $B_1 \in \mathscr{B}_1 \Rightarrow B_1 \subseteq A$, which is a contradiction.

(iii) $|\mathscr{C}| = |\{C : C = A_i \cup A_j';\ A_i, A_j \in \mathscr{A}_1\}| = |\mathscr{A}_1 - \mathscr{A}_1| \geqslant |\mathscr{A}_1| = |\mathscr{B}_1|$.
(iv) $|\mathscr{A}| + |\mathscr{B}| = |\mathscr{A}| + |\mathscr{B}_2| + |\mathscr{B}_1| \leqslant |\mathscr{A}| + |\mathscr{B}_2| + |\mathscr{C}| = |\mathscr{D}| \leqslant 2^{n-1}$.

6.14 If $1 \leqslant h < n$ let $\mathscr{A}$ consist of all subsets of $\{1, \ldots, n\}$ of size $\leqslant h$ and containing 1, and let $\mathscr{B}$ consist of all subsets of size $\geqslant h$ and not containing 1. Then

$$|\mathscr{A}| + |\mathscr{B}| = \sum_{i=0}^{h-1} \binom{n-1}{i} + \sum_{i=h}^{n-1} \binom{n-1}{i} = 2^{n-1}.$$

6.15 Take $\mathscr{B} = \mathscr{A}'$. The conditions of 6.13 become $A_1 \not\subseteq A_2'$ and $A_2' \not\subseteq A_1$ whenever $A_1, A_2 \in \mathscr{A}$, i.e. $A_1 \cap A_2 \neq \varnothing$, $A_1 \cup A_2 \neq S$. The conclusion becomes $2|\mathscr{A}| \leqslant 2^{n-1}$.

6.16 Assume that all the A_i and B_i are subsets of $\{1, \ldots, r\}$ and use induction on r. Consider the induction step. Put $\mathscr{A} = \{A_1, \ldots, A_m\}$, $\mathscr{B} = \{B_1, \ldots, B_n\}$, $\mathscr{C} = \{A - \{r\} : A - \{r\} \in \mathscr{A}$ and $A \cup \{r\} \in \mathscr{A}\}$, $\mathscr{D} = \{A - \{r\} : A \in \mathscr{A}\}$, $\mathscr{E} = \{B : B \in \mathscr{B}, r \notin B\}$, $\mathscr{F} = \{B - \{r\} : B \in \mathscr{B}\}$. Show that $|\mathscr{C}| + |\mathscr{D}| = m$.

Apply the induction hypothesis to $\mathscr{C}$ and $\mathscr{E}$ and to $\mathscr{D}$ and $\mathscr{F}$ to obtain $|\mathscr{C}| \leqslant |\mathscr{C} - \mathscr{E}|$ and $|\mathscr{D}| \leqslant |\mathscr{D} - \mathscr{F}|$. Then show that if $E \in \mathscr{C} - \mathscr{E}$ then $E - \{r\}$ and $E \cup \{r\}$ are in $\mathscr{A} - \mathscr{B}$. Show also that if $E \in \mathscr{D} - \mathscr{F}$ then either $E - \{r\}$ or $E \cup \{r\}$ is in $\mathscr{A} - \mathscr{B}$. Hence $|\mathscr{A} - \mathscr{B}| \geqslant |\mathscr{C} - \mathscr{E}| + |\mathscr{D} - \mathscr{F}|$ so that $m = |\mathscr{C}| + |\mathscr{D}| \leqslant |\mathscr{C} - \mathscr{E}| + |\mathscr{D} - \mathscr{F}| \leqslant |\mathscr{A} - \mathscr{B}|$. (Ahlswede and Daykin 1979)

6.17 Let $\mathscr{B}$ be a maximum-sized intersecting family in the ideal $\mathscr{A}$. Extend $\mathscr{B}$ upwards to an upset $\mathscr{C}$. Then $\mathscr{B} = \mathscr{C} \cap \mathscr{A}$ and Kleitman gives $2^n|\mathscr{B}| = 2^n|\mathscr{C} \cap \mathscr{A}| \leqslant |\mathscr{C}| . |\mathscr{A}| \leqslant 2^{n-1}|\mathscr{A}|$ by Theorem 1.1.1.

6.18 Suppose $\mathscr{A}$ has two bases A_1, A_2. If $A_1 \cap A_2 \neq \varnothing$, the result follows from Theorem 6.4.4. If $A_1 \cap A_2 = \varnothing$, every intersecting family is in $\mathscr{P}(A_1)$ or $\mathscr{P}(A_2)$; therefore the maximum size of an intersecting family is $\max(2^{|A_1|-1}, 2^{|A_2|-1})$. But there is a star with this size.

6.19 (ii) $E \in \mathscr{E} \Leftrightarrow E \in \mathscr{A}^*, E \notin \mathscr{B}^* \Leftrightarrow E' \notin \mathscr{A}^*, E' \in \mathscr{B}^* \Leftrightarrow E' \in \mathscr{F}$.

(iii) $\mathscr{D}$ is an upset so $\mathscr{D}'$ is an ideal, so that there is a permutation σ such that $D_i' \in D_{\sigma(i)}$ for each $D_i \in \mathscr{D}$, i.e. $D_i \cup D_{\sigma(i)} = S$.

(iv) By (ii) the sets in $\mathscr{E}$ and $\mathscr{F}$ are in complementary pairs, the union of each pair being S. Together with the pairs in (iii) they give a pairing for $\mathscr{A}^* \cup \mathscr{B}^*$. Finally, each pair contributes at most one to $|\mathscr{A}| + |\mathscr{B}|$.

6.20 (i) By maximality of $|\mathcal{B}|$, $\mathcal{B}$ is an upset, so $\mathcal{B}'$ is a downset.
(ii) Extend $\mathcal{A}$ to an intersecting family $\mathcal{B}$ of size 2^{n-1}. The condition $A \cup B \neq S$ is equivalent to $A' \cap B' \notin \emptyset$, so that $\mathcal{A}'$ is an intersecting family in the ideal $\mathcal{B}'$. Hence $|\mathcal{A}| = |\mathcal{A}'| \leqslant \frac{1}{2}|\mathcal{B}'| = \frac{1}{2}|\mathcal{B}| = 2^{n-2}$.

6.21 $((A \cup E)' \cup E)' = (A \cup E) \cap E' = A \cap E' = A$ since $A \cap E = \emptyset$. Therefore $B = (A \cup E)'$ is also in $\mathcal{D}$. We also have $A \cap B = A \cap (A \cup E)' = A \cap A' \cap E' = \emptyset$, and $A \cup B = A \cup (A \cup E)' = A \cup (A' \cap E') = S - E$.

6.22 Take $A \in \mathcal{D}$ and define $D_1 = M_1 \cup (A \cap M')$, $D_2 = D_1' \cap E'$.

6.23 Let $\mathcal{B} = \mathcal{A} - \mathcal{A}$. $\mathcal{A} \cap \mathcal{B} = \emptyset$, so that $|\mathcal{A} \cup \mathcal{B}| \geqslant 2|\mathcal{A}|$. No two sets in $\mathcal{A} \cup \mathcal{B}$ have union S, so no set and its complement can both be in $\mathcal{A} \cup \mathcal{B}$; therefore $|\mathcal{A} \cup \mathcal{B}| \leqslant 2^{n-1}$. Thus $2|\mathcal{A}| \leqslant 2^{n-1}$.

Exercises 7

7.1 (a) $23 = \binom{6}{4} + \binom{4}{3} + \binom{3}{2} + \binom{1}{1}$; $27 = \binom{6}{4} + \binom{5}{3} + \binom{2}{2} + \binom{1}{1}$;

$37 = \binom{7}{4} + \binom{3}{3} + \binom{2}{2}$.

(b) $\binom{6}{3} + \binom{4}{2} + \binom{3}{1} + \binom{1}{0} = 30$.

7.2 $25 = \binom{6}{3} + \binom{3}{2} + \binom{2}{1}$ so the 25th set is $\{7, 4, 2\}$.

7.3 $\{8, 6, 5, 3, 1\}$ is the mth set where $m = \binom{7}{5} + \binom{5}{4} + \binom{4}{3} + \binom{2}{2} + \binom{1}{1} = 32$

$\{8, 7, 5, 4, 3\}$ is the mth set where $m = \binom{7}{5} + \binom{6}{4} + \binom{5}{3} = 46$.

7.4 (i) $A < B \Leftrightarrow$ largest member of $A + B$ $(= A' + B')$ is in $B \Leftrightarrow$ largest member of $A' + B'$ is not in $B' \Leftrightarrow B' < A'$.
(ii) and (iii) are now immediate.

7.5 If A is the mth set where m is given by (7.1), then A is given by (7.3). So $a = a_t$ and $b = t - 1$. The suggested B is therefore $\{1 + a_l, \ldots, 1 + a_t; 1, \ldots, t - 1\}$ if $t > 1$; this is the nth set where $n = \binom{a_l}{l} + \ldots + \binom{a_t}{t} + \binom{t-1}{t-1} = m + 1$. If $t = 1$, $B = \{1 + a_l, \ldots, 1 + a_1\}$. If $1 + a_1 < a_2$ this is the mth set where

$n = \binom{a_l}{l} + \ldots + \binom{a_2}{2} + \binom{1+a_1}{1} = m+1$. If $1+a_1 = a_2$ let s be the smallest i such that $1+a_i < a_{i+1}$. Then B is the nth set where $n = \binom{a_l}{l} + \ldots + \binom{a_{s+1}}{s+1} + \binom{1+a_s}{s} = m+1$ on using (7.2).

7.6 (a) $\binom{a_k}{k} + \ldots + \binom{a_s}{s} \leqslant m$. Also $M+1 = \binom{a_k}{k+1} + \ldots + \binom{a_s}{s+1} + \binom{s}{s}$, so that

$$|\Delta F_{k+1}(M+1)| = \binom{a_k}{k} + \ldots + \binom{a_s}{s} + s > m.$$

(b) $M + m = \binom{1+a_k}{k+1} + \ldots + \binom{1+a_s}{s+1} + \binom{s}{s} + \ldots + \binom{t+1}{t+1}$ so that

$$|\Delta F_{k+1}(M+m)| = \binom{1+a_k}{k} + \ldots + \binom{1+a_s}{s}$$

$$+ \binom{s}{s-1} + \ldots + \binom{t+1}{t} = m + |\Delta F_k(m)|.$$

7.7

$$\Delta\mathscr{A} = \Delta F_k(r)^{\hat{1}} \cup F_{k-1}(s)^{\hat{1}} \cup (\{1\} \cup \Delta F_{k-1}(s)^{\hat{1}}).$$

If $s \leqslant |\Delta F_k(r)|$ then $F_{k-1}(s)^{\hat{1}} \subseteq \Delta F_k(r)^{\hat{1}}$ so that

$$|\Delta\mathscr{A}| = |\Delta F_k(r)^{\hat{1}}| + |\{1\} \cup \Delta F_{k-1}(s)^{\hat{1}}| = |\Delta F_k(r)| + |\Delta F_{k-1}(s)|.$$

Similarly, if $s > |\Delta F_k(r)|$, $|\Delta\mathscr{A}| = s + |\Delta F_{k-1}(s)|$. But $|\Delta F_k(r+s)| \leqslant |\Delta\mathscr{A}|$.

7.8 First suppose $a_t > t$. Then

$$r + s = m = \left\{\binom{a_k - 1}{k} + \ldots + \binom{a_t - 1}{t}\right\}$$

$$+ \left\{\binom{a_k - 1}{k-1} + \ldots + \binom{a_t - 1}{t-1}\right\}.$$

Thus

$$s \leqslant \binom{a_k - 1}{k-1} + \ldots + \binom{a_t - 1}{t-1} \Rightarrow r \geqslant \binom{a_k - 1}{k} + \ldots + \binom{a_t - 1}{t}$$

$$\Rightarrow |\Delta F_k(r)| \geqslant s.$$

Next suppose $a_t = t$. Let $m = \binom{a_k}{k} + \ldots + \binom{a_h}{h} + \binom{h-1}{h-1} + \ldots + \binom{t}{t}$ where $a_h > h$. Then

$$m = \left\{\binom{a_k - 1}{k} + \ldots + \binom{a_h - 1}{h}\right\} + \left\{\binom{a_k - 1}{k-1} + \ldots + \binom{a_h - 1}{h-1} + \binom{h-2}{h-2} + \ldots + \binom{t-1}{t-1}\right\}.$$

Thus

$$s < \binom{a_k - 1}{k-1} + \ldots + \binom{a_t - 1}{t-1}$$
$$\Rightarrow r > \binom{a_k - 1}{k} + \ldots + \binom{a_h - 1}{h}$$
$$\Rightarrow r \geqslant \binom{a_k - 1}{k} + \ldots + \binom{a_h - 1}{h} + \binom{h-1}{h-1} \Rightarrow |\Delta F_k(r)| \geqslant s.$$

Similarly if $s \geqslant \binom{a_k - 1}{k-1} + \ldots + \binom{a_t - 1}{t-1}$. Adapt the argument if $t = 1$ or if $m \leqslant k$.

7.9 (a) (i) Immediate from Kruskal–Katona on writing v in its $(k-1)$-binomial representation.

(ii) If $m_2 = \binom{n}{k}$ then the right-hand side is $\binom{n}{k-1} + |\Delta F_k(m_1)| = |\Delta F_k(m_2)| + |\Delta F_k(m_1)|$. So suppose now $0 < m_1 \leqslant m_2 < \binom{n}{k}$. Use (i) to show that $\left|\Delta F_{k+1}\left(\binom{n}{k+1} + m_1\right)\right| > m_2$ and

$$\left|\Delta F_{k+1}\left(\binom{n+1}{k+1} + m_1 + m_2 - \binom{n}{k}\right)\right|$$
$$= \binom{n+1}{k} + \left|\Delta F_k\left(m_1 + m_2 - \binom{n}{k}\right)\right|$$

whence

$$\binom{n+1}{k} + \left|\Delta F_k\left(m_1 + m_2 - \binom{n}{k}\right)\right| = \left|\Delta F_{k+1}\left(\binom{n}{k+1} + m_1 + m_2\right)\right|$$
$$\leqslant \left|\Delta F_{k+1}\left(\binom{n}{k+1} + m_1\right)\right| + |\Delta F_k(m_2)|$$

(by Exercise 7.7) $= \binom{n}{k} + |\Delta F_k(m_1)| + \Delta F_k(m_2)|$.

(b) Let m_i be the number of $A \in \mathscr{A}$ such that $A \in S_i$. Then $m_1 + m_2 = |\mathscr{A}|$ and $|\Delta\mathscr{A}| \geqslant |\Delta F_k(m_1)| + |\Delta F_k(m_2)|$, so use (ii).

7.10 Take $k = l$; every A is a B, so $m \leqslant u$. If $m > \binom{n-1}{k-1}$ then $u > \binom{n-1}{k-1}$ so $m < \binom{n-1}{k-1}$, which is a contradiction.

7.11. Take m as in (7.1), so that the mth set is as in (7.3). Use the representations of $m+1$ as in the solution to Exercise 7.5 and the Kruskal–Katona theorem to obtain expressions for $|\Delta F_k(m)|$ and $|\Delta F_k(m+1)|$. For example, if $t > 1$, the difference is $t - 1 \leqslant k - 1$.

7.12 (i) Take $\mathscr{A}_0$ to consist of the last m subsets of S of size $\frac{1}{2}n$. Then by Exercise 7.4 $|\mathscr{A}_0 \cap \mathscr{A}_0'| = 2\,|\mathscr{A}_0| - |\mathscr{A}_0 \cup \mathscr{A}_0'| = 2m - \binom{n}{n/2}$. But if $t(\mathscr{A})$ is the number of disjoint pairs of subsets in $\mathscr{A}$ then $t(\mathscr{A}_0) = \frac{1}{2}\,|\mathscr{A}_0 \cap \mathscr{A}_0'|$.

(ii) First replace sets of size $<\frac{1}{2}n$ by sets of size $\frac{1}{2}n$. So now suppose $|A| = \frac{1}{2}n$ for all $A \in \mathscr{A}$. $t(\mathscr{A}) = \frac{1}{2}\,|\mathscr{A} \cap \mathscr{A}'|$, where $|\mathscr{A} \cap \mathscr{A}'| = |\mathscr{A}| + |\mathscr{A}'| - |\mathscr{A} \cup \mathscr{A}'| \geqslant 2m - \binom{n}{n/2}$.

(iii) Note that $\frac{1}{2}\binom{n}{n/2} = \binom{n-1}{n/2-1}$.

7.13 $|\nabla\mathscr{A}| = |\Delta\mathscr{A}'| \geqslant |\Delta\mathscr{C}|$ where $\mathscr{C}$ consists of the first $|\mathscr{A}|$ $(n-k)$-subsets of S. But $|\Delta\mathscr{C}| = |\nabla\mathscr{C}'|$ where $\mathscr{C}' = \mathscr{B}$.

7.14 $|\nabla\mathscr{A}| = |\Delta\mathscr{A}'|$, so apply Corollary 7.3.6 to the collection $\mathscr{A}'$ of $(n-k)$-subsets. Since $|\mathscr{A}'| \leqslant \binom{n-1}{k-1} = \binom{n-1}{n-k}$, take $u = n-1$ to obtain

$$|\nabla\mathscr{A}| = |\Delta\mathscr{A}'| \geqslant \frac{\binom{n-1}{n-k-1}}{\binom{n-1}{n-k}}|\mathscr{A}'| = \frac{\binom{n-1}{k}}{\binom{n-1}{k-1}}|\mathscr{A}|.$$

7.15 The required position is that of $\{21 - a_i\} = \{20, 19, 17, 16, 13, 11, 10, 9, 8, 6, 5, 3\}$ in the antilexicographic ordering, i.e. the rth-*last* position in the antilexicographic order where r is its position in the squashed ordering. But

$$r = \binom{19}{12} + \binom{18}{11} + \binom{16}{10} + \binom{15}{9} + \binom{12}{8} + \binom{10}{7}$$
$$\binom{9}{6} + \binom{8}{5} + \binom{7}{4} + \binom{5}{3} + \binom{4}{2} + \binom{3}{1} = 96034.$$

So the answer is $\binom{20}{12}+1-r=125971-96034=29937$.

7.16 Note that $|\mathscr{A}|=\binom{n-2}{n-k}+\binom{n-3}{n-k-1}+\ldots+\binom{n-k-1}{n-2k+1}+\binom{n-2k}{n-2k}$ and deduce that $|\Delta^{(k)}\mathscr{A}|\geqslant\binom{n-1}{k}$. But $m\leqslant\binom{n-1}{k-1}$.

Exercises 8

8.1 (a) $S\mathscr{A}=\mathrm{AL}\mathscr{A}=\mathscr{A}$.
(b) $S\mathscr{A}$ consists of $\{1,2,3\}$, $\{1,2,4\}$, $\{3,4\}$, $\{1,5\}$.
$\mathrm{AL}\mathscr{A}$ consists of $\{4,5\}$, $\{3,5\}$, $\{1,2,5\}$, $\{2,3,4\}$.

8.2 $\mathscr{B}=\mathscr{A}\cup\mathscr{A}'$ is an antichain with $p_i=p_{n-i}$, so apply 8.2.2 to obtain $\tau(\mathscr{B})+\xi(\mathscr{B})\leqslant 2$. If $n=2m$ this gives

$$\sum_{i<n/2}\frac{p_i}{\binom{n-1}{i-1}}+\sum_{i<n/2}\frac{p_i}{\binom{n-1}{n-i}}+\frac{2p_m}{\binom{n-1}{m}}\leqslant 2,$$

whence

$$\sum_{i\leqslant n/2}\frac{p_i}{\binom{n-1}{i-1}}\leqslant 1.$$

The case n odd is easier.

8.3 Suppose that $m\geqslant\binom{n-1}{k-1}$. If $\mathscr{B}=\{B_i\}$, then $\mathscr{A}'\cup\mathscr{B}$ is an antichain (note that $A'\subseteq B\Rightarrow A\cup B=\{1,\ldots,n\}$, contradicting the given conditions). Thus by 8.1.5, $|\mathscr{B}|+\delta_{l+1}(\delta_{l+2}(\ldots\delta_{n-k}(|\mathscr{A}'|))\ldots)\leqslant\binom{n}{l}$.

But

$$|\mathscr{A}'|\geqslant\binom{n-1}{k-1}=\binom{n-1}{n-k},$$

so

$$|\mathscr{B}|\leqslant\binom{n}{l}-\binom{n-1}{l}=\binom{n-1}{l-1}.$$

8.4 The following theorem follows from 8.1.5 and 8.3.4. Let $p_1,\ldots,p_k$ be given non-negative integers, $k\leqslant\frac{1}{2}n$. Then there exists an intersecting antichain with parameters p_i if and only if

$$p_1+\delta_1(p_2+\delta_2(p_3+\ldots)))\leqslant\binom{n-1}{k-1}.$$

8.5 Apply the ordinary LYM inequality to the corresponding antichain on $\{1, \ldots, n-1\}$.

8.6 (i) Each construction puts into $\mathcal{B}$ pairs of complements in $\{1, \ldots, n+1\}$. Note that for each i exactly one construction applies.
(ii) Check all the possibilities. For example, $A_i \subset A_j' \cup \{n+1\}$ in $\mathcal{B}$ would require $A_i \subset A_j'$ so that by (a) $A_j' \cup \{n+1\}$ is *not* in $\mathcal{B}$, which is a contradiction.
(iii) is obvious. Applying 5.2.3 gives

$$|\mathcal{A}| \leqslant \tfrac{1}{2}|\mathcal{B}| \leqslant \binom{n}{[(n+1)/2]-1} = \binom{n}{[n/2]+1}.$$

8.7 Apply 8.1.5 and divide by $\binom{n}{k}$. This is stronger than the LYM inequality since by the normalized matching property

$$\frac{\delta_{k+1}(p_{k+1})}{\binom{n}{k}} \geqslant \frac{p_{k+1}}{\binom{n}{k+1}}.$$

8.8 (i) Yes: e.g. 3-sets containing 1 and 4-sets not containing 1.
(ii) No: ruled out by LYM.
(iii) No: ruled out by 8.1.5.

8.9 (i) By Theorem 8.2.4 we can have equality only when all non-zero p_is correspond to the largest denominator, i.e. $i = m-1$, m or $m+1$.
(ii) Considered the squashed antichain $\mathcal{B}$ with non-zero parameters $q_m = p_m$, $q_{m-1} = p_{m-1} + p_{m+1}$. Since $\mathcal{A}$ is complement-free, $p_m \leqslant \frac{1}{2}\binom{2m}{m}$ so the sets of size m in $\mathcal{B}$ are really subsets of a $(2m-1)$-set; thus $|\Delta(m)\mathcal{B}| \geqslant p_m$. Thus $\binom{2m}{m-1} \geqslant p_{m-1} + p_{m+1} + |\Delta(m)\mathcal{B}| \geqslant p_{m-1} + p_{m+1} + p_m = \binom{2m}{m-1}$, so that $|\Delta(m)\mathcal{B}| = p_m$. Let $p_m = \binom{u}{m}$, $u \leqslant 2m-1$; then $\binom{u}{m} = |\Delta(m)\mathcal{B}| \geqslant \binom{u}{m-1}$ whence $u \geqslant 2m-1$. Thus $p_m = 0$ or $\binom{2m-1}{m}$.
(iii)

$$\{1, \ldots, m-1\} \in \mathcal{B} \Leftrightarrow \{m+2, \ldots, 2m\} \in \mathcal{B}$$
$$\Rightarrow \{1, \ldots, m-2; m\} \in \mathcal{B},$$

so continue to change one element each time to transform into any other $(m-1)$-set. If $p_m = 0$, $p_{m-1} + p_{m+1} = \binom{2m}{m-1}$, show that $|\nabla^2(m-1)\mathscr{A}| = p_{m-1}$ so that $\nabla^2(m-1)\mathscr{A} = ((m-1)\mathscr{A})'$. Take $\mathscr{B} = (m-1)\mathscr{A}$: for $A_1 \cap A_2 = \varnothing$, $A_1 \in (m-1)\mathscr{A} \Rightarrow A_1 \subseteq A_2' \Rightarrow A_2' \in \nabla^2((m-1)\mathscr{A}) \Rightarrow A_2' \in ((m-1)\mathscr{A})' \Rightarrow A_2 \in (m-1)\mathscr{A}$.

Exercises 9

9.1 $\mathscr{A}$: (1, 0, 2, 1), (0, 1, 1, 2), (2, 0, 2, 0).
$C\mathscr{A}$: (0, 0, 0, 4), (0, 0, 1, 3), (0, 0, 2, 2).
$\Delta\mathscr{A}$: (0, 0, 2, 1), (1, 0, 1, 1), (1, 0, 2, 0), (0, 0, 1, 2), (0, 1, 0, 2), (0, 1, 1, 1), (2, 0, 1, 0).
$\Delta C\mathscr{A}$: (0, 0, 0, 3), (0, 0, 1, 2), (0, 0, 2, 1).
$C\Delta\mathscr{A}$: (0, 0, 0, 3), (0, 0, 1, 2), (0, 0, 2, 1), (0, 0, 3, 0), (0, 1, 0, 2), (0, 1, 1, 1), (0, 1, 2, 0).

9.2 Suppose $k \leqslant k_1$. If $\mathscr{B} = \{(0, k), (1, k-1), \ldots, (h, k-h)\}$ then $\Delta\mathscr{B} = \{(0, k-1), (1, k-2), \ldots, (h-1, k-h), (h, k-h-1)\}$ if $h < k$, so that $|\Delta\mathscr{B}| - |\mathscr{B}| = 0$, while if $h = k$ then $|\Delta\mathscr{B}| - |\mathscr{B}| = -1$. For any non-initial block (i.e. a block not containing $(0, k)$), $|\Delta\mathscr{B}| - |\mathscr{B}| \geqslant 0$. Hence if $\mathscr{A} = \cup \mathscr{B}_i$ where the $\mathscr{B}_i$ are blocks, $|\Delta\mathscr{A}| - |\mathscr{A}| = \Sigma\,\{|\Delta\mathscr{B}_i| - |\mathscr{B}_i|\} \geqslant |\Delta C\mathscr{A}| - |C\mathscr{A}|$ since $C\mathscr{A}$ is an initial block. Thus $|\Delta\mathscr{A}| \geqslant |\Delta C\mathscr{A}|$. But $\Delta C\mathscr{A}$ is compressed, so $\Delta C\mathscr{A} \subseteq C\Delta\mathscr{A}$. Next consider the cases $k_1 \leqslant k \leqslant k_2$ and $k > k_2$ similarly.

9.3 Start with $\mathscr{A}_0 = F_{l+1}(m)$. If $\Delta\mathscr{A}_0 \neq (l)S$ then the last member of $(l)S$ is not in $\Delta\mathscr{A}_0$, so remove the last member from $\mathscr{A}_0$ and replace it by the last member of $(l)S$; call the resulting antichain $\mathscr{A}_1$. If $\Delta(l+1)\mathscr{A}_1 \cup (l)\mathscr{A}_1 \neq (l)S$, remove the last member from $(l+1)\mathscr{A}_1$ and replace it by the second-last member of $(l)S$ to obtain the antichain $\mathscr{A}_2$. Eventually we must obtain $\Delta(l+1)\mathscr{A}_i \cup (l)\mathscr{A}_i = (l)S$ as required; otherwise the process will result in some $\mathscr{A}_i = (l)S \cup F_{l+1}(m - N_l)$, which is not an antichain.

9.4 Yes.

9.5 Suppose that $\boldsymbol{y} \in T'$ and $\boldsymbol{x}$ is such that $x_i \leqslant y_i$ for each i. Then $(y_2, \ldots, y_n)$ is one of the first a_{y_1} members of $S(k_2, \ldots, k_n)$. Thus $(x_2, \ldots, x_n)$ is one of the first a_{y_1} members, and hence one of the first a_{x_1} members. Thus $\boldsymbol{x} \in T'$.

9.6 Since $|\Delta\mathscr{A}| \geq |\Delta C\mathscr{A}|$ we might as well assume that $\mathscr{A}$ is compressed. $\mathscr{A}$ is then the first $|\mathscr{A}| \leq \binom{n-1}{k}$ k-sets, and none of them contain n. Considering them as k-subsets of an $(n-1)$-set; the normalized matching property gives

$$\frac{|\Delta\mathscr{A}|}{\binom{n-1}{k-1}} \geq \frac{|\mathscr{A}|}{\binom{n-1}{k}}.$$

9.7 (i) The required conclusion will follow if we can show that the shadow of any subcollection $\{X_{i_1}, \ldots, X_{i_r}\}$ contains at least r sets. But $r \leq t$, so $|\Delta\{X_{i_1}, \ldots, X_{i_r}\}| \geq |\Delta C\{X_{i_1}, \ldots, X_{i_r}\}| = |\Delta F_l(r)| \geq r$.
(ii) Let $t > M$, and take $\mathscr{A} \supseteq \{X_1, \ldots, X_{M+1}\} = F_l(M+1)$. Then there are not enough sets in the shadow of $\mathscr{A}$ to represent $X_1, \ldots, X_{M+1}$ since $|\Delta F_l(M+1)| < M+1$.
(iii) Let M be the largest integer $\leq |(l)S|$ such that $t \leq M$ implies $|\Delta^{l-k}F_l(t)| \geq t$. Then, if $X_1, \ldots, X_t$ are $t \leq M$ l-subsets, there exist t k-subsets $Z_1, \ldots, Z_t$ such that $Z_i \subset X_i$ for each i.

Exercises 10

10.1 (i) $f(k_1, \ldots, k_n) = \sum_{n \in J} w(J)$.
(ii) Take $\mathscr{A}$ to consist of all vectors in $S(k_1, \ldots, k_n)$ with $1 \leq x_n < k_n$.

10.2 If $|J| = i$, then $w(J) = k^i$.

10.3 If k_i and k_j are odd, take all vectors x with $x_i < \frac{1}{2}k_i$ and $x_j > \frac{1}{2}k_j$.

10.4 $|\mathscr{A}| = \frac{1}{2}\sum_J g(J)$ where $g(J) = |\boldsymbol{x} \in \mathscr{A} : s(\boldsymbol{x}) = J| + |\boldsymbol{x} \in \mathscr{A} : s(\boldsymbol{x}) = J'|$. Suppose there is a J and $\boldsymbol{x}, \boldsymbol{y} \in \mathscr{A}$ with $s(\boldsymbol{x}) = J$ and $s(\boldsymbol{y}) = J'$. Then we must have $x_i = k_i$ for $i \in J$, $x_i = 0$ otherwise, and $y_i = k_i$ for $i \notin J$ and $y_i = 0$ otherwise. Thus $g(J) = 2$ for such J. Thus $|\mathscr{A}| \leq \frac{1}{2}\sum_J \max(w(J), w(J'), 2)$. But $k_i > 1$ for some i, so that $w(J)w(J') > 1$, whence $\max(w(J), w(J'), 2) = \max(w(J), w(J'))$. (Geréb-Graus 1983)

10.5 We can look on $\mathscr{D}_0$ as a set of subsets of an $(n-1)$-set. Thus $|(n-i)\mathscr{D}_0| = \binom{n-1}{n-i} = \binom{n-1}{i-1}$ and $|(i)\mathscr{D}_0| = \binom{n-1}{i}$.

10.6 $\mathscr{D}_j$ is isomorphic to $S' = S([\frac{1}{2}(k_{j+1}-1)], k_{j+2}, \ldots, k_n)$, with rank i in S' corresponding to rank $i + \frac{1}{2}(k_1 + \ldots + k_j)$ in $\mathscr{D}_j$.

The rank of S' is $K' = [\frac{1}{2}(k_{j+1} - 1)] + k_{j+2} + \ldots + k_n$. Thus

$$|(\mathrm{i})\ \mathscr{D}_j| \leq |([\tfrac{1}{2}(K+1)])\mathscr{D}_j|$$
$$\Leftrightarrow |(i - \tfrac{1}{2}k_1 - \ldots - \tfrac{1}{2}k_j)S'| \leq |([\tfrac{1}{2}(k_{j+1} + \ldots + k_n + 1)])S'|.$$

Verify that $\frac{1}{2}K' \leq \frac{1}{2}(k_{j+1} + \ldots + k_n + 1) \leq i - \frac{1}{2}(k_1 + \ldots + k_j)$, and apply Theorem 3.1.2 to S'.

10.7 If K is odd, $|\mathscr{A}| = 2|\mathscr{A}^+|$. By Theorem 10.2.6 and Exercise 10.6, $\sum_i q_{i,j} \leq |(\frac{1}{2}(K+1))\mathscr{D}_j|$ for each j, so $|\mathscr{A}| \leq 2\sum_{j=0}^{t} |(\frac{1}{2}(K+1))\mathscr{D}_j|$. If K is even, $|\mathscr{A}| = 2|\mathscr{A}^+|$ or $2|\mathscr{A}^+| + 1$, where $|\mathscr{A}^+| \leq \sum_{j=0}^{t} |(\frac{1}{2}K)\mathscr{D}_j|$, and $2|\mathscr{A}^+| = |\mathscr{A}^+| + |\mathscr{A}^-| \leq$ number of $\frac{1}{2}K$-vectors other than $(\frac{1}{2}k_1, \ldots, \frac{1}{2}k_n)$.
(ii) For K even take all $\frac{1}{2}K$-vectors. For K odd, take $Y \cup Y^c$ where $Y = \bigcup_{i=0}^{t} (\frac{1}{2}(K+1))\mathscr{D}_i$.

10.8 $\mathscr{D}_j$ is isomorphic to $S([\frac{1}{2}(k_{j+1} - 1)], k_{j+2}, \ldots, k_n)$, which has the LYM property.

10.9 The same conclusion holds. We have

$$\sum_{K-k \leq i \leq K - \frac{1}{2}\sum_{l=1}^{j+1} k_l} \frac{q_{i,j}}{|(i)\mathscr{D}_j|} \leq 1$$

for each j, so that $\sum_i q_{i,j} \leq |(K-k)\mathscr{D}_j|$ since, as in Exercise 10.6, $|(i)\mathscr{D}_j| \leq (K-k)\mathscr{D}_j|$. Thus

$$\sum_{i,j} q_{i,j} \leq \sum_j |(K-k)\mathscr{D}_j| = \sum_j |(k)\mathscr{B}_{j+1}|.$$

10.10 Clearly $\mathscr{A}$ will include all divisors of rank $\leq l$ if $|\mathscr{A}|$ is maximum. For each divisor of m of rank $l+1$, $\mathscr{A}$ can contain at most one multiple of it; thus $\mathscr{A}$ can contain at most N_{l+1} divisors of rank greater than l. In terms of vectors $\boldsymbol{x} = (x_1, \ldots, x_n)$, we obtain the following: if for all $\boldsymbol{x}, \boldsymbol{y} \in \mathscr{A}$, $\sum_i \min(x_i, y_i) \leq l$, then $|\mathscr{A}| \leq \sum_{i=0}^{l+1} N_i$.

10.11 If $|\boldsymbol{y}| = |\boldsymbol{z}|$ then the result is immediate. Suppose if possible that $|\boldsymbol{z}| < |\boldsymbol{y}|$ and $\boldsymbol{z} < \boldsymbol{y}$. Follow the argument of the last paragraph of the proof of Corollary 9.2.5 with $\boldsymbol{z}$ in place of $\boldsymbol{a}$ and $\boldsymbol{y}$ in place of $\boldsymbol{b}$ to obtain a contradiction. You have to replace '$\in \mathscr{A}$' by 'is a subset of a set in $(\mathscr{A}^+)^*$'.

Exercises 11

11.1

$$|\mathscr{F}| = 2\left\{\binom{n-2}{\frac{1}{2}(n-3)} + \binom{n-2}{\frac{1}{2}(n-1)}\right\} = 2\binom{n-1}{\frac{1}{2}(n-1)}.$$

Check all possibilities. Note: where does the condition given after Example 11.2.1 fail to hold?

11.2 When partitioned into chains as in Chapter 3, the rectangle gives $\min(p, q)$ chains, each containing one set of size $[\frac{1}{2}n]$.

11.3 You must verify that $\boldsymbol{w}_i - \boldsymbol{x}_n$ and $\boldsymbol{w}_j + \boldsymbol{x}_n$ $(i \neq j)$ are at least 2 apart. But their difference in the $\boldsymbol{x}_n$ direction is at least 2.

11.4 Let $S = X_1 \cup X_2$ be the colour decomposition of S. Partition X_1, X_2 into symmetric chains and combine them to obtain symmetric rectangles as in the proof of Theorem 11.2.1. Partition each rectangle into $\min(p, q)$ rows or columns. Each resulting chain is monochromatic and each contains one set of size $[\frac{1}{2}n]$.

11.5 Colour all elements of S differently. The set of subsets of S of even size shows that $f(n, n) \geq 2^{n-1}$. But if $\mathscr{A}$ is a collection of $f(n, n)$ sets with given property, then, for any $x \in S$, $A \subseteq S - \{x\}$, $A \in \mathscr{A} \Rightarrow A \cup \{x\} \notin \mathscr{A}$. So at most one of $A, A \cup \{x\}$ is in $\mathscr{A}$, so that $|\mathscr{A}| \leq 2^{n-1}$.

11.6 Order the elements of the blocks B_1, B_2 of the first colouring so that in each block the second colouring corresponds to oddness and evenness. Partition B_1, B_2 into symmetric chains and form rectangles as in proof of Theorem 11.2.1; then follow that proof.

11.7 (ii) Imitate the growth of block sizes as in proof of Theorem 11.3.2. But this time start with all of $\mathbf{0}$, $\boldsymbol{x}_1$, $\boldsymbol{x}_2$, $\boldsymbol{x}_1 + \boldsymbol{x}_2$ in the one block. Thus at $n = 2$ we have one block of size 4 instead of blocks of sizes 1 and 3. The chain construction has blocks of sizes 4,2,2 at stage $n = 3$. Since a 2-block grows into $\binom{n}{[n/2]}$ blocks after $n-1$ steps, i.e. into $\binom{n-1}{[\frac{1}{2}(n-1)]}$ blocks after $n-2$ steps, the 4-block will, in $n-2$ steps, give $\binom{n+1}{[\frac{1}{2}(n+1)]} - 2\binom{n-1}{[\frac{1}{2}(n-1)]}$ blocks as required. At most one of the sums can be in any given block.

Exercises 12

12.1 Take all subsets of S containing a fixed element x, together with the set $S - \{x\}$.

12.2 (i) Let $u=(u_1 \ldots u_i)^{1/i}$. Then $1>u_1 \ldots u_{i+1} \geqslant u^{i+1}$ so that

$$f(i+1) \geqslant \frac{i+1}{1-u^{i+1}} = \frac{i+1}{(1-u)(1+u+\ldots+u^i)}$$

$$> \frac{i}{(1-u)(1+u+\ldots+u^{i-1})} = \frac{i}{1-u^i} = f(i).$$

Similarly for g.

(ii) $a_i = \dfrac{n-i+1}{i}$.

(iii) Take $\quad u_i = \dfrac{1}{a_{k+1}} \quad$ and $\quad v_i = a_{k+1-i}$.

12.3 $\Sigma_i x_i \leqslant 1$ by LYM; $x_j^* \geqslant \Sigma_{i \geqslant j} x_i$ by the normalized matching property. (Compare Exercise 2.2.)

12.4 $-u+\binom{n}{k}v = m_k$ and $-u+\binom{n}{i}v \leqslant m_i$ yield $m_k - m_i \leqslant v\left(\binom{n}{k}-\binom{n}{i}\right)$. If we can show that

$$\frac{m_k - m_j}{\binom{n}{k}-\binom{n}{j}} \leqslant \frac{\binom{n}{k}}{\binom{n}{k}-\binom{n}{k-1}} \leqslant \frac{\binom{n}{k+1}}{\binom{n}{k+1}-\binom{n}{k}} \leqslant \frac{m_i - m_k}{\binom{n}{i}-\binom{n}{k}}$$

then we can choose v in the middle and then u. The middle inequality is precisely the log concavity condition; the first inequality reduces to

$$\binom{n}{k}\left\{\binom{n}{k-2}+\ldots+\binom{n}{j}\right\} \leqslant \binom{n}{k-1}\left\{\binom{n}{k-1}+\ldots+\binom{n}{j+1}\right\}$$

which is also true by log concavity.

12.5 Replace $\binom{n}{i}$ by N_i, and imitate the argument. The LYM and normalized matching properties, and log concavity, all hold. Therefore if $\mathscr{A}$ is an antichain in a multiset with $|\mathscr{A}| \geqslant N_k$, $k \leqslant \frac{1}{2}K$, then $|\mathscr{A}^*| \geqslant \Sigma_{i \leqslant k} N_i$.

12.6 (iv) $y_1 + \ldots + y_{[n/2]}$

$$= \binom{n}{[n/2]} + \theta\left(\binom{n}{0} + \tfrac{1}{2}\binom{n}{1} + \tfrac{1}{3}\binom{n}{2} + \ldots + \left(\frac{1}{n+1}\binom{n}{n}\right)\right.$$

$$\left. + \mathrm{O}\left(\binom{n}{[n/4]}\right)(0<\theta<1)\right.$$

$$= \binom{n}{[n/2]} + \frac{\theta 2^n}{n+1} + \mathrm{O}\left(\binom{n}{[n/4]}\right)$$

on integrating $1 + \binom{n}{1}x + \ldots + \binom{n}{n}x^n = (1+x)^n$ and putting $x = 1$. You must verify that the y_i satisfy the constraints (12.13). If $j \leqslant \frac{1}{2}n$, $\sum_{j/2 \leqslant k \leqslant j} ky_k = j\binom{n}{j} - l\binom{n}{l} + \sum' iy_i$ where l is the greatest integer less than $\frac{1}{2}j$ and where $\sum'$ is over all i between l and $\frac{1}{2}l$. This sum is the same form as the first, so repeat the argument.

12.7 Let Q denote the poset of all non-empty subsets of T and let R denote the poset of all subsets of $S - T$. Then both Q and R are LYM posets and have log concave rank numbers. Their product is $C(n, k)$, and so the result follows from Theorem 12.1.3.

12.8 (i) If $X \subset A_r'$ and $Y \subset A_r$ and $Y \in P$, then $X \cup Y \in P$ since $X \cap A_i = \varnothing$ for each i.
(ii) Induction on r and $|S|$. If $A_r \neq S$, the result follows by induction. If $A_r = S$, let P'' be the subposet of $\mathcal{P}(A_r)$ obtained by dropping the requirement $a_r \leqslant |X \cap A_r| \leqslant b_r$. By induction P'' is LYM with log concave rank numbers. P' is obtained from P'' by deleting the bottom a_r and top $n - b_r$ ranks since $X \cap A_r = X \cap S = X$. But such deletion does not affect LYM or log concavity.
(iii) Use Exercise 4.7 and Theorem 12.1.3.

Exercises 13

13.1 Suppose $\max(\mathcal{F}) = \mathcal{A}$. If $X \in \mathcal{F}$ then $X \leqslant Y$ for some $Y \in \max(\mathcal{F}) = \mathcal{A}$, and so $X \in \bar{\mathcal{A}}$. If $X \in \bar{\mathcal{A}}$, $X \leqslant Y$ for some $Y \in \mathcal{A} = \max(\mathcal{F})$ and so $X \leqslant Y$ where $Y \in \mathcal{F}$, so that $X \in \mathcal{F}$. Thus $\mathcal{F} = \bar{\mathcal{A}}$.

Suppose $\mathcal{F} = \bar{\mathcal{A}}$. If $X \in \max(\mathcal{F})$ then $X \in \mathcal{F}$ so $X \in \bar{\mathcal{A}}$ and so

$X \leqslant Y$ for some $Y \in \mathscr{A}$; but then $Y \in \bar{\mathscr{A}} = \mathscr{F}$, so $X = Y$ by maximality and so $X \in \mathscr{A}$. If $X \in \mathscr{A}$ then $X \in \bar{\mathscr{A}} = \mathscr{F}$, and $X \in \max(\mathscr{F})$, since if $X < Y$ where $Y \in \mathscr{F} = \bar{\mathscr{A}}$ then $X < Y \leqslant Z$ for some $Z \in \mathscr{A}$, contradicting the antichain property; therefore $X \in \max(\mathscr{F})$. Thus $\mathscr{A} = \max(\mathscr{F})$.

13.2 Let $X \in \max(\mathscr{A} \cup \mathscr{B})$. Certainly $X \in \bar{\mathscr{A}} \cup \bar{\mathscr{B}}$. If $X < Y$ where $Y \in \bar{\mathscr{A}} \cup \bar{\mathscr{B}}$, then $X < Y \in \bar{\mathscr{A}}$, say, so that $X < Y \leqslant Z$, $Z \in \mathscr{A}$ and so $X \notin \max(\mathscr{A} \cup \mathscr{B})$. Therefore $X \in \max(\bar{\mathscr{A}} \cup \bar{\mathscr{B}})$, and $\max(\mathscr{A} \cup \mathscr{B}) \subseteq \max(\bar{\mathscr{A}} \cup \bar{\mathscr{B}})$. The converse is similar.

13.3 Consider X in category (i) or (ii) as in the text. You must show that $X \in \max(\bar{\mathscr{A}} \cap \bar{\mathscr{B}})$. If $X \in \mathscr{A} \cap \mathscr{B}$ then $X \in \bar{\mathscr{A}} \cap \bar{\mathscr{B}}$, and if $X \leqslant Y \in \bar{\mathscr{A}} \cap \bar{\mathscr{B}}$ then $Y \leqslant A$ for some $A \in \mathscr{A}$, so that $X \leqslant A$ with $A, X \in \mathscr{A}$ and so $A = X = Y$ and $X \in \max(\bar{\mathscr{A}} \cap \bar{\mathscr{B}})$. If $X \in \mathscr{A} \cup \mathscr{B}$ and $X \subseteq W$ for some $W \in \mathscr{A} \cup \mathscr{B}$ then, say, $X \in \mathscr{A}$, $W \in \mathscr{B}$; therefore $X \in \bar{\mathscr{A}}$ and $X \in \bar{\mathscr{B}}$ so that $X \in \bar{\mathscr{A}} \cap \bar{\mathscr{B}}$. If $X \leqslant Y \in \bar{\mathscr{A}} \cap \bar{\mathscr{B}}$, proceed as before.

13.4 Take $P = \{a, b, c\}$ with $c < a, c < b$. If $\mathscr{A} = \{a\}$ and $\mathscr{B} = \{b\}$ then $\mathscr{A} \wedge \mathscr{B} = \{c\}$ and $\mathscr{A} \nabla \mathscr{B} = \{a, b\}$.

13.5 (i) First $\Leftrightarrow$ by definition. Now prove $\mathscr{A} \preccurlyeq \mathscr{B} \Leftrightarrow \mathscr{B}_{\mathscr{A}} = \mathscr{B}$. Suppose $\mathscr{A} \preccurlyeq \mathscr{B}$. If $\mathscr{B}_{\mathscr{A}} \subset \mathscr{B}$ (strict inclusion) there exists $B \in \mathscr{B}$ such that $A \nleqslant B$ for all $A \in \mathscr{A}$. If there exists $A \in \mathscr{A}$ such that $B \leqslant A$ for this B, then $B \leqslant A \leqslant B'$ for some $B' \in \mathscr{B}$ since $\mathscr{A} \preccurlyeq \mathscr{B}$; therefore $B = B'$ and $A \leqslant B$. Therefore this B is incomparable with every $A \in \mathscr{A}$ and could therefore be added to $\mathscr{A}$, contradicting maximality. Thus $\mathscr{B}_{\mathscr{A}} = \mathscr{B}$. The converse is similar.
(ii) If $\mathscr{A}_{\mathscr{B}} \cup \mathscr{A}^{\mathscr{B}} \subset \mathscr{A}$ strictly, there is an $A \in \mathscr{A}$ which can be added to $\mathscr{B}$ to give a larger antichain.
(iii) Certainly $\mathscr{A}_{\mathscr{B}} \cap \mathscr{B}_{\mathscr{A}} \subseteq \mathscr{A} \cap \mathscr{B}$. But if $X \in \mathscr{A} \cap \mathscr{B}$ then $X \in \mathscr{A}_{\mathscr{B}}$ (since $X \geqslant X$) and $X \in \mathscr{B}_{\mathscr{A}}$; so $X \in \mathscr{A}_{\mathscr{B}} \cap \mathscr{B}_{\mathscr{A}}$.
(iv) $A \in \mathscr{A}_{\mathscr{B}} \cap \mathscr{A}^{\mathscr{B}} \Rightarrow B' \leqslant A \leqslant B$ for some $B, B' \Rightarrow B' = A = B \Rightarrow A \in \mathscr{A} \cap \mathscr{B}$. The converse is easy.

13.6 $\Sigma_i \min(|C_i|, k) = k\alpha_k + \Sigma_{|C_i|<k} |C_i| = k\alpha_k + |S_k|$.

13.7 For a k-saturated chain partition $\mathscr{C}$, $d_k(P) = |S_k| + k\alpha_k$. But $\alpha_k \geqslant d_1(P - S_k)$, so that $d_k(P) \geqslant |S_k| + kd_1(P - S_k)$. On the other hand, $d_k(P) \leqslant |S| + kd_1(P - S)$ for all $S \subseteq P$.

13.8 Consider the top and bottom members $\mathscr{A}^*, \mathscr{A}_*$ of $\mathscr{S}(P)$. If they have a member A in common, then A is in every member $\mathscr{B}$ of $\mathscr{S}(P)$. For $\mathscr{A}_* \preccurlyeq \mathscr{B}$, so $A \leqslant B$ for some $B \in \mathscr{B}$, but then

$B \leqslant C$ for some $C \in \mathscr{A}^*$ since $\mathscr{B} \leqslant \mathscr{A}^*$; therefore $A \leqslant C$ where $A, C \in \mathscr{A}^*$, so $A = C = B$ and $A \in \mathscr{B}$.

13.9 $d_k(P) = |S_k| + k\alpha_k$, $d_{k+1}(P) \leqslant |S_k| + (k+1)\alpha_k$, and $d_{k-1}(P) \leqslant |S_k| + (k-1)\alpha_k$. The first two give $\Delta_{k+1}(P) \leqslant \alpha_k$. The first and third give $\Delta_k \geqslant \alpha_k$.

13.10 If $\Delta_k(P) = \Delta_{k+1}(P)$ then, by Exercise 13.9, both are equal to α_k. Also $d_k(P) = |S_k| + k\alpha_k$. Thus $d_{k+1}(P) = d_k(P) + \Delta_{k+1}(P) = d_k(P) + \alpha_k = |S_k| + (k+1)\alpha_k$.

References

Ahlswede, R. and Daykin, D. E. (1978). An inequality for the weights of two families of sets, their unions and intersections. *Z. Wahrsch. V. Geb.* **43,** 183–5.

—— and —— (1979). The number of values of combinatorial functions. *Bull. Lond. math. Soc.* **11,** 49–51.

Aigner, M. (1973). Lexicographic matching in Boolean algebras. *J. combinat. Theory B* **14,** 187–94.

Anderson, I. (1967*a*). On primitive sequences. *J. Lond. math. Soc.* **42,** 137–48.

—— (1967*b*). *Some problems in combinatorial number theory.* Ph.D. Thesis, University of Nottingham.

—— (1968). On the divisors of a number. *J. Lond. math. Soc.* **43,** 410–8.

—— (1969). A variance method in combinatorial number theory. *Glasgow math. J.* **10,** 126–9.

—— (1976). Intersection theorems and a lemma of Kleitman. *Discrete Math.* **16,** 181–5.

—— (1985). Counting common representatives and symmetric chain decompositions. *Proc. R. Soc. Edin. A* **100,** 151–5.

Baker, K. (1969). A generalization of Sperner's lemma. *J. combinat. Theory* **6,** 224–5.

Baumert, L. D., McEliece, R. J., Rodemich, E. R., and Rumsey, H. (1980). A probabilistic version of Sperner's theorem. *Ars Combinat.* **9,** 91–100.

Berge, C. (1976). A theorem related to the Chvátal conjecture. *Proc. 5th Br. Combinatorial Conf.,* 1975 (eds. C. St. J. A. Nash-Williams and J. Sheehan), pp. 35–40. Utilitas Math. Publishing Inc., Winnipeg.

Bollobás, B. (1965). On generalized graphs. *Acta math. Acad. Sci. Hung.* **16,** 447–452.

—— (1973). Sperner systems consisting of pairs of complementary subsets. *J. combinat Theory A* **15,** 363–6.

Bondy, J. A. (1972). Induced subsets. *J. combinat Theory B* **12,** 201–2.

Brace, A. and Daykin, D. E. (1972). Sperner type theorems for finite sets. *Proc. Br. Combinatorial Conf., Oxford,* 1972, pp. 18–37. Institute of Mathematics and its Applications, Southend-on-Sea.

de Bruijn, N. G., Tengbergen, C, and Kruyswijk, D. (1951). On the set of divisors of a number. *Nieuw Arch. Wiskd.* **23,** 191–3.

Cai, M. (1984). On a problem of Katona on minimal completely separating systems with restrictions. *Discrete Math.* **48,** 121–3.

Canfield, E. R. (1978*a*). On a problem of Rota. *Adv. Math.* **29,** 1–10.

—— (1978*b*). On the location of the maximum Stirling number of the second kind. *Stud. appl. Math.* **59,** 83–93.

Clements, G. F. (1970). On existence of distinct representative sets for subsets of a finite set. *Can. J. Math.* **22,** 1284–92.
—— (1971). Sets of lattice points which contain a maximal number of edges. *Proc. Am. math. Soc.* **27,** 13–15.
—— (1973). A minimization problem concerning subsets of a finite set. *Discrete Math.* **4,** 123–8.
—— (1974). Inequalities concerning subsets of a finite set. *J. combinat. Theory A* **17,** 227–244.
—— (1976). Intersection theorems for sets of subsets of a finite set. *Q. J. Math. Oxf.* (2) **27,** 325–337.
—— (1977*a*). An existence theorem for antichains. *J. combinat. Theory A* **22,** 368–71.
—— (1977*b*). More on the generalized Macaulay theorem. *Discrete Math.* **18,** 253–64.
—— (1984*a*). Antichains in the set of subsets of a multiset. *Discrete Math.* **48,** 23–45.
—— (1984*b*). On uniqueness of maximal antichains of subsets of a multiset. *Period. math. Hung.* **15,** 307–13.
—— and Gronau, H. D. O. F. (1981). On maximal antichains containing no set and its complement. *Discrete Math.* **33,** 239–47.
—— and Lindström, B. (1969). A generalization of a combinatorial theorem of Macaulay. *J. combinat. Theory* **7,** 230–8.
Daykin, D. E. (1974*a*). A simple proof of Katona's theorem. *J. combinat. Theory A* **17,** 252–3.
—— (1974*b*). Erdös–Ko–Rado from Kruskal–Katona. *J. combinat. Theory A* **17,** 254–5.
—— (1977). A lattice is distributive iff $|A| . |B| \leq |A \vee B| . |A \wedge B|$. *Nanta Math.* **10,** 58–60.
—— (1984). Antichains of subsets of a finite set. In *Graph theory and combinatorics* (ed. B. Bollobás). Academic Press, London.
——, Frankl, P., Greene, C., and Hilton, A. J. W. (1981). A generalization of Sperner's theorem. *J. Aust. Math. Soc.* **3,** 481–5.
——, Godfrey, J., and Hilton, A. J. W. (1974). Existence theorems for Sperner families. *J. combinat. Theory A* **17,** 245–51.
——, Hilton, A. J. W. and Miklos, D. (1983). Pairings from downsets and upsets in distributive lattices. *J. combinat. Theory A* **34,** 215–30.
——, Kleitman, D. J. and West, D. B. (1979). The number of meets between two subsets of a lattice. *J. combinat. Theory A* **26,** 135–56.
—— and Lovász, L. (1976). The number of values of a Boolean function. *J. Lond. math. Soc.* **12,** 225–30.
Dilworth, R. P. (1950). A decomposition theorem for partially ordered sets. *Ann. Math.* **51,** 161–5.
—— (1960). Some combinatorial problems on partially ordered sets. *Combinatorial Analysis, Proc. Symp. appl. Math.*, pp. 85–90. American Mathematical Society, Providence, Rhode Island.
—— and Greene, C. (1971). A counterexample to the generalization of Sperner's theorem. *J. combinat Theory* **10,** 18–21.
Ein, L. M. H., Richman, D. R., Kleitman, D. J., Shearer, J. and Sturtevant, D. (1981). Some results on systems of finite sets that satisfy a certain intersection condition. *Stud. appl. Math.* **65,** 269–74.

Engel, K. and Gronau, H. D. O. F. (1985). An intersection–union theorem for integer sequences. *Discrete Math.* **54,** 153–9.
Erdös, P. (1945). On a lemma of Littlewood and Offord. *Bull. Am. Math. Soc.* **51,** 898–902.
——, Herzog, M., and Schönheim, J. (1970). An extremal problem on the set of noncoprime divisors of a number. *Isr. J. Math.* **8,** 408–12.
—— and Hindman, N. (1984). Enumeration of intersecting families. *Discrete Math.* **48,** 61–65.
——, Ko C. and Rado R. (1961). Intersection theorems for systems of finite sets. *Q. J. Math. Oxf.* (2) **12,** 313–20.
—— and Schönheim, J. (1970). On the set of non-pairwise coprime divisors of a number. *Combinatorial theory and its applications, Balatonfured,* 1969, *Colloq. Janos Bolyai* **4,** 369–75. North-Holland, Amsterdam.
Ford, L. R. and Fulkerson, D. R. (1956). Maximal flow through a network. *Can. J. Math.* **8,** 399–404.
—— and —— (1958). Network flow and systems of representatives. *Can. J. Math.* **10,** 78–85.
Fortuin, C. M., Kasteleyn, P. W., and Ginibre, J. (1971). Correlation inequalities on some partially ordered sets. *Commun. math. Phys.* **22,** 89–103.
Frankl, P. (1976*a*). On Sperner families satisfying an additional condition. *J. combinat. Theory A* **20,** 1–11.
—— (1976*b*). The Erdös–Ko–Rado theorem is true for $n = ckt$. *Combinatorics, Colloq. Janos Bolyai* **18,** 365–75.
—— (1982). An extremal problem for two families of sets. *Eur. J. Combinat.* **3,** 125–7.
—— (1983). On the trace of finite sets. *J. combinat. Theory A* **34,** 41–5.
—— (1984). A new short proof for the Kruskal–Katona theorem. *Discrete Math.* **48,** 327–9.
—— and Furedi, Z. (1980). The Erdös–Ko–Rado theorem for integer sequences. *SIAM J. Algebraic discrete Meth.* **1,** 376–81.
—— and Hilton, A. J. W. (1977). On the minimum number of sets comparable with some members of a finite set. *J. Lond. math. Soc.* **15,** 16–8.
Freese, R. (1974). An application of Dilworth's lattice of maximal antichains. *Discrete Math.* **7,** 107–9.
Geréb-Graus, M. (1983). An extremal theorem on divisors of a number. *Discrete Math.* **46,** 241–8.
Graham, R. L., Yao, A. C., and Yao, F. F. (1980). Some monotonicity properties of partial orders. *SIAM J. Algebraic discrete Meth.* **1,** 251–8.
Greene, C. and Hilton, A. J. W. (1979). Some results on Sperner families. *J. combinat. Theory A* **26,** 202–9.
——, Katona, G., and Kleitman, D. J. (1976). Extensions of the Erdös–Ko–Rado theorem. *Stud. appl. Math.* **55,** 1–8.
—— and Kleitman, D. J. (1976*a*). The structure of Sperner k-families. *J. combinat. Theory A* **20,** 41–68.
—— and —— (1976*b*). Strong versions of Sperner's theorem. *J. combinat. Theory A* **20,** 80–8.
—— and —— (1978). Proof techniques in the theory of finite sets. In *Studies in combinatorics* (ed. G. C. Rota). *M.A.A. Stud. Math.* **17,** 22–79.

Griggs, J. R. (1977) Sufficient conditions for a symmetric chain order. *SIAM J. appl. Math.* **32,** 807–9.
—— (1980*a*). On chains and Sperner *k*-families in ranked posets. *J. combinat. Theory A* **28,** 156–68.
—— (1980*b*). The Littlewood–Offord problem: tightest packing and an *M*-part Sperner theorem. *Eur. J. Combinat.* **1**, 225–34.
—— (1982). Collections of subsets with the Sperner property. *Trans. Am. math. Soc.* **269,** 575–91.
—— (1984). Maximum antichains in the product of chains. *Order* **1,** 21–8.
—— and Kleitman, D. J. (1977). A three-part Sperner theorem. *Discrete Math.* **17,** 281–9.
——, Odlyzko, A. M., and Shearer, J. B. *k*-colour Sperner theorems. *J. combinat. Theory A*, to be published.
——, Stahl, J., and Trotter, W. T. (1984). A Sperner theorem on unrelated chains of subsets. *J. combinat. Theory A* **36,** 124–7.
——, Sturtevant, D., and Saks, M. (1980). On chains and Sperner *k*-families in ranked posets. *J. combinat. Theory* **29,** 391–4.
Grimmett, G. R. (1973). A generalization of a theorem of Kleitman and Milner. *Bull. Lond. math. Soc.,***5,** 157–8.
Gronau, H. D. O. F. (1980). On maximal antichains consisting of sets and their complements. *J. combinat. Theory A* **29,** 370–5.
Hall, M. (1948). Distinct representatives of subsets. *Bull. Am. math. Soc.* **54,** 922–6.
Hall, P. (1935). On representatives of subsets. *J. Lond. math. Soc.* **10,** 26–30.
Halmos, P. R. and Vaughan, H. E. (1950). The marriage problem. *Am. J. Math.* **72,** 214–5.
Hansel, G. (1966). Sur le nombre des fonctions Booléennes monotones de *n* variables. *C. R. Acad. Sci. Paris* **262,** 1088–90.
Harper, L. H. (1967). Stirling behaviour is asymptotically normal. *Ann. Math. Statist.* **38,** 410–4.
—— (1974). The morphology of partially ordered sets. *J. combinat. Theory A* **17,** 44–58.
Hilton, A. J. W. (1974). Analogues of a theorem of Erdös, Ko and Rado on a family of finite sets. *Q. J. Math. Oxf.* (2) **25,** 19–28.
—— (1976). A theorem on finite sets. *Q. J. Math. Oxf.* (2) **27,** 33–6.
—— (1977). An intersection theorem for a collection of families of subsets of a finite set. *J. Lond. math. Soc.* **15,** 369–76.
—— (1978). Some intersection and union theorems for several families of finite sets. *Mathematika* **25,** 125–8.
—— (1979). A simple proof of the Kruskal–Katona theorem and of some associated binomial inequalities. *Period. math. Hung.* **10,** 25–30.
—— and Milner, E. C. (1967). Some intersection theorems for systems of finite sets. *Q. J. Math. Oxf.* (2) **18,** 369–84.
Hoggar, S. G. (1974). Chromatic polynomials and log concavity. *J. combinat. Theory B* **16,** 248–54.
Holley, R. (1974). Remarks on the FKG inequalities. *Commun. math. Phys.* **36,** 227–31.
Hsieh, W. N. (1975). Intersection theorems for systems of finite vector spaces. *Discrete Math.* **12,** 1–16.

—— and Kleitman, D. J. (1973). Normalized matching in direct products of partial orders. *Stud. appl. Math.* **52,** 285–9.

Jichang, S. and Kleitman, D. J. (1984). Superantichains in the lattice of partitions of a set. *Stud. appl. Math.* **71,** 207–41.

Kalai, G. (1984). Intersection patterns of convex sets. *Isr. J. Math.* **48,** 161–74.

Katona, G. O. H. (1964). Intersection theorems for systems of finite sets. *Acta math. Acad. Sci. Hung.* **15,** 329–37.

—— (1966a). A theorem on finite sets. In *Theory of graphs. Proc. Colloq. Tihany,* 1966, pp. 187–207. Akademiai Kiado. Academic Press, New York.

—— (1966*b*). On a conjecture of Erdös and a stronger form of Sperner's theorem. *Stud. Sci. Math. Hung.* **1,** 59–63.

—— (1972*a*). Families of subsets having no subset containing another with small difference. *Nieuw Arch. Wiskd.* (3) **20,** 54–67.

—— (1972b). A simple proof of the Erdös–Ko–Rado theorem. *J. combinat. Theory B* **13,** 183–4.

—— (1973). Two applications of Sperner-type theorems (for search theory and truth functions). *Period math. Hung.* **4,** 19–23.

—— (1974). Solution of a problem of A. Ehrenfeucht and J. Mycielski. *J. combinat. Theory A* **17,** 265–6.

Kleitman, D. J. (1965). On a lemma of Littlewood and Offord on the distribution of certain sums. *Math. Z.* **90,** 251–9.

—— (1966). Families of non-disjoint subsets. *J. combinat. Theory* **1,** 153–5.

—— (1968). On a conjecture of Milner on k-graphs with non-disjoint edges. *J. combinat. Theory* **5,** 153–6.

—— (1970). On a lemma of Littlewood and Offord on the distribution of linear combinations of vectors. *Adv. Math.* **5,** 1–3.

—— (1974). On an extremal property of antichains in partial orders. In *Combinatorics* (eds. M. Hall and J. H. van Lint), *Math. Centre Tracts* **55,** 77–90. Mathematics Centre, Amsterdam.

—— (1976*a*). Some new results on the Littlewood–Offord problem. *J. combinat. Theory A* **20,** 89–113.

—— (1976*b*). Extremal properties of collections of subsets containing no two sets and their union. *J. combinat. Theory A* **20,** 390–2.

—— and Markowsky, G. (1974). On Dedekind's problem: the number of monotone Boolean functions II. *Trans. Am. math. Soc.* **45,** 373–89.

—— and Milner, E. C. (1973). On the average size of sets in a Sperner family. *Discrete Math.* **6,** 141–7.

—— and Spencer, J. (1973). Families of k-independent sets. *Discrete Math.* **6,** 255–62.

König, D. (1931). Graphok es matriscok. *Mat. Fiz. Lapok* **38,** 116–9.

Korshunov, A. D. (1981). The number of monotone Boolean functions. *Probl. Kibernet.* **38,** 5–108.

Kruskal, J. B. (1963). The number of simplices in a complex. In *Mathematical optimization techniques* (ed. R. Bellman), pp. 251–78. University of California Press, Berkeley.

Leeb, unpublished data.

Lehmer, D. H. (1964). The machine tools of combinatorics. In *Applied combinatorial mathematics* (ed. E. Beckenbach), pp. 5–31. Wiley, New York, 5–31.

Li, S.-Y. R. (1977). An extremal problem among subsets of a set. *J. combinat. Theory A* **23,** 341–3.

—— (1978). Extremal theorems on divisors of a number. *Discrete Math.* **24,** 37–46.

Liggett, T. M. (1977). Extensions of the Erdös–Ko–Rado theorem and a statistical application. *J. combinat. Theory A* **23,** 15–21.

Lih, K. W. (1980). Sperner families over a subset. *J. combinat. Theory A* **29,** 182–5.

Lindsey, J. H. (1964). Assignment of numbers to vertices. *Am. math. Mon.* **71,** 508–16.

Lindström, B. (1980). A partition of $L(3, n)$ into saturated symmetric chains. *Eur. J. Combinat.* **1,** 61–63.

Lipski, W. (1978). On strings containing all subsets as substrings. *Discrete Math.* **21,** 253–9.

Littlewood, J. and Offord, C. (1943). On the number of real roots of a random algebraic equation III. *Mat. Sb.* **12,** 277–85.

Livingston, M. L. (1979). An ordered version of the Erdös–Ko–Rado theorem. *J. combinat. Theory A* **26,** 162–5.

Lovász, L. (1977). Flats in matroids and geometric graphs. In *Combinatorial surveys* (ed. P. J. Cameron), pp. 45–86. Academic Press, New York.

—— (1979). *Combinatorial problems and exercises.* North-Holland, Amsterdam.

Lubell, D. (1966). A short proof of Sperner's theorem. *J. combinat. Theory* **1,** 299.

Macaulay, F. S. (1927). Some properties of enumeration in the theory of modular systems. *Proc. Lond. math. Soc.* **26,** 531–55.

Marica, J. (1971). Orthogonal families of sets. *Can. math. Bull.* **14,** 573–6.

—— and Schönheim, J. (1969). Differences of sets and a problem of Graham. *Can. math. Bull.* **12,** 635–7.

Meschalkin, L. D. (1963). A generalization of Sperner's theorem on the number of subsets of a finite set. *Theor. Probl. Appl.* **8,** 203–4.

Miklós, D. (1984). Great intersecting families of edges in hereditary hypergraphs. *Discrete Math.* **48,** 95–9.

Milner, E. C. (1968). A combinatorial theorem on systems of sets. *J. Lond. math. Soc.* **43,** 204–6.

Mirsky, L. (1971). A dual of Dilworth's decomposition theorem. *Am. Math. Mon.* **78,** 876–7.

Moon, A. (1982). An analogue of the Erdös–Ko–Rado theorem for the Hamming schemes $H(n, q)$. *J. combinat. Theory A* **32,** 386–90.

Mörs, M. (1985). A generalization of a theorem of Kruskal. *Graphs Combinat.* **1,** 167–83.

Perfect, H. (1969). Remark on a criterion for common transversals. *Glasgow math. J.* **10,** 66–7.

—— (1984). Addendum to a paper of M. Saks. *Glasgow math. J.* **25,** 31–3.

Perles, M. A. (1963). A proof of Dilworth's decomposition theorem for partially ordered sets. *Isr. J. Math.* **1,** 105–7.

Purdy, G. (1977). A result on Sperner collections. *Util. Math.* **12,** 95–9.

Rands, B. M. I. (1982). An extension of the Erdös–Ko–Rado theorem to t-designs. *J. combinat. Theory A* **32,** 391–5.
Rota, G.-C. (1967). A generalization of Sperner's theorem. *J. combinat. Theory* **2,** 104.
Saks, M. (1979). A short proof of the existence of k-saturated partitions of partially ordered sets. *Adv. Math.* **33,** 207–11.
Sali, A. (1983). Stronger form of an M-part Sperner theorem. *Eur. J. Combinat.* **4,** 179–83.
Schönheim, J. (1974*a*). On a problem of Daykin concerning intersecting families of sets. *Proc. Br. Combinatorics Conf. Aberystwyth* 1973 (ed. T. P. McDonough and V. C. Mavron). *London. math. Soc. Lect. Notes* **13,** 139–140.
—— (1974*b*). On a problem of Purdy related to Sperner systems. *Can. math. Bull.* **17,** 135–6.
—— (1976). Hereditary systems and Chvátal's conjecture. *Proc. 5th Br. Combinatorial Conf., Aberdeen,* 1975 (ed. C. St. J. A. Nash-Williams and J. Sheehan), pp. 537–40. Utilitas Math. Publishing Inc., Winnipeg.
Seymour, P. D. (1973). On incomparable collections of sets. *Mathematika* **20,** 208–9.
—— (1977). The matroids with the max-flow min-cut property. *J. combinat. Theory B* **23,** 189–222.
—— and Welsh, D. J. A. (1975). Combinatorial applications of an inequality from statistical mechanics. *Math. Proc. Camb. phil. Soc.* **77,** 485–95.
Shearer, J. B. (1979). A simple counterexample to a conjecture of Rota. *Discrete Math.* **28,** 327–30.
—— and Kleitman, D. J. (1979). Probabilities of independent choices being ordered. *Stud. appl. Math.* **60,** 271–6.
Shepp, L. A. (1980). The FKG inequality and some monotonicity properties of partial orders. *SIAM J. Algebraic Discrete Meth.* **1,** 295–9.
—— (1982). The XYZ conjecture and the FKG inequality. *Ann. of Probab.* **10,** 824–7.
Spencer, J. H. (1974). A generalized Rota conjecture for partitions. *Stud. appl. Math.* **53,** 239–41.
Sperner, E. (1928). Ein Satz über Untermengen einer endlichen Menge. *Math. Z.* **27,** 544–8.
Stanley, R. P. (1980). Weyl groups, the hard Lefschetz theorem, and the Sperner property. *SIAM J. Algebraic Discrete Meth.* **1,** 168–84.
Stanton, D. (1980). Some Erdös–Ko–Rado theorems for Chevalley groups. *SIAM J. Algebraic Discrete Meth.* **1,** 160.
Stein, P. (1983). On Chvátal's conjecture related to a hereditary system. *Discrete Math.* **43,** 97–105.
Tarjan, T. G. (1975). Complexity of lattice-configurations. *Stud. Sci. Math. Hung.* **10,** 203–11.
Tuza, Z. (1984). Helly-type hypergraphs and Sperner families. *Eur. J. Combinat.* **5,** 185–7.
West, D. B. (1980). A symmetric chain decomposition of $L(4, n)$. *Eur. J. Combinat.* **1,** 379–83.
—— (1982). Extremal problems in partially ordered sets. In *Ordered sets: Proc. NATO Advanced Study Institute* 1981 (ed. I. Rival), pp. 473–521. Reidel, Dordrecht.

——, Harper, L. H. and Daykin, D. E. (1983). Some remarks on normalized matching. *J. combinat. Theory A* **35,** 301–8.

White, D. E. and Williamson, S. G. (1977). Recursive matching algorithms and linear orders on the subset lattice. *J. combinat. Theory A* **23,** 117–27.

Wilson, R. M. (1984). The exact bound in the Erdös-Ko-Rado theorem. *Combinatorica* **4,** 247–57.

Woodall, D. R. (1978). Minimax theorems in graph theory. *Selected topics in graph theory* (eds. L. W. Beineke and R. J. Wilson), pp. 237–70. Academic Press, London.

Yamamoto, K. (1954). Logarithmic order of free distributive lattices. *J. math. Soc. Jpn.* **6,** 343–53.

Zionts, S. (1974). *Linear and integer programming*. Prentice-Hall, Englewood Cliffs, NJ.

Index

A CATALOG OF SELECTED

DOVER BOOKS

IN SCIENCE AND MATHEMATICS

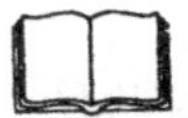

Astronomy

CHARIOTS FOR APOLLO: The NASA History of Manned Lunar Spacecraft to 1969, Courtney G. Brooks, James M. Grimwood, and Loyd S. Swenson, Jr. This illustrated history by a trio of experts is the definitive reference on the Apollo spacecraft and lunar modules. It traces the vehicles' design, development, and operation in space. More than 100 photographs and illustrations. 576pp. 6 3/4 x 9 1/4. 0-486-46756-2

EXPLORING THE MOON THROUGH BINOCULARS AND SMALL TELESCOPES, Ernest H. Cherrington, Jr. Informative, profusely illustrated guide to locating and identifying craters, rills, seas, mountains, other lunar features. Newly revised and updated with special section of new photos. Over 100 photos and diagrams. 240pp. 8 1/4 x 11. 0-486-24491-1

WHERE NO MAN HAS GONE BEFORE: A History of NASA's Apollo Lunar Expeditions, William David Compton. Introduction by Paul Dickson. This official NASA history traces behind-the-scenes conflicts and cooperation between scientists and engineers. The first half concerns preparations for the Moon landings, and the second half documents the flights that followed Apollo 11. 1989 edition. 432pp. 7 x 10. 0-486-47888-2

APOLLO EXPEDITIONS TO THE MOON: The NASA History, Edited by Edgar M. Cortright. Official NASA publication marks the 40th anniversary of the first lunar landing and features essays by project participants recalling engineering and administrative challenges. Accessible, jargon-free accounts, highlighted by numerous illustrations. 336pp. 8 3/8 x 10 7/8. 0-486-47175-6

ON MARS: Exploration of the Red Planet, 1958-1978--The NASA History, Edward Clinton Ezell and Linda Neuman Ezell. NASA's official history chronicles the start of our explorations of our planetary neighbor. It recounts cooperation among government, industry, and academia, and it features dozens of photos from Viking cameras. 560pp. 6 3/4 x 9 1/4. 0-486-46757-0

ARISTARCHUS OF SAMOS: The Ancient Copernicus, Sir Thomas Heath. Heath's history of astronomy ranges from Homer and Hesiod to Aristarchus and includes quotes from numerous thinkers, compilers, and scholasticists from Thales and Anaximander through Pythagoras, Plato, Aristotle, and Heraclides. 34 figures. 448pp. 5 3/8 x 8 1/2. 0-486-43886-4

AN INTRODUCTION TO CELESTIAL MECHANICS, Forest Ray Moulton. Classic text still unsurpassed in presentation of fundamental principles. Covers rectilinear motion, central forces, problems of two and three bodies, much more. Includes over 200 problems, some with answers. 437pp. 5 3/8 x 8 1/2. 0-486-64687-4

BEYOND THE ATMOSPHERE: Early Years of Space Science, Homer E. Newell. This exciting survey is the work of a top NASA administrator who chronicles technological advances, the relationship of space science to general science, and the space program's social, political, and economic contexts. 528pp. 6 3/4 x 9 1/4. 0-486-47464-X

STAR LORE: Myths, Legends, and Facts, William Tyler Olcott. Captivating retellings of the origins and histories of ancient star groups include Pegasus, Ursa Major, Pleiades, signs of the zodiac, and other constellations. "Classic." – *Sky & Telescope.* 58 illustrations. 544pp. 5 3/8 x 8 1/2. 0-486-43581-4

A COMPLETE MANUAL OF AMATEUR ASTRONOMY: Tools and Techniques for Astronomical Observations, P. Clay Sherrod with Thomas L. Koed. Concise, highly readable book discusses the selection, set-up, and maintenance of a telescope; amateur studies of the sun; lunar topography and occultations; and more. 124 figures. 26 halftones. 37 tables. 335pp. 6 1/2 x 9 1/4. 0-486-42820-6

Chemistry

MOLECULAR COLLISION THEORY, M. S. Child. This high-level monograph offers an analytical treatment of classical scattering by a central force, quantum scattering by a central force, elastic scattering phase shifts, and semi-classical elastic scattering. 1974 edition. 310pp. 5 3/8 x 8 1/2. 0-486-69437-2

HANDBOOK OF COMPUTATIONAL QUANTUM CHEMISTRY, David B. Cook. This comprehensive text provides upper-level undergraduates and graduate students with an accessible introduction to the implementation of quantum ideas in molecular modeling, exploring practical applications alongside theoretical explanations. 1998 edition. 832pp. 5 3/8 x 8 1/2. 0-486-44307-8

RADIOACTIVE SUBSTANCES, Marie Curie. The celebrated scientist's thesis, which directly preceded her 1903 Nobel Prize, discusses establishing atomic character of radioactivity; extraction from pitchblende of polonium and radium; isolation of pure radium chloride; more. 96pp. 5 3/8 x 8 1/2. 0-486-42550-9

CHEMICAL MAGIC, Leonard A. Ford. Classic guide provides intriguing entertainment while elucidating sound scientific principles, with more than 100 unusual stunts: cold fire, dust explosions, a nylon rope trick, a disappearing beaker, much more. 128pp. 5 3/8 x 8 1/2. 0-486-67628-5

ALCHEMY, E. J. Holmyard. Classic study by noted authority covers 2,000 years of alchemical history: religious, mystical overtones; apparatus; signs, symbols, and secret terms; advent of scientific method, much more. Illustrated. 320pp. 5 3/8 x 8 1/2. 0-486-26298-7

CHEMICAL KINETICS AND REACTION DYNAMICS, Paul L. Houston. This text teaches the principles underlying modern chemical kinetics in a clear, direct fashion, using several examples to enhance basic understanding. Solutions to selected problems. 2001 edition. 352pp. 8 3/8 x 11. 0-486-45334-0

PROBLEMS AND SOLUTIONS IN QUANTUM CHEMISTRY AND PHYSICS, Charles S. Johnson and Lee G. Pedersen. Unusually varied problems, with detailed solutions, cover of quantum mechanics, wave mechanics, angular momentum, molecular spectroscopy, scattering theory, more. 280 problems, plus 139 supplementary exercises. 430pp. 6 1/2 x 9 1/4. 0-486-65236-X

ELEMENTS OF CHEMISTRY, Antoine Lavoisier. Monumental classic by the founder of modern chemistry features first explicit statement of law of conservation of matter in chemical change, and more. Facsimile reprint of original (1790) Kerr translation. 539pp. 5 3/8 x 8 1/2. 0-486-64624-6

MAGNETISM AND TRANSITION METAL COMPLEXES, F. E. Mabbs and D. J. Machin. A detailed view of the calculation methods involved in the magnetic properties of transition metal complexes, this volume offers sufficient background for original work in the field. 1973 edition. 240pp. 5 3/8 x 8 1/2. 0-486-46284-6

GENERAL CHEMISTRY, Linus Pauling. Revised third edition of classic first-year text by Nobel laureate. Atomic and molecular structure, quantum mechanics, statistical mechanics, thermodynamics correlated with descriptive chemistry. Problems. 992pp. 5 3/8 x 8 1/2. 0-486-65622-5

ELECTROLYTE SOLUTIONS: Second Revised Edition, R. A. Robinson and R. H. Stokes. Classic text deals primarily with measurement, interpretation of conductance, chemical potential, and diffusion in electrolyte solutions. Detailed theoretical interpretations, plus extensive tables of thermodynamic and transport properties. 1970 edition. 590pp. 5 3/8 x 8 1/2. 0-486-42225-9

Engineering

FUNDAMENTALS OF ASTRODYNAMICS, Roger R. Bate, Donald D. Mueller, and Jerry E. White. Teaching text developed by U.S. Air Force Academy develops the basic two-body and n-body equations of motion; orbit determination; classical orbital elements, coordinate transformations; differential correction; more. 1971 edition. 455pp. 5 3/8 x 8 1/2. 0-486-60061-0

INTRODUCTION TO CONTINUUM MECHANICS FOR ENGINEERS: Revised Edition, Ray M. Bowen. This self-contained text introduces classical continuum models within a modern framework. Its numerous exercises illustrate the governing principles, linearizations, and other approximations that constitute classical continuum models. 2007 edition. 320pp. 6 1/8 x 9 1/4. 0-486-47460-7

ENGINEERING MECHANICS FOR STRUCTURES, Louis L. Bucciarelli. This text explores the mechanics of solids and statics as well as the strength of materials and elasticity theory. Its many design exercises encourage creative initiative and systems thinking. 2009 edition. 320pp. 6 1/8 x 9 1/4. 0-486-46855-0

FEEDBACK CONTROL THEORY, John C. Doyle, Bruce A. Francis and Allen R. Tannenbaum. This excellent introduction to feedback control system design offers a theoretical approach that captures the essential issues and can be applied to a wide range of practical problems. 1992 edition. 224pp. 6 1/2 x 9 1/4. 0-486-46933-6

THE FORCES OF MATTER, Michael Faraday. These lectures by a famous inventor offer an easy-to-understand introduction to the interactions of the universe's physical forces. Six essays explore gravitation, cohesion, chemical affinity, heat, magnetism, and electricity. 1993 edition. 96pp. 5 3/8 x 8 1/2. 0-486-47482-8

DYNAMICS, Lawrence E. Goodman and William H. Warner. Beginning engineering text introduces calculus of vectors, particle motion, dynamics of particle systems and plane rigid bodies, technical applications in plane motions, and more. Exercises and answers in every chapter. 619pp. 5 3/8 x 8 1/2. 0-486-42006-X

ADAPTIVE FILTERING PREDICTION AND CONTROL, Graham C. Goodwin and Kwai Sang Sin. This unified survey focuses on linear discrete-time systems and explores natural extensions to nonlinear systems. It emphasizes discrete-time systems, summarizing theoretical and practical aspects of a large class of adaptive algorithms. 1984 edition. 560pp. 6 1/2 x 9 1/4. 0-486-46932-8

INDUCTANCE CALCULATIONS, Frederick W. Grover. This authoritative reference enables the design of virtually every type of inductor. It features a single simple formula for each type of inductor, together with tables containing essential numerical factors. 1946 edition. 304pp. 5 3/8 x 8 1/2. 0-486-47440-2

THERMODYNAMICS: Foundations and Applications, Elias P. Gyftopoulos and Gian Paolo Beretta. Designed by two MIT professors, this authoritative text discusses basic concepts and applications in detail, emphasizing generality, definitions, and logical consistency. More than 300 solved problems cover realistic energy systems and processes. 800pp. 6 1/8 x 9 1/4. 0-486-43932-1

THE FINITE ELEMENT METHOD: Linear Static and Dynamic Finite Element Analysis, Thomas J. R. Hughes. Text for students without in-depth mathematical training, this text includes a comprehensive presentation and analysis of algorithms of time-dependent phenomena plus beam, plate, and shell theories. Solution guide available upon request. 672pp. 6 1/2 x 9 1/4. 0-486-41181-8

HELICOPTER THEORY, Wayne Johnson. Monumental engineering text covers vertical flight, forward flight, performance, mathematics of rotating systems, rotary wing dynamics and aerodynamics, aeroelasticity, stability and control, stall, noise, and more. 189 illustrations. 1980 edition. 1089pp. 5 5/8 x 8 1/4. 0-486-68230-7

MATHEMATICAL HANDBOOK FOR SCIENTISTS AND ENGINEERS: Definitions, Theorems, and Formulas for Reference and Review, Granino A. Korn and Theresa M. Korn. Convenient access to information from every area of mathematics: Fourier transforms, Z transforms, linear and nonlinear programming, calculus of variations, random-process theory, special functions, combinatorial analysis, game theory, much more. 1152pp. 5 3/8 x 8 1/2. 0-486-41147-8

A HEAT TRANSFER TEXTBOOK: Fourth Edition, John H. Lienhard V and John H. Lienhard IV. This introduction to heat and mass transfer for engineering students features worked examples and end-of-chapter exercises. Worked examples and end-of-chapter exercises appear throughout the book, along with well-drawn, illuminating figures. 768pp. 7 x 9 1/4. 0-486-47931-5

BASIC ELECTRICITY, U.S. Bureau of Naval Personnel. Originally a training course; best nontechnical coverage. Topics include batteries, circuits, conductors, AC and DC, inductance and capacitance, generators, motors, transformers, amplifiers, etc. Many questions with answers. 349 illustrations. 1969 edition. 448pp. 6 1/2 x 9 1/4. 0-486-20973-3

BASIC ELECTRONICS, U.S. Bureau of Naval Personnel. Clear, well-illustrated introduction to electronic equipment covers numerous essential topics: electron tubes, semiconductors, electronic power supplies, tuned circuits, amplifiers, receivers, ranging and navigation systems, computers, antennas, more. 560 illustrations. 567pp. 6 1/2 x 9 1/4. 0-486-21076-6

BASIC WING AND AIRFOIL THEORY, Alan Pope. This self-contained treatment by a pioneer in the study of wind effects covers flow functions, airfoil construction and pressure distribution, finite and monoplane wings, and many other subjects. 1951 edition. 320pp. 5 3/8 x 8 1/2. 0-486-47188-8

SYNTHETIC FUELS, Ronald F. Probstein and R. Edwin Hicks. This unified presentation examines the methods and processes for converting coal, oil, shale, tar sands, and various forms of biomass into liquid, gaseous, and clean solid fuels. 1982 edition. 512pp. 6 1/8 x 9 1/4. 0-486-44977-7

THEORY OF ELASTIC STABILITY, Stephen P. Timoshenko and James M. Gere. Written by world-renowned authorities on mechanics, this classic ranges from theoretical explanations of 2- and 3-D stress and strain to practical applications such as torsion, bending, and thermal stress. 1961 edition. 560pp. 5 3/8 x 8 1/2. 0-486-47207-8

PRINCIPLES OF DIGITAL COMMUNICATION AND CODING, Andrew J. Viterbi and Jim K. Omura. This classic by two digital communications experts is geared toward students of communications theory and to designers of channels, links, terminals, modems, or networks used to transmit and receive digital messages. 1979 edition. 576pp. 6 1/8 x 9 1/4. 0-486-46901-8

LINEAR SYSTEM THEORY: The State Space Approach, Lotfi A. Zadeh and Charles A. Desoer. Written by two pioneers in the field, this exploration of the state space approach focuses on problems of stability and control, plus connections between this approach and classical techniques. 1963 edition. 656pp. 6 1/8 x 9 1/4. 0-486-46663-9